Current Progress in Agricultural Genomics and Allied Sciences

Current Progress in Agricultural Genomics and Allied Sciences

Edited by **Harvey Parker**

SYRAWOOD
PUBLISHING HOUSE
New York

Published by Syrawood Publishing House,
750 Third Avenue, 9th Floor,
New York, NY 10017, USA
www.syrawoodpublishinghouse.com

Current Progress in Agricultural Genomics and Allied Sciences
Edited by Harvey Parker

International Standard Book Number: 978-1-68286-296-4 (Hardback)

Printed in China.

Contents

Preface

Genetic engineering is a rapidly developing branch of biotechnology and is specifically useful for the development of agriculture. This book consists of contributions made by international experts on genetic engineering and biotechnology. This book traces the progress of this field and highlights some of its key concepts and applications through discussions on topics like genetics and plant breeding, agricultural chemistry, seed science, etc. This book is most suitable for students and academicians pursuing biotechnology and allied sciences. As this field is emerging at a rapid pace, the contents of this book will help the readers understand the modern concepts and applications of the subject.

The researches compiled throughout the book are authentic and of high quality, combining several disciplines and from very diverse regions from around the world. Drawing on the contributions of many researchers from diverse countries, the book's objective is to provide the readers with the latest achievements in the area of research. This book will surely be a source of knowledge to all interested and researching the field.

In the end, I would like to express my deep sense of gratitude to all the authors for meeting the set deadlines in completing and submitting their research chapters. I would also like to thank the publisher for the support offered to us throughout the course of the book. Finally, I extend my sincere thanks to my family for being a constant source of inspiration and encouragement.

Editor

Assessment of Factors Associated with Low Yield of Cashew among Farmers in Growing Areas of Nigeria

A. E. Agbongiarhuoyi[1*], E. O. Uwagboe[1], O. S. Ibiremo[1], F. O. Olasupo[1] and E. O. Aigbekaen[1]

[1]Cocoa Research Institute of Nigeria (CRIN) P.M.B. 5244 Ibadan. Nigeria.

Authors' contributions

The research work was carried out in collaboration with authors from different areas of specializations. Author AEA conceptualized the study. All authors were involved in field data collection while authors AEA and EOU performed the statistical analyses. Author AEA wrote the manuscript and authors OSI, EOA and FOO reviewed it. All authors read and approved the final manuscript.

Editor(s):
(1) Marco Aurelio Cristancho, National Center for Coffee Research, CENICAFÉ, Colombia.
Reviewers:
(1) Anonymous, Iran.
(2) Anonymous, Ghana.
(3) Anonymous, Malawi.

ABSTRACT

The study assessed the factors affecting low yield of cashew among farmers in Kwara and Edo States between March and April, 2014. It examined the socio-economic characteristics, factors contributing to cashew low yield, sources of cashew planting materials and sources of information on cashew cultivation among farmers in the study areas. Data collection was through structured questionnaire administered on 160 respondents selected through multistage sampling procedure. Descriptive statistics and chi-square were used in data analysis. Results showed that majority of the farmers were ageing with a mean age of 52 years and 60% of them had more than 16 years of cashew farming experience. Most (69%) cashew farms were less than 15 years indicating that the cashew trees were of average age and are still in their productive years. The mean yield of respondents was 10 bags using 80kg bag which was low because this number is not economical considering other inputs of production. Majority of respondents obtained cashew planting materials from fellow farmers and their own farms and got information on cashew cultivation from fellow farmers. Chi-square results revealed that there was significant relationship ($p<0.05$) between insect pest attack, poor cashew price, lack of improved variety, farm abandonment, high cost of farm

Corresponding author: E-mail: toniagbons@yahoo.com;

labour, inadequate farm maintenance, no government support and yield of cashew. Efforts should be made by stakeholders in the cashew value chain to address relevant factors affecting low yield through a holistic government intervention programme such as Agricultural Transformation Agenda (ATA).This will develop a sustainable marketing strategy that will guarantee good prices of cashew nuts.

Keywords: Cashew low yield; farmers; socio-economic characteristics.

1. INTRODUCTION

Cashew plant is native to Brazil, from where the Portuguese explorers introduced it to Nigeria in the 15th and 16th century [1]. It is held with great esteem and value in many customs and cultures. It is known by many names across some parts of the world. In Mozambique, the Maconde tribe refer to it as the 'Devil's Nut'. In Nigeria (English) is called cashew, Potuguese-*caju,* French-*cajou,* Spanish-*maranon,* Italian-*anacardio,* Dutch-*acajou* and Indonesia-*jambu mente.* According to [2], Cashew nuts are offered at wedding ceremonies as a token of fertility and are considered by many to have aphrodisiac properties.

Cashew is a major product for millions of small-scale farmers in Sub-Sahara Africa. Globally, the annual production is about 2.1 million tons of raw nuts with an estimated value of US$ 1.5 to 2 billion [3]. The main cashew production regions are Africa, Vietnam, India and Brazil.

Cashew grows almost everywhere in Nigeria with production spanning across 27 out of 36 States in all the geo-political zones. Irrespective of the predominance of this crop in many states, yield per hectare is not encouraging. Nigerian yield in terms of nuts with shell is 22,855.19 kg/ha in 2012. When compared to yields from other countries within the same period like Peru (51,136.36 kg/ha), Philippines (46,807.81 kg/ha), Vietnam (38,944.84 kg/ha) and Mexico (30,088.44 kg/ha) according to [4] the yield in Nigeria is relatively low. Its product potentials had not been well harnessed due to varied factors ranging from low yielding plantations to inadequate farm management, processing facilities and marketing problems. Most cashew trees start bearing fruit in the third or fourth year, and are likely to reach their mature yield by the seventh year with favourable condition. The average nut yield of a mature tree is in the range of 7-11 kg per annum [2].

Nigeria is rated among the top five cashew producing countries. It is now ranked world's 2nd largest producer after Vietnam which is the highest producer with 1,190,900 tonnes. Nigeria currently produces 836,500 tonnes of cashew [4]. Three main cashew products are traded on the international market: raw nuts, cashew kernels and cashew nut shell liquid (CNSL).

In terms of annual trade, cashew earned Nigeria ₦24b in 2012 and provides about 600,000 jobs to people engaged in its production [5]. In the area of health benefits, cashew apple contains five times more vitamin C than an orange which makes it unique among other fruits [11]. The apple is also used traditionally as a curative for scurvy and stomach ailments like dysentery and diarrhoea. Fresh or distilled, it is a potent diuretic, possessing antiscorbutic properties and is useful for kidney troubles and in advanced cases of cholera [7]. Clinically, cashew nut consumption has also been proven to improve sperm count and reduce infertility among couples [8]. Cashew nut contains 47% fat, 21% protein and 22% carbohydrate according to a study carried out by [9].

As a way of reducing wastage of fleshy cashew apple on farmers' farms, the Cocoa Research Institute of Nigeria (CRIN) which has mandate for research into cocoa, cashew, coffee, kola and tea crops has so far developed various products from the cashew apple and these include: Top quality thermally stable chocolate from cashew, evaluated and acknowledged internationally. Top quality sweet (8-12% v/v alcohol), dry (17-18% v/v alcohol) wines and improved technique for processing cashew apples into juice, drink and jam have also been developed.

A locally fabricated juice extractor was developed by CRIN to produce cashew apple juice adaptable for use on a cottage industry scale. Also, CRIN is now producing cashew nuts coated with honey. These efforts are already adding value to the socio-economic life of consumers who are having high demand for them within the country. Despite these contributions to the national economy, the issue of declining production has remained a serious challenge to all stakeholders in the cashew value chain.

Cashew production has not received adequate attention from the Federal Government since the advent of oil in the 1960s. There is abandonment and significant decline in yield of cashew in most growing areas of Nigeria [10]. For instance, it was reported that a farm having 400ha cultivated to cashew in Ogun State needs the intervention of CRIN scientists in rehabilitating and putting it into productive venture. However, the Agricultural Transformation Agenda (ATA) programme of government has added cashew to the existing value chain crops. CRIN in collaboration with the Federal Government distributed 10,000kg of Jumbo cashew nuts to farmers in 22 States of Nigeria from 23rd to 31st July 2014. This support was part of the transformation process in the cashew sub-sector.

It is envisaged that this intervention will boost the yield of cashew in the country. In view of this, the study was designed to address the inherent challenges associated with cashew field production in order to proffer possible solutions.

The main objective of the study was to assess the factors contributing to low yield of cashew among farmers in Kwara and Edo States. The specific objectives were to describe the socio-economic characteristics, ascertain the sources of cashew planting materials and the sources of information on cashew cultivation among cashew farmers.

It was hypothesized that there is no significant relationship between yield and factors affecting low cashew production.

2. METHODOLOGY

A multi-stage procedure was used to select respondents for the study. Stage1: Two States Kwara and Edo were randomly selected across two geo-political zones in Nigeria, South South and North Central. Stage 2: In each State, two Local Government Areas (LGAs) noted for cashew production were chosen. These include Isin and Irepodun in Kwara, Owan East and Etsako East in Edo States. Two villages per LGA where cashew is well produced were selected. Stage 3: Twenty farmers were selected in each village to make eighty per state using simple random sampling technique bringing the total number to 160 farmers. A list of cashew farmers in the two states selected was obtained from the Tree Crop Units (TCU) in the State ministry of Agriculture. A structured interview schedule instrument was used for field data collection from cashew farmers in the study areas. Descriptive statistics and chi-square were used in data analysis. The variables were measured at 0.05 Level of probability.

3. RESULTS AND DISCUSSION

3.1 Socio-economic Characteristics of Cashew farmers

The results showed that 69% of the respondents were males while 31% were females (Table 1). It means that more men were involved in cashew production than the women in the study areas. This is consistent with the results of [11,12] who reported similar findings. A number of the respondents (29%) were between 55-64 years with a mean age of 52 years indicating that the farmers were aging. This will have negative implication on their ability to increase production of cashew and invariably on yield. Majority (97%) of them were married. In terms of educational status, 42%, 21%, 24% had primary, secondary and tertiary education, respectively. It means that most of the respondents had one form of education or the other which will assist in adoption of technology in cashew production. Generally, education is thought to create a favourable mental aptitude for the acceptance of new practices [13]. The findings in Table 1 further revealed that many respondents (61%) cultivated 1-4 hectares of cashew farms. This indicates the small-scale nature of most of their farm holdings in the study areas. [14] also found that majority of cashew farmers in Oyo State were small holders. Most (69%) cashew farms in Kwara and Edo States were less than 15 years. It implies that the cashew trees were of average age and are still in their productive years. A reasonable percentage of the respondents (60%) had more than 16 years of farming experience. It means that they are experienced in cashew cultivation and this will improve production practices.

3.2 Cashew yield

In Table 2, the mean yield of respondents was 10 bags using 80kg bag. The respondents that obtained 1-5 bags were more than those having 5 and above. This quantity is rather low because a farmer cannot break-even with this number considering what it will amount to in monetary terms. A bag of 80kg of dried cashew nuts sells for between N4,800 to N5,250 ($29.09 to $31.81) in Kwara and Edo States of Nigeria. It implies that farmers' productivity needs to be improved so as to have more bags that will generate more income and enhanced livelihood.

The production can be correlated with the hectarage of the farmers in which most of them have less than 4 hectares. This result is in consonant with the work of [14] who reported similar finding.

3.3 Factors that Contribute to Low Yield of Cashew

Table 3 shows that poor price of cashew (73%), insect pest attack (51%) (Stem girdler, Stem borer, termites and grasshopper), inadequate government support to cashew production (44%), high cost of farm labour (41%) and farm abandonment (35%) were considered to be very serious factors contributing to low yield of cashew while soil problem, farm location, none soil test before planting, lack of fertilizer use,

Table 1. Distribution of socio-economic characteristics of cashew farmers

Variables	Frequency	Percentage	Mean
Sex			
Male	110	68.8	
Female	50	31.2	
Age of farmers (Years)			
15-24	3	1.9	52.4
25-34	10	6.3	
35-44	32	20.0	
45-54	37	23.1	
55-64	47	29.3	
65 and above	31	19.4	
Marital status			
Married	155	96.9	
Single	5	3.1	
Educational status			
No formal Education	9	5.6	
Primary	67	41.9	
Secondary	34	21.2	
Tertiary	38	23.8	
Adult Education	12	7.5	
Farm size (Hectares)			
1-2	52	32.5	6.7
3-4	45	28.1	
5-6	16	10.0	
7-8	16	10.0	
9-10	9	5.6	
11 and above	22	13.8	
Age of cashew farms (Years)			
1-5	15	9.3	14.7
6-10	51	31.9	
11-15	44	27.5	
16-20	20	12.5	
21-25	12	7.5	
26 and above	18	11.3	
Experience (Years)			
3-9			14.7
10-16	39	24.4	
17-23	65	40.6	
24-30	31	19.4	
31 and above	21	13.1	
	4	2.5	

Source: Field survey, 2014 (n=160)

old age of cashew tree were not seen as serious factors. In addition, inadequate farm maintenance (40%) was also serious. It means that the very serious and serious factors constitute important variables to low yield of cashew in the study areas. It therefore implies that for increased cashew productivity, the identified factors must be properly addressed through an articulate programme that will make cashew economy to be more attractive. This could be done via formation of farmers' cooperatives and ATA which will make farmers' voice to be held by the government as well as relevant private partners in the cashew industry. In addition, appropriate integrated pest management scheme will be necessary to combat insect pest infestation in cashew farms.

3.4 Sources of Cashew Planting Materials

In Table 4, most of the respondents got cashew planting materials from fellow farmers and their own farms. Only 1.3% and 11% collected cashew planting materials from CRIN and Ministry of Agriculture in Edo and Kwara States respectively. It implies that majority of cashew in the fields of farmers were from the old varieties planted in their farms. CRIN as the institute with the national mandate on improvement of cashew in Nigeria needs to intensify efforts in making improved cashew varieties available to farmers so that the much desired yield would be achieved. Also, the State ministry of Agriculture in cashew producing States need to supply farmers with good planting materials so that they will get better seeds or seedlings as input for their farms. The implication of obtaining unrecommended genotypes for planting could be grave as they will not only lead to low productivity due to the inbreeding depression but will also break down the genetic potential of the crop.

3.5 Sources of Information on Cashew Cultivation

The results in Table 5 indicates that majority (65%) of the respondents obtained information on cashew cultivation from fellow farmers while others got theirs from other sources. This finding is in line with the work of [15]. They reported that farmers rely on neighbours, friends and relatives in obtaining information on cashew production. The extension agents that were supposed to be farmers' main source of information were quite low (about18%). There is need for increased extension agents' farmer contact in order to disseminate appropriate cultivation practices on cashew to them. This will go a long way in improving better access to information source that will increase production.

3.6 Association between Yield and Factors Affecting Low Cashew Production

The results in Table 6 reveal that there was significant relationship ($P<0.05$) between insect pest attack, poor cashew price, lack of improved variety, farm abandonment, high cost of farm labour, inadequate farm maintenance, no government support and yield of cashew. It implies that these were important factors contributing to low cashew yield in the study areas. All stakeholders in the cashew value chain should note these and work out appropriate interventions in reducing their effect on cashew yield.

Table 2. Distribution of cashew yield/80kg bag by respondents

Yield/80kg bag	Frequency	Percentage
1-5	36	22.5
6-10	31	19.4
11-15	29	18.1
16-20	13	8.1
21-25	2	1.3
26-30	24	15.0
31 and above	25	15.6

Source: Field survey, 2014 (n=160)

Table 3. Distribution of factors contributing to low yield of cashew

Factors	Frequency	Percentage
Insect pest attack		
Very serious	81	50.6
Serious	56	35.0
Not serious	23	14.4
Disease infection		
Very serious	42	26.3
Serious	60	37.5
Not serious	58	36.2
Soil problem		
Very serious	7	4.4
Serious	16	10.0
Not serious	137	85.6
Lack of fertilizer use		
Very serious	16	10.0
Serious	19	11.9
Not serious	125	78.1
Farm location		
Very serious	6	3.8
Serious	20	12.5
Not serious	134	83.7
Soil test before planting		
Very serious	9	5.6
Serious	20	12.5
Not serious	131	81.9
Inadequate farm maintenance		
Very serious	33	20.6
Serious	64	40.0
Not serious	63	39.4
No improved variety		
Very serious	45	28.1
Serious	45	28.1
Not serious	70	43.8
Old age of cashew tree		
Very serious	17	10.6
Serious	36	22.5
Not serious	107	66.9
Farm abandonment		
Very serious	56	35.0
Serious	31	19.4
Not serious	73	45.6
Poor price of cashew		
Very serious	116	72.5
Serious	30	18.8
Not serious	14	8.7
High cost of farm labour		
Very serious	65	40.6
Serious	57	35.6
Not serious	38	23.8
Climate variability		
Very serious	42	26.3
Serious	58	36.2
Not serious	60	37.5
Fire outbreak		
Very serious	46	28.7

Factors	Frequency	Percentage
Serious	55	34.4
Not serious	59	36.9
Inadequate government support		
Very serious	70	43.8
Serious	55	34.4
Not serious	35	21.8

Source: Field survey, 2014 (n=160)

Table 4. Distribution of respondents' sources of cashew planting materials

Sources	Frequency	Percentage
Own farm		
Yes	56	35.0
No	104	65.0
Fellow farmer		
Yes	94	58.8
No	66	41.2
Ministry of Agriculture		
Yes	17	10.6
No	143	89.4
Open market		
Yes	9	5.6
No	151	94.4
CRIN		
Yes	2	1.3
No	158	98.7

Source: Field survey, 2014 (n=160)

Table 5. Distribution of respondents' sources of information on cashew cultivation

Sources	Obtained information	Did not obtain information
Agric. Extension agents	28 (17.5)	132 (82.5)
Fellow farmer	104 (65.0)	56 (35.0)
Radio	34 (21.3)	126 (78.7)
Television	38 (23.8)	122 (76.2)
Newspaper	15 (9.4)	145 (90.6)
GSM phone	8 (5.0)	152 (95.0)
Farmer association	31 (19.4)	129 (80.6)
Non Governmental Oganisation (NGO)	3 (1.9)	157 (98.1)

Source: Field survey, 2014 (n=160) Figures in parentheses are percentages

Table 6. Chi square analysis of the association between yield and factors affecting low cashew production

Variables	df	χ^2 Value	p Value
Insect attack	2	31.737	0.000
Disease infestation	2	3.650	0.161
Poor cashew price	2	112.850	0.000
No improved variety	2	7.812	0.020
Farm abandonment	2	16.738	0.000
Fire outbreak	2	1.662	0.436
High cost of farm labour	2	7.212	0.027
Climate variability	2	3.650	0.161
Inadequate farm maintenance	2	11.638	0.003
Inadequate government support	2	11.563	0.003

Source: Field survey, 2014 df Degree of freedom, χ^2 is Chi-square, $p<0.05$

4. CONCLUSION

The study concludes that insect pest attack (stem girdler, stem borer, termites grasshopper), poor price of cashew nuts, inadequate government support to cashew production, high cost of farm labour, farm abandonment, inadequate farm maintenance and lack of improved variety were considered to be important factors contributing to low yield of cashew in producing areas of Edo and Kwara States of Nigeria. Majority of respondents obtained cashew planting materials from fellow farmers and their own farms. They got information on cashew cultivation from fellow farmers. Few respondents received information from extension agents that were supposed to be farmers' main source. Socio-economic variables indicated that men were more involved in cashew production than the women. The farmers were ageing and most of them had long years of farming experience. From the foregoing, efforts should be made by stakeholders in the cashew value chain to address relevant factors militating against low yield through an integrated pest management system, government intervention programme such as Agricultural Transformation Agenda (ATA) and a formidable market strategy that will guarantee good prices of cashew nuts. This will enhance farmers' livelihood and boost production.

COMPETING INTERESTS

Authors have declared that no competing interests exist.

REFERENCES

1. Asogwa EU, Hammed LA, Ndubuaku TCN. Integrated production and protection practices of cashew (*Anacardium occidentale*) in Nigeria. African Journal of Biotechnology. 2008;7(25):4868-4873.
2. Azam-Ali SH, Judge EC. Small-scale cashew nut processing. Published by FAO; 2004. Accessed 06/06/2014. Available:http://www.fao.org/inpho_archive/content/documents/vlibrary/ac306e/ac306e00.htm
3. Food and Agriculture organization of the United Nations, top 5 facts. Published by FAO; 2010. accessed 04/07/2014. Available: http www.fao.org
4. FAO, Food and agricultural organization statistical production report on cashew nuts with shell. Published by FAO; 2014. Accessed 04/07/2014. Available:http://faostat3.fao.org/faostat-gateway/go/to/download/Q/*/E
5. The African commodity report, cashew. Published by Delmas; 2013. Accessed 04/07/2014. Available:http://www.delmas.com/static/communication
6. Pradeepkumar T, Jyothibhaskar SB, Satheesa KN. Management of hoticultural crops. Published by New Delhi. 2008;500.
7. Mathew J, Sobhana A, Mini C. Opportunities for income enhancement from cashew plantations through cashew apple processing. In: Masawe PAL, Esegu JFO, Kasuga LJF, Mneney EE, Mujuni D. (Eds). Proceedings of the second international cashew conference, Kmpala Uganda, CABI International, Wallingford, UK. 2013;143-149.
8. Bakare W. Cashew cultivation: Local and international prospects; 2012. Accessed 07/06/2014. Available:http://backtofarmnigeria.wordpress.com/
9. Pragnya GS, Padinjarathil HB. Effect of gamma irradiation on shelf life of cashew apple. A bachelor's degree dissertation submitted to Vinayaka Mission's Kirupanada Variyar Engineering college, Vinayaka Mission's University, Salem. 2010;1-2.
10. Hammed LA, Anikwe JC, Adedeji AR. Cashew nuts and production development in Nigeria. American-Eurasian Journal of Scientific Research. 2008;3(1):54-61.
11. Wongnaa CA. Analysis of factors affecting the production of cashew in Wenchi Municipality, Ghana. Journal of Agricultural Sciences. 2013;8(1):8-1.
12. Agbongiarhuoyi AE. Aigbekaen EO, Akinbile LA. Awareness of cashew products potentials and market information among farmers in Kogi State, Nigeria. ARPN Journal of Agricultural and Biological Science. 2008;3(4):10-15.
13. Caswell M, Fuglie K, Ingram C, Jans S, Kascak C. Adoption of agricultural production practices: Lessons learned from the US. Department of agriculture area studies project. Washington DC. US department of agriculture. Resource Economics Division, Economic Research service, Agriculture Economic Report No. 2001;792.

14. Uwagboe EO, Adeogun SO, Odebode SO. Constraints of farmers in cashew production: A case study of Orire LGA of Oyo State, Nigeria. ARPN Journal of Agricultural and Biological Science. 2010;5(4):27-31.
15. Masawe PAL, Kapinga FA. Towards a sustainable cashew industry in Tanzania. In: Masawe PAL, Esegu JFO, Kasuga LJF, Mneney EE, Mujuni D. (Eds). Proceedings of the second international cashew conference, Kmpala Uganda, CABI International, Wallingford, UK. 2010;112-118.

2

Increase of Spinach Growth Through the use of Larger Plug Cell Volume and an Exogenous BAP Spray

J. Di Matteo[1], J. Rattin[1] and A. Di Benedetto[1,2*]

[1]Faculty of Agricultural Sciences, National University of Mar del Plata, (B7620ZAA), Balcarce, Province of Buenos Aires, Argentina.
[2]Faculty of Agronomy, University of Buenos Aires, Avenida San Martín 4453 (C1417DSE), Buenos Aires, Argentina.

Authors' contributions

This work was carried out in collaboration between all authors. All authors read and approved the final manuscript.

Editor(s):
(1) Lanzhuang Chen, Laboratory of Plant Biotechnology, Minami Kyushu University, Japan.
Reviewers:
(1) Vincent Ezin, University of Abomey-Calavi 01 BP 526 Cotonou, Benin, AVRDC- The World Vegetable Center in Taiwan, Taiwan.
(2) Ferdaous Mani, Department of horticol and in vitro sciences, High Agronomic Institute, Tunisia.

ABSTRACT

Aims: In spinach, transplanting has replaced direct seeding but small containers may cause root restriction effects that would be decreased post-transplant yield. Although a single hormonal regulator (BAP) spray is effective in overcoming root restriction, the mechanisms involved are unknown. The aim of this work was to analyze spinach growth changes by the use of different plug cell volumes and BAP as foliar spray.
Study Design: A randomized complete block factorial design with three blocks was used.
Place and Duration of Study: The experiments were conducted at the INTA Balcarce Experimental Station campus, Argentina (37° 45′ S, 58° 18′ W and altitude 130 m) from 5th March to 24th May 2008 and repeated once from 15th March to 19th May 2009.
Methodology: We analyzed growth changes by the use of different plug cell volumes and BAP as foliar spray on dry weight accumulation and partitioning in *Spinacea oleracea* plants in two field experiments.
Results: The use of large plug cell volumes and BAP sprays at the pre-transplant stage increased

**Corresponding author: E-mail: dibenede@agro.uba.ar;*

plant growth, through an increase in the relative growth rate (RGR) and the rate of leaf area expansion (RLAE). We found a close direct relationship between RGR and net assimilation rate (NAR) but an inverse relationship between RGR and leaf area ratio (LAR). In addition, we found that NAR significantly increased as leaf thickness, intercellular spaces and stomatal density increased. A change in dry weight partitioning toward shoots was found as well.

Conclusion: From a grower's point of view, the use of large plug cell volume would increase spinach growth. On the other hand, a promising approach to increasing crop productivity is the use of plant growth and development regulators such as BAP.

Keywords: Leaf traits; nursery; photoassimilate partitioning; productivity; root restriction.

1. INTRODUCTION

Technological advances in transplanting have contributed to the growth of the industry by reducing costs and increasing the reliability of production. However, small containers may cause root restriction effects, i.e. a physical stress imposed on a root system which leads to a significant decrease in post-transplant growth [1-3]. Surprisingly, the root restriction associated with container volume seems to have received little consideration in the scientific literature [4].

We have previously shown that a single exogenous application of cytokinin 6-benzylaminopurine (BAP) to plants grown in small pots at the pre-transplant stage may override the shoot growth limitation due to root restriction [2,5]. We have also shown that the pre-transplant application of BAP increases both lettuce and celery yield [1,3] through higher shoot fresh weight accumulation and the rate of leaf area expansion (RLAE) as a result of higher relative growth rate (RGR) and the rate of leaf appearance (RLA).

In addition, it has been shown that an endogenous decrease in cytokinins levels decreased the development of the aerial part of plants [6-9]). Root and shoot growth are closely coordinated, presumably by a cytokinin signaling pathway which is largely hormonal in nature with a major site of control located in the root system [10,11]. Thus, increasing root growth may lead to a corresponding increase in the synthesis of cytokinins [12].

Spinach (*Spinacea oleracea* L.), a leaf vegetable with a high nutritive value, is of considerable economic importance. Since transplanting has almost replaced direct seeding, the objective of this work was to analyze spinach growth changes by the use of different plug cell volumes and the hormonal regulator BAP through the physiological mechanisms involved.

2. MATERIALS AND METHODS

2.1 Plant Material, Treatments and Experiments

The experiments were conducted at the INTA Balcarce Experimental Station campus, Argentina (37° 45′ S, 58° 18′ W and altitude 130 m) from 5th March to 24th May 2008 and repeated once from 15th March to 19th May 2009.

Spinach (cv. 'Ballet') seeds (a flowering-resistant genotype) were sown in 288- (6.18 cm^3 $cell^{-1}$) or 200- (13.90 cm^3 $cell^{-1}$) cell trays and were transplanted to a Typic Argiudol soil. Soil containing clay-loam-sand at 17.5, 47.5 and 35% (v/v) respectively with pH=6.09; EC=0.16 dS/m; N=0.26% and P=0.91 mmol L^{-1}. Seedlings were sprayed with BAP (6-benzylaminopurine) (SIGMA EC 214-927-5) (Sigma-Aldrich Co., St. Louis, MO, USA) solutions (0, 5, 50, 100 and 200 mg L^{-1}) either when first true leaf pairs appeared (pre-transplant application) or one week after transplant (post-transplant application) [1]. BAP was previously diluted in alcohol 80%.

Plants were grown under greenhouse facilities and were irrigated as needed (to compensate 80% relative evapo-transpiration) with high quality tap water; a weekly fertilization (Stage 2: 50 mg L^{-1} N; Stage 3-4: 100 mg L^{-1} N; post-transplant: 150 mg L^{-1} N) [13] was used.

Half hourly averages air temperature was measured using a HOBO H08-001-02 data logger (Onset Computer Corporation, MA, U.S.A) protected from direct radiation by aluminum foil shades. Mean temperatures during the experiments ranged from 8.9 to 18.2°C.

2.2 Growth Evaluations

For destructive measurements, plants (10 per treatment and block) were harvested at the transplant stage and both in the middle and at the end of the experiments. The numbers of new leaves that had appeared were also recorded on the same dates and the rate of leaf appearance (RLA) was calculated as the slope of the number of new (visible) leaves *vs.* time (in weeks). The fresh weight (FW) of roots and shoots was determined separately for each plant. Separate plant parts were dried at 80°C for 48 hours and weighed to obtain the dry weight of shoots (including leaf blades and stems) and roots. Leaves were oven-dried at 80°C for 48 hours. Leaf area was determined with a LI-COR 3000A automatic leaf area meter (LI-COR BioSciences, Lincoln, NE, USA).

The relative growth rate (RGR) was calculated as the slope of the regression of the natural logarithm (ln) of the whole plant on a dry weight basis *vs.* time (in days), while the rate of leaf area expansion (RLAE) was calculated as the slope of the regression of the ln of total leaf area *vs.* time (in days). Mean net assimilation rate (NAR) and leaf area ratio (LAR) were calculated according with Potter and Jones [14] as follows:

$$NAR = (k_w\, W_0\, e^{k_w t}) / (A_0\, e^{k_a t})$$
$$LAP = (k_a\, A_0\, e^{k_a t}) / (k_w\, W_0\, e^{k_w t})$$

Where: k_w: RGR ($days^{-1}$); W_0: extrapolated value of total dry weight at time zero (g); A_0: extrapolated value of leaf area at time zero (cm^2); k_a: RLAER ($days^{-1}$); t: time (in days) at the midpoint of the experimental period and e: base of natural logarithms.

The specific leaf area on a FW basis (SLA_{FW}) was calculated as the ratio between area of the new individual leaf and leaf FW. The leaf weight rate (LWR) values measured biomass allocated to leaves *vs.* biomass allocated to other plant parts. The allometric coefficients between roots and shoots were calculated as the slope of the straight-line regression of the ln of root dry weight *vs.* the ln of shoot dry weight.

Tissue from the middle region of the lamina was fixed in formalin-acetic-alcohol. The size and densities of epidermal stomata size were recorded on this tissues discolored with ethyl alcohol (96°) at boiling temperature. Leaf thickness was determined from leaf lamina tissues embedded in paraffin, sectioned at 20 µm on a rotary microtome. Leaf samples were stained with safranin-crystal violet-fast green. Data are the mean of three leaves per treatment and block from ten leaf cross-sections per leaf. An image analysis system (Image Pro Express v 6.0, Media Cybernetics, USA) facilitated quantitative anatomical measurements.

2.3 Statistical Analysis

We used a randomized complete factorial design with three blocks of each treatment: Cell size, BAP concentration (BAP_c) and BAP application time (BAP_{at}). Since we found no significant differences between the two experiments, we considered them together. Data were subjected to two way analysis of variance and means were separated by Tukey's test ($P<0.05$); STATISTICA 8 (StatSoft) software was used. Straight-line regression slopes were tested for parallelism (test for equal slope) [15].

3. RESULTS

3.1 Leaf area and biomass accumulation

The shoot fresh weight at the end of the experiments was higher for control plants grown in 200-cell trays than for those grown in 288-cell ones and for those sprayed at the pre-transplant stage with 5 to 100 mg L^{-1} BAP. The higher the BAP dose, the higher the shoot fresh weight. On the other hand, a high (200 mg L^{-1}) pre-transplant or post-transplant dose decreased aerial fresh weight (plants grown in 200-cell trays) or showed little changes (Fig. 1).

Spinach plants from 200-cell trays showed higher total leaf area than those from 288-cell trays at the end of the experiment. In general, total leaf area in control plants was increased by a single BAP spray, but was highest when BAP was sprayed at the pre-transplant stage. The higher leaf area was the result of an increase in both the individual leaf area and RLA (Table 1).

RLAE was higher in control plants from 200-cell than from 288-cell trays and was significantly increased with a BAP spray at almost all doses tested at both pre- and post-transplant stages. RGR showed similar results except when BAP was applied at the post-transplant stage in plants grown in 200-cell trays. NAR was higher in plants grown in 200-cell trays than in those grown in 288-cell ones. A single BAP spray (at all doses) increased NAR plants grown in 288-cell trays both at the pre- and at the post-transplant stage; BAP failed to increase NAR at the post-transplant stage in 200-cell grown plants. Results for LAR were opposite. Specific leaf area (SLA_{FW}) was higher in control plants grown in 288-cell trays than in those grown in 200-cell ones, but in both cases was significantly decreased by a BAP spray. Leaf weight ratio (LWR) increased when plants were sprayed with BAP but even was higher in plants grown in 200-cell plug trays than in those grown in 288-cell ones (Table 2).

3.2 Dry Weight Partitioning

The allometries from Table 3 show that controls plants grown in 288- and 200-cell trays partitioned a similar photo-assimilate proportion to roots and that a single BAP application significantly changed photo-assimilate partitioning towards shoots.

3.3 Anatomical Traits

Leaf thickness was higher in plants grown in 200-cells trays than in those grown in 288-cell ones. Leaf thickness was significantly increased by a BAP spray, but the response was different according to the BAP dose applied and application time; the same response was found when size and density of stomata were recorded (Table 4).

3.4 Growth Rates and Anatomical Changes Ratios

When plotting the data from all treatments, we found a close direct relationship (r^2=0.822) between RGR and NAR (Fig. 2a) and an inverse relationship between RGR and LAR (r^2=0.761) (Fig. 2b). Nevertheless, control plants grown in 200-cell trays showed higher NAR and lower LAR values than those grown in 288-cell trays.

On the other hand, NAR significantly increased according to an increase in leaf thickness (r^2=0.694) (Fig. 3a), intercellular space proportion (r^2=0.727) (Fig. 3b) and stomata density (r^2=0.495) (Fig. 3c). NAR relationships were higher in control plants grown in 200-cell trays than in those grown in 288-cells ones.

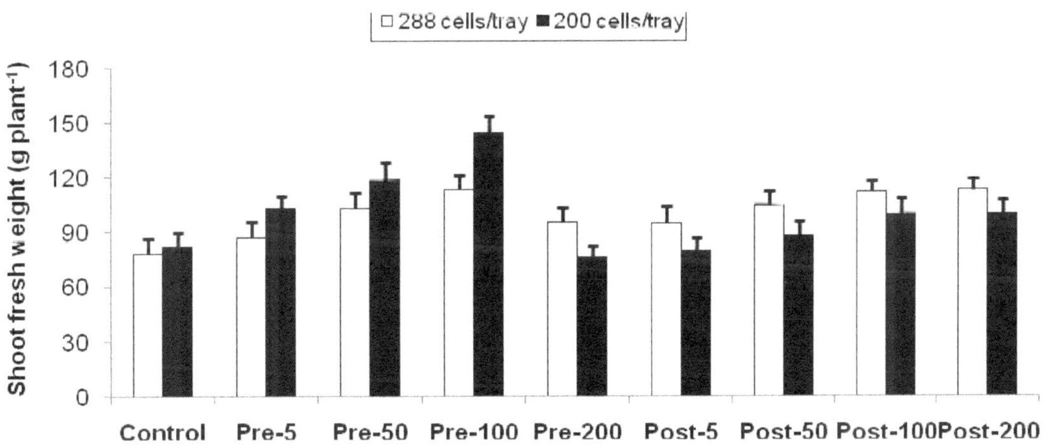

ANOVA	
Significantly	
PCV	***
BAP_c	***
BAP_{at}	***
PCV x BAP_c	***
PCV x BAP_{at}	***
PCV x BAP_c x BAP_{at}	***

*Significance *** 0.001*

Fig. 1. The effects of two plug cell volume (288- and 200-cells tray^{-1}) (PCV) and a pre-transplant or post-transplant BAP sprays (5, 50,100 or 200 mg L^{-1}) on spinach shoot fresh weight at the end of the experiment (n=6). Both, standard errors and the analysis of variance (ANOVA) are indicated

Table 1. Changes in the total leaf area, the individual leaf area and the rate of leaf appearance (RLA) at the end of the experiment in spinach plants grown at two plug cell volume (PCV) and sprayed with different concentrations of BAP solutions at the pre- or post-transplant stage (data are the mean of two experiment) (n=6)±SE. Different lower-case letters indicate statistically significant differences (*P*=.05) between BAP treatments for each cell volume while different capital letters indicate statistically significant differences (*P*=.05) for each BAP treatment between different cell volumes. The probability of the RLA slope being zero was *P* = .001. The significance of interactions (ANOVA) has been indicated as well

PCV	Total leaf area (cm^2 $plant^{-1}$)	Individual leaf area (cm^2 $leaf^{-1}$)	RLA (leaves $week^{-1}$)
288 cells $tray^{-1}$			
Control	499.37^{cB}	34.07^{cB}	1.122^{cB}
Pre-5	598.74^{bB}	42.71^{bB}	1.262^{aB}
Pre-50	681.25^{aB}	45.18^{bA}	1.233^{aB}
Pre-100	670.18^{aB}	53.72^{aA}	1.292^{aB}
Pre-200	656.61^{aA}	41.50^{bB}	1.142^{bB}
Post-5	595.80^{bA}	41.48^{bA}	1.108^{bB}
Post-50	598.93^{bA}	42.43^{bA}	1.157^{bB}
Post-100	610.34^{bA}	38.90^{bA}	1.131^{bB}
Post-200	638.42^{bA}	43.02^{bA}	1.147^{bB}
200 cells $tray^{-1}$			
Control	562.04^{cA}	39.21^{cA}	1.219^{cA}
Pre-5	743.23^{aA}	50.94^{aA}	1.358^{aA}
Pre-50	768.69^{aA}	46.04^{aA}	1.384^{aA}
Pre-100	773.18^{aA}	51.30^{aA}	1.350^{aA}
Pre-200	649.18^{bA}	49.12^{aA}	1.202^{bA}
Post-5	604.75^{bA}	45.76^{bA}	1.272^{bA}
Post-50	642.65^{bA}	48.80^{bA}	1.224^{cA}
Post-100	639.43^{bA}	45.58^{bA}	1.212^{cA}
Post-200	654.23^{bA}	42.58^{bA}	1.225^{cA}

ANOVA			
Significantly	**Total leaf area**	**Individual leaf area**	**RLA**
PCV	***	***	***
BAP_c	***	***	***
BAP_{at}	*	*	***
PCV x BAP_c	***	***	***
PCV x BAP_{at}	**	**	***
PCV x BAP_c x BAP_{at}	**	**	***

*Significance *** 0.001 ** 0.01 * 0.05*

4. DISCUSSION

Leaf expansion and synthesis of photo-assimilates are essential for plant growth. Our results showed that a higher total leaf area in spinach (Table 1) as a result of higher RLAE (Table 2) was closely related to plug cell volume and both BAP spray concentration and BAP application time (pre- or post-transplant stage). Previous reports have shown similar increases in total leaf area in other leafy vegetables such as lettuce and celery [1,3], but the fact that this result would be associated with a higher individual leaf area and RLA (Table 1) has not been previously published in vegetables. We have recently suggested that an increase in plug cell size or exogenous BAP would increase leaf size or leaf appearance as an expression of meristem growth and phyllochron decrease respectively in ornamental foliage plants [16,5,17]. Cytokinins control growth mainly through the regulation of cell division and differentiation in the stem apical meristem [18]. Similarly, the size of the apical meristem is in turn related to the rate of leaf appearance, as shown in experiments with cytokinins deficient *Arabidopsis* mutants [19].

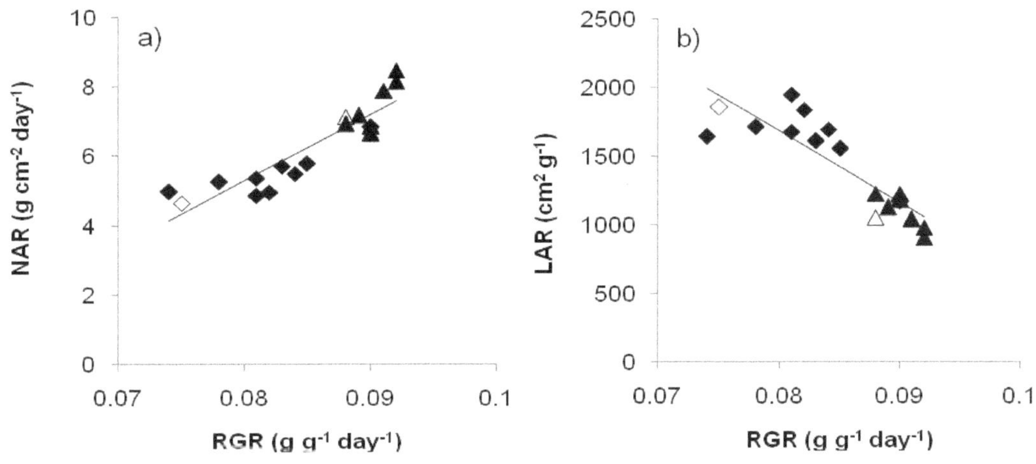

Fig. 2. The net assimilation rate (NAR) (a) and the leaf area ratio (LAR) (b) related to the relative growth rate (RGR). The straight-line regressions were: NAR = 192.11 RGR - 10.07 (r^2= 0.822) and LAR = - 52,175.00 RGR + 5,856.20 (r^2= 0.761). ◊: 288-grown controls; Δ: 200-grown controls

Table 2. Changes in the rate of leaf area expansion (RLAE), the relative growth rate (RGR), the net assimilation rate (NAR), the leaf area ratio (LAR), the specific leaf area (SLA_{FW}) and the leaf weight rate (LWR) in spinach plants grown at two plug cell volume (PCV) and sprayed with different concentrations of BAP solutions at the pre- or post-transplant stage (n = 6) ± SE. Different lower-case letters indicate statistically significant differences between BAP treatments for each cell volume while different capital letters indicate statistically significant differences for each BAP treatment between different cell volumes at *P* = .05 by Tukey's test (SLA_{FW} and LWR) or according the test for equal slope (testing for parallelism) (RLAE, RGR, NAR and LAR); the probability of these slopes being zero were *P* = .001

PCV	RLAE (cm^2 cm^{-2} day^{-1})	RGR (g g^{-1} day^{-1})	NAR (g cm^{-2} day^{-1}) x 10^5	LAR (cm^2 g^{-1})	SLA_{FW} (cm^2 g^{-1})	LWR (g g^{-1})
288 cells tray^{-1}						
Control	0.028^{bB}	0.076^{bB}	4.65^{bB}	1856.20^{aA}	3.53^{aA}	1.389^{dB}
Pre-5	0.036^{aA}	0.084^{aB}	5.97^{aB}	1540.71^{bA}	3.11^{bA}	1.532^{cB}
Pre-50	0.036^{aA}	0.082^{aB}	5.96^{aB}	1535.37^{bA}	2.61^{bA}	2.822^{bA}
Pre-100	0.037^{aA}	0.083^{aB}	5.71^{aB}	1511.63^{bA}	2.86^{bA}	2.531^{bB}
Pre-200	0.038^{aA}	0.075^{bB}	5.78^{aB}	1556.20^{bA}	2.76^{bA}	2.838^{aB}
Post-5	0.037^{aA}	0.074^{bB}	5.49^{aB}	1594.26^{bA}	2.73^{bA}	1.450^{cA}
Post-50	0.037^{aA}	0.071^{bB}	5.37^{aB}	1577.40^{bA}	2.61^{bA}	1.542^{cA}
Post-100	0.036^{aA}	0.071^{bB}	5.27^{aB}	1548.86^{bA}	2.77^{bA}	1.790^{cA}
Post-200	0.035^{aA}	0.070^{bB}	5.26^{aB}	1511.29^{aB}	2.87^{bA}	1.587^{cA}
200 cells tray^{-1}						
Control	0.032^{bA}	0.082^{bA}	7.15^{bA}	1248.38^{aB}	3.13^{aB}	1.590^{dA}
Pre-5	0.038^{aA}	0.092^{aA}	8.16^{aA}	906.87^{bB}	2.66^{bB}	2.005^{cA}
Pre-50	0.040^{aA}	0.091^{aA}	7.90^{aA}	1007.95^{bB}	2.64^{bA}	2.507^{bA}
Pre-100	0.041^{aA}	0.092^{aA}	8.49^{aA}	977.50^{bB}	2.84^{bA}	3.021^{aA}
Pre-200	0.035^{aA}	0.084^{bA}	6.94^{bA}	1224.15^{aB}	2.80^{bA}	3.008^{aA}
Post-5	0.036^{aA}	0.080^{bA}	6.86^{bA}	1224.15^{aB}	2.88^{bA}	1.754^{cA}
Post-50	0.036^{aA}	0.080^{bA}	6.85^{bA}	1182.04^{aB}	2.71^{bA}	1.776^{cA}
Post-100	0.034^{bA}	0.079^{bA}	7.20^{bA}	1124.39^{aB}	2.42^{bB}	1.729^{cA}
Post-200	0.033^{bA}	0.080^{bA}	6.67^{bA}	1170.28^{aB}	2.57^{bB}	1.753^{cA}

ANOVA						
Significantly	**RLAE**	**RGR**	**NAR**	**LAR**	**SLA_{FW}**	**LWR**
PCV	***	***	***	***	***	***
BAP_c	***	***	***	***	***	**
BAP_{at}	ns	**	**	**	***	**
PCV x BAP_c	***	***	***	***	***	**
PCV x BAP_{at}	**	**	**	**	***	**
PCV x BAP_c x BAP_{at}	**	**	**	**	***	**

*Significance *** 0.001 ** 0.01 'ns' No significant*

Table 3. Changes in the allometric relationships between shoots and roots for spinach plants grown at two plug cell volume (PCV) and sprayed with different BAP solutions at the pre- or post-transplant stage using a straight-line regression analysis between the natural logarithm of the roots dry weight and the natural logarithm to the shoot dry weight (ln Root dry weight = $\alpha + \beta$ x ln Shoot dry weight). Mean values (n=6) $\pm$ SE. The coefficients of determination (r^2) are indicated. Different lower-case letters indicate statistically significant differences between BAP treatments for each cell volume while different capital letters indicate statistically significant differences for each BAP treatment between different cell volumes according the test for equal slope (testing for parallelism). The probability of the slope being zero was *P* = .001

PCV	α	β	r^2
288 cells tray^{-1}			
Control	-1.77	0.69^{aA}	0.675
Pre-5	-1.97	0.58^{bA}	0.788
Pre-50	-2.25	0.55^{bA}	0.807
Pre-100	-2.17	0.60^{bA}	0.748
Pre-200	-2.33	0.59^{bA}	0.661
Post-5	-1.77	0.60^{bA}	0.782
Post-50	-2.04	0.53^{bA}	0.722
Post-100	-1.92	0.60^{bA}	0.836
Post-200	-1.76	0.61^{bA}	0.822
200 cells tray^{-1}			
Control	-1.65	0.65^{aA}	0.771
Pre-5	-1.76	0.50^{bB}	0.730
Pre-50	-1.93	0.60^{bA}	0.663
Pre-100	-2.01	0.55^{bB}	0.750
Pre-200	-1.91	0.55^{bA}	0.821
Post-5	-1.50	0.60^{bA}	0.854
Post-50	-1.70	0.56^{bA}	0.841
Post-100	-1.78	0.58^{bA}	0.841
Post-200	-1.81	0.52^{bB}	0.836

ANOVA	
Significantly	
PCV	***
BAP_c	***
BAP_{at}	**
PCV x BAP_c	**
PCV x BAP_{at}	**
PCV x BAP_c x BAP_{at}	**

*Significance *** 0.001 ** 0.01*

The productivity of most typical field grown leafy vegetables is closely associated with a high both fresh and dry weight per plant. The possibility to reach a higher plant yield was strongly restricted by the plug cell volume used, perhaps related to insufficient cytokinins supply [1,3]. Fig. 1 shows that the higher spinach fresh weight was found in control plants transplanted from 200-cell trays and those which were sprayed with a single 100 mg L^{-1} BAP at the pre-transplant stage. These results are in agreement with the fact that availability of hormones synthesized in the root apex and reallocated to shoots would be reduced when the vertical cell root growth is impeded by the container base [20].

Since there were no significant differences in dry weight content for control or treated plants (data not shown), it is possible to describe the photo-assimilate acquisition and partition rates on a dry weight base. Thus, RGR increased in plants grown in 200-cell trays and when a 5 to 100 mg L^{-1} BAP spray was applied at the pre-transplant stage (Table 2). RGR can be described as the product between NAR or 'physiological component', and LAR or 'morphological component'. In our experiment, the higher spinach RGR would be explained by an increase in NAR (Table 2) and by the close positive relationship between NAR and RGR (r^2=0.822) (Fig. 2a). An increased efficiency of carbon fixation would be associated with higher NAR values, since this variable is largely the net result of carbon gain and carbon loss [21].

Both direct additions of cytokinins [22] and over expression in transgenic plants [23] have been reported to increase leaf thickness, which may further enhance carbon assimilation per unit leaf area. On the other hand, fresh biomass per unit leaf area is well correlated to leaf thickness [24].

Table 4. Changes in leaf thickness, stomata density and stomata size in the last expanded leaf at the end of the experiments in spinach plants grown at two plug cell volume (PCV) and sprayed with different concentrations of BAP solutions at the pre- or post-transplant stage. Mean values (n=6) ± SE. Different lower-case letters indicate statistically significant differences (*P* = .05) between BAP treatments for each cell volume while different capital letters indicate statistically significant differences (*P* = .05) by Tukey's test for each BAP treatment between different cell volumes

	Leaf thickness ($\mu m\ leaf^{-1}$)	Stomata density ($stomata\ mm^{-2}$)	Stomata size ($\mu m\ stomata^{-1}$)
288 cells $tray^{-1}$			
Control	418.18^{cB}	55.73^{cB}	904.80^{aB}
Pre-5	535.15^{aB}	87.58^{aB}	723.76^{bB}
Pre-50	542.42^{aB}	79.62^{aB}	837.33^{bA}
Pre-100	538.18^{aB}	79.85^{aB}	842.00^{bA}
Pre-200	445.45^{bB}	79.78^{aB}	659.40^{cB}
Post-5	433.33^{bB}	85.80^{aB}	586.13^{cA}
Post-50	448.48^{bB}	79.83^{aB}	628.00^{cA}
Post-100	425.76^{cB}	85.54^{aB}	732.67^{bA}
Post-200	415.15^{cB}	71.66^{bB}	779.20^{bA}
200 cells $tray^{-1}$			
Control	469.70^{dA}	63.69^{cA}	1014.10^{aA}
Pre-5	562.12^{aA}	87.58^{aA}	921.07^{bA}
Pre-50	566.97^{aA}	111.46^{aA}	805.93^{cA}
Pre-100	568.18^{aA}	108.50^{aA}	853.67^{cA}
Pre-200	513.64^{bA}	87.80^{bA}	879.20^{bA}
Post-5	483.33^{cA}	99.62^{aA}	565.20^{eA}
Post-50	457.88^{cA}	87.58^{bA}	733.70^{cA}
Post-100	475.76^{cA}	95.54^{aA}	659.40^{eA}
Post-200	439.39^{dA}	88.51^{bA}	742.30^{cA}

ANOVA			
Significantly	**Leaf thickness**	**Stomata density**	**Stomata size**
PCV	***	***	***
BAP_c	***	***	***
BAP_{at}	***	***	***
PCV x BAP_c	***	***	***
PCV x BAP_{at}	***	***	***
PCV x BAP_c x BAP_{at}	***	***	***

*Significance *** 0.001*

Resistances encountered in the mesophyll from the intercellular spaces to the chloroplasts may relevantly decrease CO_2 concentration in the chloroplast and may contribute to limiting photosynthesis [25]. Aalto and Juurola [26] showed that large gradients in CO_2 concentrations can be observed in the stomatal capillary tube and in the mesophyll cells and that the concentration changes less than 1% across the airspaces of the mesophyll. This suggests that the gas phase diffusion is fast enough to provide a plentiful CO_2 supply, even in the narrowest passages between the cells [27]. Airspace effects may be larger in some species. It is profitable to have a large amount of absorbing surface because CO_2 transport is fast in air and slow in solution. The concentration decreased faster in the mesophyll cells than in the airspaces, because the diffusion coefficient is four orders of magnitude smaller in liquids than in gases. Due to the fast gas phase diffusion and equal distance from cell walls, chloroplasts in different parts of the leaf obtain almost equal quantities of CO_2 for photosynthesis. SLA and internal leaf anatomy often affect net gas exchange because of their effects on internal CO_2 conductance to the site of carboxylation, internal shading and competition for CO_2 among carboxylation sites. On the other hand, Magyar-Tábori et al. [28] have indicated that cytokinins can influence the leaf structure. Although a negative relationship between LAR and RGR (r^2= 0.761) (Fig. 2b) was found in the present study, there were statistically significant changes in the 'morphological component' of growth (LAR) through different leaf attributes such as leaf thickness, stomata density and stomata size related to cell size or BAP concentration and time of application (pre- or post-transplant) (Table 4). We also found a change in specific leaf area on a fresh weight basis (SLA_{FW}) (Table 2). A significant proportion of NAR values would be explained by the positive relationship between leaf thickness (r^2=0.694) (Fig. 3a), intercellular spaces (r^2=0.727) (Fig. 3b) and stomata density (r^2=0.495) (Fig. 3c). The BAP-mediated increase in NAR might be the consequence of different possible effects of cytokinin on the efficiency of carbon fixation. Another possibility would be that BAP could promote leaf blade thickening, thus leading to an increase in the levels of photosynthetic machinery per unit leaf area [29]. Mediavilla et al. [30] and Gandolfo et al. [31] showed that changes in leaf dry mass per unit area are caused by variations in internal anatomy and leaf tissue density, and are not simply a consequence of changes in leaf thickness. Our results are in agreement with these previous findings (Table 4, Fig. 3).

Photo-assimilates from leaves are used for root and stem growth. In addition, the size of the different plant sinks determines the photo-assimilate partitioning among the plant organs. Our results showed that a decrease in root restriction through the use of larger plug cell and an exogenous BAP application increased dry weight partitioning to shoots as indicated by both the allometric relationships between roots and shoots (Table 3) and the leaf weight rate (LWR), which measures the allocation of biomass to leaves *vs.* that to other plant parts [21] (Table 2). However, complementary information about sugar contents in different organs of the plant is needed.

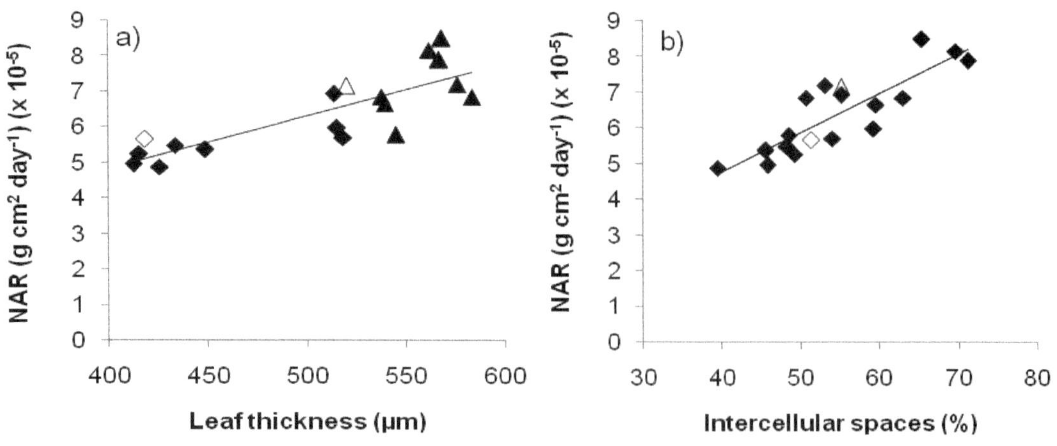

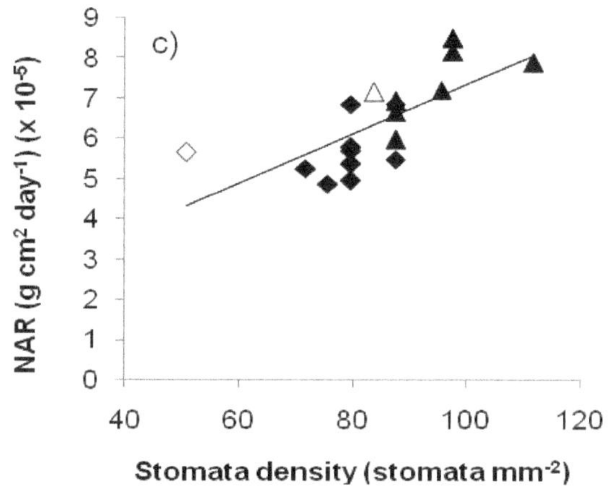

Fig. 3. The net assimilation rate (NAR) related to leaf thickness (a), Intercellular spaces (b) and stomata density (c). The straight-line regressions were: NAR = 0.015 leaf thickness -0.95 (r^2= 0.694), NAR = 0.110 intercellular spaces + 0.40 (r^2= 0.727) and NAR = 0.061 stomata density + 1.22 (r^2= 0.495) respectively. ◊: 288-grown controls; Δ: 200-grown controls

Changes in the allometric slopes reflect variations in carbon partitioning [32], and carbohydrate partitioning between competing sites is fundamental to plant growth; especially because plants are capable of modifying their resource allocation to favor the growth of their growing parts [33]. Plant tissues and organs rich in cytokinins are known to attract assimilate translocation. Ron'zhina (2004) [34] showed that the sink-promoting effect of cytokinins in detached leaves was related to the source-to-sink transformation of the portion of the mature leaf and manifests itself by changing the direction of the phloem assimilate translocation. Incorporation of cytokinin-producing bacteria into the root zone of lettuce plants has been shown to double the speed of accumulation of shoot biomass at the normal level of water supply [35].

5. CONCLUSION

In summary, spinach growth is limited by root restriction during the plug tray propagation stages mainly associated with a decrease in dry weight accumulation. The use of larger plug cells and BAP sprays at the pre-transplant stage increased plant growth (expressed as RGR or RLAE) through an increase in NAR as a result of higher leaf thickness, stomata size, stomata density and intercellular spaces and a change in dry weight partitioning toward shoots.

From a grower's point of view, the use of large plug cell volume would increase spinach growth. On the other hand, a promising approach to increasing crop productivity is the use of plant growth and development regulators such as BAP. This compound, even at very low doses, is capable of changing the intensity and direction of plant physiological processes. Because BAP is not phytotoxic, it can be used in the nursery plant industry on a large scale.

ACKNOWLEDGEMENTS

This work was supported by the University of Buenos Aires Science Program under Grant G054 and W193 and the University of Mar del Plata Science Program under AGR 317/010 and AGR 379/12.

COMPETING INTERESTS

Authors have declared that no competing interests exist.

REFERENCES

1. Araki A, Rattin J, Di Benedetto A, Mirave P. Temperature and cytokinin relationships on lettuce (*Lactuca sativa* L.) and celery (*Apium graveolens* L.) nursery growth and yield. Int. J. Agric. Res. 2007;2(8):725-730.
2. Di Benedetto A. Root restriction and post-transplant effects for bedding pot plants. In: Aquino JC. Editor. Ornamental plants: Types, cultivation and nutrition. Nova Science Publishers, Inc. NY, USA. 2011;47-79.
3. Coro M, Araki A, Rattin J, Miravé P, Di Benedetto A. Lettuce and celery responses to both BAP and PBZ related to the plug

cell volume. Amer. J. Exp. Agric. 2014;4(10):1103-1119.
4. Poorter H, Bühler J, Van Dusschoten D, Climent J, Postma JA. Pot size matters: A meta-analysis of the effects of rooting volume on plant growth. Funct. Plant Biol. 2012;39(11):839-850.
5. Di Benedetto A, Pagani A. Dry weight accumulation in the *Impatiens walleriana* pot plant in responses to different pre-transplant plug cell volume. Eur. J. Hortic. Sci. 2013;78(2):76-85.
6. Werner T, Motyka V, Laucou V, Smets R, Van Onckelen H. Cytokinin-deficient transgenic *Arabidopsis* plants show multiple developmental alterations indicating opposite functions of cytokinins in the regulation of shoot and root meristem activity. Plant Cell. 2003;15(11):2532-2550.
7. Shani E, Yanai O, Ori N. The role of hormones in shoot apical meristem function. Curr. Op. Plant Biol. 2006;9(5):484-489.
8. Kurepin LV, Emery RJN, Pharis RP, Reid DM. Uncoupling light quality from light irradiance effects in *Helianthus annuus* shoots: Putative roles for plant hormones in leaf and internode growth. J. Exp. Bot. 2007;58(8):2145-2157.
9. Kyozuka J. Control of shoot and root meristem function by cytokinin. Curr. Op. Plant Biol. 2007;10:442-446.
10. Dodd IC. Root-to shoot signalling: Assessing the roles of 'up' in the up and down world of long-distance signalling in planta. Plant Soil. 2005;274(1-2):251-270.
11. Hirose N, Takei K, Kuroha T, Kamada-Nobusada T, Hayashi H, Sakakibara H. Regulation of cytokinin biosynthesis, compartmentalization and translocation. J. Exp. Bot. 2008;59(1):75-83.
12. O'Hare TJ, Turnbull CGN. Root growth, cytokinin and shoot dormancy in lychee (*Litchi chinensis* Sonn.). Sci. Hortic-Amsterdam. 2004;102(2):257-266.
13. Styer RC, Koranski DS. Plug and Transplant Production. A grower's Guide. Ball Publishing, Batavia, Illinois, USA; 1997.
14. Potter J, Jones JW. Leaf area partitioning as an important factor in growth. Plant Physiol. 1977;59(1):10-14.
15. Motulsky H, Christopoulus A. Fitting models to biological data using lineal and nonlinear regression. A practical guide to curve fitting. Graph Pad Software Inc. San Diego, CA, USA. 2003;351.
16. Di Benedetto A, Galmarini C, Tognetti J. Changes in leaf size and in the rate of leaf production contribute to cytokinin-mediated growth promotion in *Epipremnum aureum* L. cuttings. J. Hortic. Sci. Biotechnol. 2013;88(2):179-186.
17. De Lojo J, Di Benedetto A. Biomass accumulation and leaf shape can be modulated by an exogenous BAP spray in the ornamental foliage plant *Monstera deliciosa* (Liebm.). J. Hortic. Sci. Biotechnol. 2014; 89(2):136-140.
18. Jaillais Y, Chory J. Unraveling the paradoxes of plant hormone signaling integration. Nat. Structur. Mol. Biol. 2010;17(6):642-645.
19. Francis D, Halford NG. Nutrient sensing in plant meristems. Plant Mol. Biol. 2006;60(6):981-993.
20. Di Benedetto A, Molinari J, Rattin J. The effect of transplant in sweet maize (*Zea mays* L.). II. Container root restriction. Int. J. Agric. Res. 2006;1(6):555-563.
21. Shipley B. Net assimilation rate, specific leaf area and leaf mass ratio: Which is most closely correlated with relative growth rate? A meta-analysis. Funct. Ecol. 2006;20(4):565-574.
22. Bosselaers JP. Cytokinin effects on leaf architecture in *Phaseolus vulgaris* L. J Exp. Bot. 1983;34(145):1007-1017.
23. Rupp HM, Frank M, Werner T, Strnad M, Schmülling T. Increased steady state mRNA levels of the STM and KNAT1 homeobox genes in cytokinin overproducing *Arabidopsis thaliana* indicate a role for cytokinins in the shoot apical meristem. Plant J. 1999;18(5):557-563.
24. Milla R, Reich PB, Niinemets Ü, Castro-Diaz P. Environmental and developmental controls on specific leaf area are little modified by leaf allometry. Funct. Ecol. 2008;22(4):565-576.
25. Flexas J, Barbour MM, Brendel O, Cabrera HM, Carriqui M, Diaz-Espejo A, Douthe C, Dreyer E, Ferrio JP, Gago J, Galle A, Galmes J, Kodama N, Medrano H, Niinemets Ü, Peguero-Pina JJ, Pou A, Ribas-Carbo M, Tomas M, Tosens T, Warren CR. Mesophyll diffusion conductance to CO_2: An unappreciated central player in photosynthesis. Plant Sci. 2012;193-194(5):70-84.
26. Aalto T, Juurola E. A three-dimensional model of CO_2 transport in airspaces and mesophyll cells of a silver birch leaf. Plant, Cell Environ. 2002;25(11):1399-1409.

27. Kondo T, Kajita R, Miyazaki A, Hokoyama M, Nakamura-Miura T, Mizuno S, Masuda Y, Irie K, Tanaka Y, Takada S, Kakimoto T, Sakagami Y. Stomatal density is controlled by a mesophyll-derived signaling molecule. Plant Cell Physiol. 2010;51(1):1-8.
28. Magyar-Tábori K, Dobránszki J, Teixeira da Silva JA, Bulley SM, Hudák I. The role of cytokinins in shoot organogenesis in apple. Plant Cell Tiss. Org. 2010;101(3):251-267.
29. Ferjani A, Yano S, Horiguchi G, Tsukaya H. Control of leaf morphogenesis by long- and short-distance signaling: Differentiation of leaves into sun or shade types and compensated cell enlargement. Plant Cell Monogr. 2008;10:47-62.
30. Mediavilla S, Escudero A, Heilmeier H. Internal leaf anatomy and photosynthetic resource-use efficiency: Interspecific and intraspecific comparisons. Tree Physiol. 2001;21(4):251–259.
31. Gandolfo E, De Lojo J, Gómez D, Pagani A, Molinari J, Di Benedetto A. Anatomical changes involved in the response of *Impatiens wallerana* to different pre-transplant plug cell volumes and BAP sprays. Eur. J. Hortic. Sci. 2014;79(4):226-232.
32. Kobe RK, Iyer M, Walters MB. Optimal partitioning theory revisited: Nonstructural carbohydrates dominate root mass responses to nitrogen. Ecology. 2010;91(1):166-179.
33. Price CA, Weitz JS. Zero-sum allocational strategies determine the allometry of specific leaf area. Am. J. Bot. 2010;97(11):1808-1815.
34. Ron'zhina ES. Structural and functional rearrangements of mesophyll as a probable basis for the cytokinin-dependent assimilate translocation in detached leaves. Russ. J. Plant Physl. 2004;51(3):333-341.
35. Arkhipova TN, Prinsen E, Veselov SU, Martinenko EV, Melentiev AI, Kudoyarova GR. Cytokinin producing bacteria enhance plant growth in drying soil. Plant Soil. 2007;292(1-2):305-315.

3

Adoption of Improved Soyabean Technologies in Processing and Utilization as Supplement with Local Foods in Bauchi Local Government Area, Bauchi State, Nigeria

E. F. Fabiyi[1*]

[1]*Agricultural Economics and Extension Department, College of Agricultural Sciences, Landmark University, Omu-Aran, P.M.B 1001, Kwara State, Nigeria.*

Author's contribution

The sole author designed, analyzed and interpreted and prepared the manuscript.

Editor(s):
(1) Masayuki Fujita, Department of Plant Sciences, Kagawa University, Japan.
Reviewers:
(1) O.P. Kolawole, Crop Processing Laboratory, Agricultural Engineering Department, Federal University of Technology Akure (FUTA), Nigeria.
(2) Ezekia Svotwa,Tobacco Research Board, Kutsaga Research Station, Zimbabwe.
(3) Kabi Prasad Pokhrel, Research Division, Tribhuvan University, Nepal.
(4) Ian McFarlane, School of Agriculture, Policy and Development, University of Reading, UK.
(5) Kanichiro Matsumura, Tokyo University of Agriculture Japan.

ABSTRACT

The study was carried out to determine the level of adoption of improved soyabean (*Glycine Max* (L) *Merril*) processing and utilization technologies as supplement with local food in Bauchi Local Government Area (LGA), Bauchi State, Nigeria. Bauchi LGA was purposively selected because there were many soyabean farmers in the area. Random sampling technique was used to select three villages from Bauchi LGA. Random cluster technique was used to select 15 farmers from each village totaling 45 farmers as sample size. Interview schedule was used to source relevant information from soyabean farmers on their socio-economic characteristics and level of adoption of the technologies. Twenty soyabean improved technologies were selected for supplementation with local foods, to assess the level of adoption of these technologies. The result revealed that both old and young, ranging from 20 to 70 years were involved in soyabean utilization and processing. The result also revealed that majority of the soyabean farmers were small-scale farmers ranging from

Corresponding author: E-mail: esfabiyi@yahoo.co.uk;

0.1 to 4 hectares (ha) and family size ranging from 1 to 21 members. The respondents had many years of processing experience (1-15years) but were using local methods which may not eliminate the anti-nutritional factors in soyabean. Majority of the farmers were illiterate having only Quranic education (58%), 33% had formal education at different levels while, 9% had no formal education. The regression analysis revealed that out of nine socio- economic characteristics used as predictors, only household size and labour were positively significant ($P<0.001$) and processing was negatively significant ($P<0.05$) while, other six were not significant. These technologies were poorly adopted. Three of them were fairly adopted namely soya "*awara*" (cheese) (73%), soya *daddawa* (36%) and soya *kunu* (36%). Ten innovations were poorly adopted while, 7 innovations had zero adoption. The poor adoption of these soyabean improved technologies may probably be mainly due to lack of adequate education and inadequate extension services. To increase the adoption of these technologies, extension workers need to reach out to people by creating awareness through mass media and practical demonstration of processing methods. This is necessary because soyabean is good for the poor and the rich, also to both old and the young for good health and vitality.

Keywords: Adoption; innovations; soyabean; processing; utilization.

1. NTRODUCTION

In Nigeria, soyabean *(Glycine Max (L) Merril)* was formerly used for animal feeds only to enhance the growth of the animals. It was not acceptable for human consumption because of some factors such as the problem of anti-nutritional factors and beany flavour. At Cornell University and the University of Illinois around late 1960s the workers there showed that heating could be used to reduce the action of lipoxygenase limiting the development of "off-flavours" produced in soya slurries [1]. Later researches conducted revealed that soaking and cooking can reduce the anti-nutritional factors in soyabean [2-4]. In Nigeria, commercial and home processing of soyabean for human consumption has increased in recent years, because of its health benefit and low cost of purchase. Soyabean can be used as supplement for many Nigerian foods to improve the protein content of human diets. Soyabean ranks as one of the richest source of protein especially when used in combination with cereals to compensate for its sulphur amino-acids deficiencies [5]. Soyabean is known to be free of cholesterol, has anti-cancer properties and its very rich in essential elements such as calcium, iron, zinc, phosphorus and various vitamins including vitamin A, B_6, thiamine, ribolaflavin and niacin (Table 1) [6].

Edible soyabeans proteins are rarely consumed directly, but rather they are used as ingredient in foods. These foods receive additional heat treatment either during processing or cooking before consumption [7].

A key problem associated with soyabean is that it contains some anti-nutritional factors, which inhibit the availability of the desirable elements such as proteins and minerals endowed in the grain [7,8]. According to [9], cooking improves texture and palatability quality of soyabean. Certain treatment such as cooking, baking, roasting and other common means of food preparation are effective in eliminating the anti-nutritional factors of raw soyabeans. However, whole soyabean have a good balance of the essential amino-acids and are an excellent dietary source of calories, minerals and vitamins. Eating soyabean products that are processed using the right methods (improved technologies) will reduce malnutrition in children [10] and promote good health in adults. Improved processing methods involved certain definite steps or pattern in order to obtain desired products suitable for human consumption. The objectionable flavour development has always discouraged soyabean use as human food in Nigeria. Europe and North America and other countries can tolerate this [8,11]. The processing is done to avoid problems of toxicity such as goitrogen associated with goiter formation [3,7]. The innovative methods of processing include soaking overnight (about 18-24 hours) [3,7], the soaking time may vary depending on the products to be produced. It is necessary for extension service to create awareness on importance of adopting improved soyabean processing technologies for human consumption and health benefit. Therefore, the main objective of this study is to assess the level of adoption of improved soyabean technologies in processing and utilization for human diet in the study area. Specific objectives are:

- To determine the socio-economic characteristics of the respondents on adoption of improved technologies in soyabean processing and utilization.

- To identify problems hindering the adoption of improved soyabean technologies.
- To create awareness through extension services about the usefulness of soyabean in human diets by supplementation with local foods.

Table 1. Nutritional composition of soyabean (per 100g)

Nutrient	Value
Protein (%)	40
Carbohydrates (%)	30
Fat (%)	20
Water (%)	10
Calcium	226mg
Phosphorus	554mg
Iron	8.4mg
Vitamin A	80iu
Vitamin B_1	1.1mg
Vitamin B_2	0.3mg
Niacin	2.2mg
Energy (cabries)	392

Osho et al. cited by [6]

2. MATERIALS AND METHODS

An investigation was carried out to assess the utilization of soyabean technologies as supplement with local foods in Bauchi Local Government Area (LGA) of Bauchi State, North eastern zone of Nigerian. The LGA falls within Nothern Guinea Savanah Vegetation Zone [12].

Bauchi LGA lies within the Western Agricultural Zone of Bauchi State. The western zone lies between 9°30' to 10°48'N and 8° 45' to 10° 15'E [13]. The study area has a tropical climate with long dry season period of seven months, ranging from October to April and a short wet season period of five months ranging from May to September. The diurnal temperature varies from an average daily minimum of 13.1°C and maximum of 31.6°C in January, while the relative humidity in August is 94% at 07.00. The LGA is made up of about 26 towns and villages which is the population for the study. There are no separate population figures for the towns and villages in the study area. The area has tropical soils of alluvial and colluvial origin. They are well drained and coarse textured with texture tending to the heavier sandy clay loams and clay loams for gravelly sandy loams [14]. Major crops grown include: sorghum, rice, groundnut, cowpea, millet, maize and soyabean. Also some vegetables and fruits are grown such as okro, *moringa*, water melon, pumpkin and carrots.

2.1 Sampling Technique

Bauchi LGA was purposively selected for this study because majority of the farmers there produce and process soyabean as food for human consumption. A simple random sampling technique was used to select three villages from the LGA. These included: *Bayara*, *Miri* and *Turum*. Random cluster sampling technique was used to select 15 soyabean farmers from each village. Pilot survey was carried out in order to have an insight into the question that was really necessary for the study and as part of training programme for the enumerators. This was done under close supervision. Forty five farmers were used for the study.

2.2 Data Collection

Data were collected through the use of interview schedule using trained enumerators. The interview schedule were designed to collect information relating to socio- economic characteristics of the respondents. These include: age, education level, household size, farm size (ha), processing experience, information sources and adoption of improved soyabean processing and utilization, social participation, labour source and awareness. Twenty soyabean improved technologies were selected for the study in relation to the local food of the people. These were: soya cheese (*awara*), soya *daddawa*, soya *kunu,* soya *yoghurt*, soya pap, soya vegetable soup, soya baby food, soya *chin-chin*, soya *kosssai*, soya bread, soya *tuwo*, soya *amala,* soya *moinmoin*, soya *danwake*, soya cake, soya *gari*, soya puff-puff and soya pancake. Hypothesis formulated stated that there is no significant relationship between socio- economic characteristics of the respondents and soyabean improved technologies adoption.

2.3 Data Analysis

The survey was carried out from 2002 to 2006. Data collected were analyzed using descriptive statistics (including frequencies, percentages and mean) to explain and show the extent of adoption of some selected soyabean technologies. Multiple regression analysis was also used to measure the influence of some socio- economic characteristics of respondents on adoption of soyabean technologies in processing and utilization (dependent variable) and the selected socio-economic characteristics were used as independent variables. In this study multiple regression analysis was used because there were more than one independent variable explaining the behaviour of the dependent variable.

3. RESULTS

Socio- Economic Characteristics of the Respondents

A total of 45 selected farmers, were interviewed from the study area. The respondents ages were ranging from 20 to70 years. Household size ranged from 1 to 21 people and farm sizes ranged from 0.1 - 4 hectres (ha). Majority of the farmers here were small-scale farmers. The processing experiencc ranged from 1 -15 years. This revealed that some of the farmers were aware of the usefulness of soyabean in human diet (Table 2a).

The educational status of the respondents in Bauchi LGA showed that majority have only Quranic education (58%), 22% had formal primary education, 4% had secondary, only 7% had tertiary education , while 9% had no formal education (Table 2b).

Table 3, shows the regression analysis result on the influence of socio-economic characteristics of respondents on adoption of 20 soyabean technologies in processing and utilization. The result revealed that household size, labour and processing experience were significant ($P<0.001$), ($P<0.001$) and ($P<0.05$) respectively, while six predictors were not significant (Table 3). Hence, the hypothesis that stated that there is no significant influence of socio- economic characteristics of farmers on adoption of improved soyabean technologies was rejected.

Table 4 shows level of adoption of the 20 selected soyabean technologies in the study area. Soya cheese had the highest of 73% adoption, soya *daddawa* and soya *kunu* had 36% adoption each, soya vegetable soup had 24% of adoption and 7 other innovations had zero adoption (these include soya yoghurt, soya feed, soya *kossai*, soya bread, soya *amala,* soya puff-puff and soya pancake).

Table 5 depicts percentage distribution of constraints or problems hindering the adoption of some soyabean improved technologies in processing and utilization. Inadequate market had the highest constraint (20%), lack of awareness had 11% while, inadequate equipment had 11% and no response had 20%.

4. DISCUSSION

The use of soyabean as supplement with local foods and level of adoption of improve soyabean technologies in processing and utilization was investigated. Twenty selected soyabean processing and utilization were used for the study in Bauchi LGA of Bauchi State, Nigeria. The selection of foods for supplementation was done according to the local meal preparation and diet. Some of these foods were common to the study area and others were common to different communities in Nigeria. Soyabean was added to kunu (a common local drink) to become soya *kunu* so also were other foods and drinks. This was done to improve the protein content of the foods and drinks in the study area to solve the problem of malnutrition. In Africa, serious mal-nutritional problems is especially acute in terms of protein deficiency [15]. Also prices of livestock products such as meat, milk and eggs have risen beyond the reach of the poor. Soyabeans are good sources of nutrients as they contain about 40% high quality protein, 20% oil as well as minerals and vitamins [15]. The fat in soyabean is the unsaturated type [16]. Soyabean contains 43grams(g) of protein per 100g, which is the highest among the pulses. In Nigeria very limited information is available on adoption of processing and utilization of soyabean foods; also on factors influence their adoption.

4.1 Socio-Economic Characteristics of Respondents on Adoption of Soyabean Technologies in the Study Area

In the study area men and women were involved in the utilization of soyabean, women are mostly involved in the processing. The descriptive analysis (Table 2) shows that the age of the respondents ranged between 20 to 70 years which revealed that both young and old were involved in the soyabean utilization. The household size ranged from 1 to 21 members. Probably people with large family members prefer to use soyabean to replace meat and fish which are very costly and soyabean products are cheaper. Processing experience ranged from 1 to 15 years. This revealed that some farmers have been producing and eating soyabean for sometimes, but may not know how to process it to remove the problem associated with anti-nutritional factors. Soyabean processing technology methods are different from local methods of processing soyabean. According to the statement of some selected women processing soyabean, their methods required soaking for 3 to 4 hours only, which is not enough to eliminate the anti-nutritional factors in soyabean. This revealed that the respondents were not aware of the danger of short time soaking and the anti-nutritional factors' problem. Their educational level and lack of social interaction may influence their level of understanding and lack of extension services for awareness. The processing technology recommendation was overnight soaking which may be up to 15 to 18 hours or even 24 hours depending on the type of soyabean products to be produced. Soaking time should be longer, while cooking time should be shorter. Soyabean is not normally cooked like other pulses because it takes several hours to become tender or cooked. Apart from removing the anti-nutritional factors in soyabean by boiling, boiling for 10 minutes in a solution of water and 0.05% of sodium bicarbonate [6,17] reduces the cooking time to as short as 30 minutes to tenderize the beans [7]. Also soaking the beans overnight before boiling will tenderize the beans and reduce the cooking time. Excessive heating may destroy certain amino acids like lysine and cystine in soyabean [3]. Cooking of soyabean is very necessary because it helps to destroy some of the anti-nutritional factors in soyabean. The soaking overnight reduces some of the anti-nutritional factors, while cooking treatments adequately eliminate the lipoxygenase and partially tenderize the beans [7].

Table 2b revealed that the majority of the respondents had only Quranic education (58%), 33% had formal education at various levels, while 9% had no formal education. It is reported that formal education enhances adoption of innovations [18]. Moreover in [19,20] it was also stated that the more education a person has the more favourable attitude he or she has towards adoption of new improved technologies. Also [21] reported that the more educated farmers are the higher there utilization of soyabean.

The regression analysis result revealed that household size and labour source were positively significant ($P<0.001$). This seems to be true because every member of a family is expected to take part in the family's work or business. Soyabean processing experience was negatively significant ($P<0.05$) (Table 3). This probably may be due to inadequate information about new improved processing methods. Other six predictors were not significant. There may be many factors responsible for this, such as inadequate extension services, lack of land to plant soyabean or lack of funds. All these constraints may affect adoption of these technologies (Table 5).

4.2 Extent of Adoption of Soyabeans Improved Technologies in the Study Area

Twenty soyabean technologies were selected and used to assess the level of adoption of soyabean improved technologies in the study area. Only three out of twenty were fairly adopted (soya *awara* 73%, soya *daddawa* 36 % and soya *kunu* 36%) (Table 4). The use of soyabean products are for alleviating malnutrition in children [10]. The goal of agricultural extension is to disseminate agricultural technologies for improving productivity, farmers' welfare and household nutritional status [22]. In the study area the few farmers planting soyabean, planted it mainly for sale, while other farmers processed it to various local foods.

Table 2a. Socio- economic characteristics of respondents in Bauchi LGA

Variables	Min.	Max	X	Sex	C.V %
Age (in years)	26	70	48.6	1.6	0.04
Household size	1.0	21	9.7	0.7	0.07
Farm size (ha)	0.1	4	0.8	0.2	0.14
Processing experience	1.0	15	4.0	0.5	0.11

Table 2b. Distribution of respondents by educational status in Bauchi LGA

Educational levels	Frequency	Percentage
Quaranic education	26	58
Adult education	4	9
Primary education	10	22
Secondary education	2	4
Tertiary education	3	7

n=45

Table 3. Regression analysis of adoption of improved soyabean technologies in processing and utilization with respondent's socio- economic characteristics in Bauchi LGA

Predicator	Coefficient	t- ratio
Constant	31.420	2.69
Age	0.024	0.28^{Ns}
Education level	-0.547	0.86^{Ns}
Household size	0.956	6.94^{***}
Farm size	0.042	-1.18^{Ns}
Processing experience	-0.365	-2.54^{*}
Information source	0.063	0.03^{Ns}
Labour source	0.056	4.17^{***}
Awareness	-1.010	-0.77^{Ns}
Social participation	-2.292	-0.90^{Ns}

*R^2 =78.1%; Ns = not significant; * = significant: P<0.005level; *** = significant: P<0.001 level*

In Benue State [23] reported that many farmers were aware of the usefulness of soyabean utilization with local diets. Here the adoption was very high ranging from 6.4 to 98.7% adoption from 12 soyabean innovations considered for the study. Also in another study [24] working in Samaru Zaria showed that 80% of women adopted soyabean supplementation of soya pap and soya soup. In many other countries of the world incorporation of soyabeans in diets were adopted and disseminated by extension agents in Sri Lanka and Zambia [8,17]. Amusat and Ademola reported that the study carried out in Oniyo community in Oyo State, Nigeria revealed that there was knowledge of the nutritional benefit of eating soyabean but do not know that soyabean consumption reduces chances of cancer, obesity, ulcer and other diseases [25]. This indicates that knowledge of soyabean benefits is yet inadequate [26] in most of the communities in Nigeria. This lack of knowledge of these health benefit caused low adoption or uses of soyabean as supplement. Obviously, awareness campaign about importance of soyabean production will probably increase its consumption and knowledge of its health benefits.

Table 4. Extent of adoption of the 20 selected soyabean innovation in Bauchi LGA

Soyabean innovations	Frequency	Percentage
Soya cheese (awara)	33	73
Soya daddawa (condiment)	16	36
Soya milk	8	18
Soya kunu	16	36
Soya yoghurt utilization technologies	0	0
Soya pap	1	2
Soya vegetable soup	11	24
Soya feed	0	0
Soya baby food	3	7
Soya chin-chin	2	4
Soya kossai	0	0
Soya bread	0	0
Soya tuwo	2	4
Soya amala	0	0
Soya moin-moin	1	2
Soya danwake	2	4
Soya cake	2	4
Soya gari	1	2
Soya puff-puff	0	0
Soya pancake	0	0

Table 5. Respondents problems hindering soyabean improved technologies adoption in processing and utilization in Bauchi LGA, Bauchi State

Problems	Frequency	Percentage
Inadequate market	9	20
Insufficient funds	7	16
Lack of fertilizer	5	11
Lack of awareness (extension services)	5	11
Insects attack	4	9
Inadequate processing Equipment	3	7
Insufficient rain	3	7
No response	9	20
Multiple response		n=45

5. CONCLUSION

This study revealed that lack of knowledge about soyabean processing technologies methods and its health benefits affected the adoption of soyabean improved technologies in the study area. Other factors such as educational level and socio- economic characteristics of the people also affected the adoption. On the other hand not everyone likes soyabean probably because of the beany flavor. This might have also contributed to the poor adoption of the supplementation with local foods.

Processing knowledge and extension communication programmes will increase adoption of soyabean processing technologies in the study area and in Nigeria as a whole. According to [24] the knowledge of the extent of adoption of soyabean utilization and major factors affecting the adoption could be utilized in designing an effective awareness through extension services or education programme on soyabean utilization.

Consequently, since soyabean is a very valuable crop and can be used to improve the protein content of many foods if proper processing methods are used, the constraints affecting utilization of soyabean needs to be addressed by government and extension workers. Even if farmers were illiterate, supporting farmers with practical training and demonstrations by extension workers will make things work.

COMPETING INTERESTS

Author has declared that no competing interests exist.

REFERENCES

1. Hosken, B. Advances in soybean processing and utilization. In 12th Australian soybean conference proceedings; 2003.
2. NAERLS (National Agricultural Research Liason Services). Soyabean production and utilization in Nigeria. Extension Bulletin No. 68, ABU, Zaria. 1994;26-36.
3. Enwere NJ. Foods of plant origin: Processing and utilization with recipes and technologies profiles. Afro–Orbis Publications Ltd., Nsukka; 1998.
4. Okoruwa A.E. Income of females and rural household expenditure in South-West Nigeria. African Rural Development Monograph series. Monograph No 29. 2002;1-19.
5. Odusanya, R. Powerful benefits of soyabean: Food, beverages and seasonings. Saturday Punch, January 12. 2002;33.
6. Osho O, Dashiell KE. Expanding soyabean production, processing and utilization in Nigeria In: R.S.B Ferris (ed.). Postharvest Technology and Commodity Marketing Proceedings of a Post Harvest Conference, 29 Nov.-1 Dec. 1995, Accra, Ghana (IITA Ibadan, Nigeria). 1998;151-156.
7. Osho SM. Development soyabean Technologies for small-scale and industrial levels. A manual workshop on small- scale and industrial level processing of soyabean. Ibadan, Nigeria; 1993.
8. Weingartner KE. Processing, nutrition of soyabean. In: Singh SB, Rachie KO, Dashiell KE (eds.) soyabean for the Tropics. John Weiley and Sons Ltd., New York. 1987;149-177.
9. Ihekorange AI, Ngoddy PO. Integrated food science and technology for the tropics. Macmillian Education Ltd., London and Oxford; 1985.
10. Okoruwa AE. Integrating crop production and utilization research to reduce malnutrition: soyabean production, processing and utilization reduce malnutrition in West Africa. Invested Symposium Paper for the South Africa Nutrition Congress 2000, Durban, South Africa; 2000.
11. Fukushima D. Recent Progress in Research and Technology for Processing and Utilization of Soyabeans. The Third International Soyabean Processing and Utilization Conference (ISPUC III): 2000 of the Innovative Era for Soyabean, October 15-20 Tsukuba, Ibaraki, Japan, 2000;11-16.
12. Keay RWJ. An outline of Nigeria vegetation, Government Printer, Nigeria; 1959.
13. BSADP. Bauchi State in brief. Prepared by Bauchi State Agricultural Development Project (BSADP); 2002.
14. BSADP. Development Statistics. Bauchi State Agricultural Development Projects; 1982.
15. Ugwu DS, Nwoke UM. Assessment of Soyabean products acceptability and consumption in Orumba South Local Government Area of Anambra State,

Nigeria. International Journal of Agricultural Science and Soil Science. 2011;1(8):314-325.
16. Adegoke GO, Gbadamosi R, Ewoerhurhoma F, Uzo-Peters PI, Falade, KO, Itiola O, Moody O, Skura B. Protection of maize and soyabeans using Aframomum danielli. Journal of European Food Research and Technology. 2002;214:408-411.
17. Nelson AI, Wei LS, Weingartner KE. Home and village preparation of soya products:In: soyabean in the tropics. Singh SB, Rachie KO, Dashiell KE. (eds.). 1987;179-184.
18. Ogunbameru BO, Pandey IM. Nigerian rural women participation in agricultural and decision-making. Focus on Adamawa and Taraba States, Nigeria a wake up call. Ed. Hodges, UNICEF, Nigeria. 1992;133-184.
19. Egbugara EA. Economic status and rural women's involvement in agriculture in Imo State. Journal of Rural Extension and Development. 1993;1:87-93.
20. Agbamu JU. Analysis of farmer's characteristics associated with adoption of soil management innovation in Ikorodu LGA of Lagos State. Nigeria Journal of Agricultural Extension.1993;7:57-66
21. Ajao O, Ogunniyi LT, Adepoju AA. Economic Efficiency of Soyabean Production in Ogo Oluwa Local Government Area of Oyo State, Nigeria. American Journal of Experimental Agriculture. 2012;2(4):667-679.
22. Savile AH. Extension in rural communities. A manual for agricultural and home extension workers. Oxford; 1978.
23. Sanginga PC, Adesina AA. Manyyong VM, Otite O, Dashiell KE. Social impact of soyabean in Nigeria's Southern Guinea Savannah, IITA, Ibadan, Nigeria; 1999.
24. Atala TK, Ajia TD, Olukosi JO. Adoption of soyabean utilization innovations among women in Samaru village of Sabon-Gari Local Government Area of Kaduna State, Nigeria. Agricultural System in Africa. 1992;2:18-24.
25. Amusat AS, Ademola AO. Utilization of Soyabean in Oniyo Community of Oyo State, Nigeria. Global Journal of Science Frontier Research Agriculture and Veterinary. 2013;13(7):6-14.
26. Olatunji SO, Etuk UR, Adesope OM. Factors Related to Continuity in Utilization of Soya Bean Products by Farm-Families in Abia State, Nigeria. Journal of Agricultural Science. 2012;3(1):15-20.

4

Contamination of Lettuce Plants Irrigated with Waste Water in Yaounde, Cameroon

Valerie Njitat Tsama[1], Godwill Mih Chewachong[2], Ives Magloire Kengne Noumsi[1], Wilfried Arsene Letah Nzouebet[1], Nkeze Nyochembeng[1] and Zachée Ambang[2*]

[1]Department of Plant Biology (Waste Water Research Unit), Faculty of Science, University of Yaounde I, P.O.Box: 8250 Yaounde, Cameroon.
[2]Department of Plant Biology (Phytopathology Research Unit), Faculty of Science, University of Yaoundé I, P.O.Box 812 Yaoundé Cameroon.

Authors' contributions

This work was carried out in collaboration between all authors. Author VNT designed the study, wrote the protocol and wrote the first draft of the manuscript. Author GMC reviewed the experimental design and all draft of the manuscript; author IMKN reviewed the experimental design and all drafts of the manuscript. Authors WALN and NN managed the analyses of the study. Author ZA identified the plants, designed the study and protocol, and reviewed all drafts. All authors read and approved the final manuscript.

Editor(s):
(1) Daniele De Wrachien, Full Professor of Irrigation and Drainage State University of Milan, Italy.
Reviewers:
(1) Anonymous, Kenya.
(2) Anonymous, Pakistan.

ABSTRACT

Aims: To evaluate the effects of various treatment combinations of water contaminated with faecal sludge on the growth and microbiological characteristics of lettuce (*Lactuca sativa* L.) cultivated around the river Avo'o discharged zone.
Study Design: We used randomized plots in which young lettuce plants obtained from a nursery were planted in four sites (S0, S1, S2 and S3) situated 3000, 810, 100 and 350 m from the raw discharge area along river Avo'o.
Place and Duration: Field studies were carried out in Nomayos (Yaounde) during the 2009-2010 growing seasons while laboratory analysis were effected at the waste water research unit of the Department of Plant Biology at the University of Yaounde I.

**Corresponding author: E-mail: zachambang@yahoo.fr;*

Methodology: Faecal streptococci, faecal coliforms and parasitic characteristics were determined in raw discharged sludge from river Avo'o water samples collected at three different points (P1, P2, P3) and in lettuce using standard protocols. Growth parameters of the lettuce plants were also evaluated.

Results: Faecal sludge, water from river Avo'o and *L. sativa* showed high levels of bacteriological and parasitological pollutants. The results also revealed high concentrations of faecal coliforms and faecal streptococci (6092 and 3390 CFU/100 ml respectively), and helminth eggs (> 273 eggs / L) in water. Treatment 3 (S3) showed the largest number of leaves with an average of 34.63 leaves in the ninth week for site 1 and 40.33 leaves for site 2. Plant height increased gradually with time and treatments. The mean values of leaf surfaces obtained after nine weeks were 11.5 cm2 and 13.99 cm2 for treatment S3 in sites 1 and 2 respectively.

Conclusion: Though the best growth of lettuce was observed in plots irrigated with water from river Avo'o, the high levels of parasite concentrations on their leaves (43 eggs/100 g) highlight the necessity for a better management of faecal sludge in Yaounde city.

Keywords: River Avo'o; faecal sludge; irrigation; growth of lettuce; microbial pollutants; Cameroon.

1. INTRODUCTION

Excreta management especially around dumping sites in urban areas of most developing countries do not respect standard hygiene and health norms. Thousands of tons of sludge from onsite sanitation systems such as traditional latrines, public toilets, and septic tanks are disposed of without prior treatment in different parts of the world every day [1]. During the last decade, an increasing number of countries explored the possibility of municipal wastewater and sludge reuse [2,3,4]. Municipal sludge utilization and the use of wastewater for irrigation is becoming widely practiced not only in countries with water deficits, but also in countries with more temperate climates. The methods and extent of reuse however varies with respect to the local infrastructure and circumstances prevailing in each country. Faced with rapidly growing populations, cities in developing countries are subjected to food insecurity and high levels of unemployment. Urban agriculture, a booming business could be a viable approach to reducing unemployment and ensuring food security, at least, for those in the cities. In effect, many people have taken refuge in the production of vegetables, especially in lowlands and swamps which are floodplains with permanent moisture, rich in organic matter [5,6,7]. Several studies have shown the effect of wastewaters on the growth and productivity of plants in various countries around the world [5,8,9,10,11,12]. A study focused on the evaluation of the biomass of several species of *Eucalyptus camaldulensis, Acacia cyanophyllus* and *Popolus nigra* grown on farmlands irrigated with wastewater revealed that *Popolus nigra* plants had the best growth in height while *Acacia cyanophyllus* manifested the highest growth in diameter [13]. In addition, *Acacia cyanophyllus* produced the greatest biomass followed by *Arundo donax*, while biomass generation was low for *Eucalyptus sp.* and lowest in *Populus nigra*. Nawaz et al. [12] assessed the amounts of heavy metals in irrigation wastewater, soils and wheat plants irrigated with contaminated wastewater from municipal sources in Abbottaba in Pakistan and reported levels of metals that were generally above the threshold limits for irrigation water and food as per international regulations. In one other study, the feasibility of using wastewater and sludge from secondary sewage treatment plants to irrigate forest plant species of *Pinus brutia* (Greek origin) and *Pinus maritime* was tested [14]. The present research was conducted in Nomayos, a peri-urban area of Yaounde, Cameroon. The aim of the study was to evaluate the effects of various treatment combinations of water contaminated with faecal sludge on the growth and microbiological characteristics of lettuce (*Lactuca sativa* L.) cultivated around the river Avo'o discharged zone by using ordinary water and water contaminated with faecal sludge. The objectives were: (a) to characterize different kinds of water used for the irrigation of *L. sativa* in the study area, (b) to evaluate the growth parameters of the plant, and (c) to evaluate the sanitary risks associated with lettuce produced by farmers in this area.

2. MATERIALS AND METHODS

2.1 Site Description

The study site is located in Nomayos, a locality situated 20 km away from Yaounde, Cameroon. The zone serves as an indiscriminate dumping site for sludge collected by vacuum trucks from onsite sanitations in Yaounde (Fig. 1). Sludge is discharged daily into the few wells available (Fig. 1a) or directly on the soils surface (Fig. 1b). The sludge deposited then leaches into rivers near the area of discharge where farmers coincidentally collect water for irrigation of their crops (mostly vegetables).

The work was carried out along the borders of two tributaries of river Avo'o. The site has a surface area of about 300 m^2 and receives approximately 1350 m^3 of untreated faecal sludge per week.

2.2 Methodology

2.2.1 Microbiological tests

Samples of faecal sludge water from river Avo'o and lettuce irrigated with water from this river were collected for different microbiological and parasitological analysis. For this, four (04) points (S0: control site located at Mbankomo, 3 km away from the faecal sludge discharge area; S1: site 1 located 810 m before the faecal sludge discharge area; S2: site 2 located 100 m near the faecal sludge discharge area and S3: site 3 located 350 m after the faecal sludge discharge area) were mapped out for sampling. The microbiological analysis involved mostly the determination of the presence of faecal streptococci and coliforms using the membrane filtration protocol described by [15]. Parasitological analysis to determine the presence of helminth eggs were carried out using the protocol described by Bailenger and modified by [15].

2.2.2 Evaluation of lettuce growth parameters

The lettuce plants used for this study were obtained from a nearby nursery and transplanted at the different sites. Plants were watered twice daily (morning and evening) by sprinkling. Plant height, number of fresh leaves as well as leaf length and width of lettuce plants in the different sites mentioned above were evaluated. Plant height was measured using a double decameter; the number of fresh leaves and number of dead plants were determined by counting while the length of leaves was measured with a caliper.

2.2.3 Data analysis

The ANOVA test was conducted to determine the effects of different treatments applied on the bacteriological, parasitological and morphological parameters of *Lactuca sativa* L. Specifically, a two way ANOVA was used to determine any correlations and interactions between the different sites (S0, S1, S2 and S3) and the different growth parameters. On the other hand, a one way ANOVA was used to determine the means of the different microbiological and parasitological parameters.

3. RESULTS AND DISCUSSION

3.1 Microbiological Characteristics of Water Used for Irrigation

Analyses of sludge water from river Avo'o and lettuce plants from the study sites revealed the presence of indicators of faecal pollution: faecal coliforms and faecal streptococci (Table 1). Mean values for these bacterial indicators in faecal sludge were 22.86 x 10^5 CFU/100 ml and 10.68 x 10^5 CFU/100 ml respectively for faecal coliforms and faecal streptococci thus indicating a significant difference (P< .05).

When water samples collected at different points (S1, S2, S3) were compared to the control S0 (1112 and 264.20 CFU/100 ml), significant differences (P< .05) were observed between the amount of faecal coliforms and faecal streptococci (Table 1). Analysis of the amount of faecal coliforms and faecal streptococci on the leaves of *L. sativa* revealed significant differences between samples from S2 and S3 when compared to those of the control (S0) as exemplified in (Table 1). In addition, a very strong Pearson's linear correlation was observed between the levels of faecal pollution indicators in the sludge water from river Avo'o and leaves of *L. sativa* (Table 1).

The data obtained reveals very high levels of faecal pathogens in sludge suggesting high levels of pollutants in the raw sludge. Similar results were obtained by [16] working on factors affecting the watering constructed wetlands using faecal sludge in tropical regions. The bacteriological characteristics of water from river Avo'o contaminated with drainage from the Nomayos sludge discharge area revealed pollution levels that were above standard restrictive values recommended for irrigation waters used in the cultivation of edible crops (< 1000 CFU for faecal bacteria and 1 helminth ova/100 g of *Lactuca sativa*) [15]. These values reflect the impact of sludge on environmental pollution. Similar observations were made by [16] who noted that the uncontrolled discharge of

Table 1. Microbiological characteristics of samples (n= 33 for each sample)

Site	CF (CFU/100 ml)		SF(CFU/100 ml)	
	Water samples	***L. sativa***	**Water samples**	***L. sativa***
Faecal sludge*	$22.86 \times 10^5 \pm 25.47 \times 10^2$		$10.68 \times 10^5 \pm 16.46 \times 10^2$	
S0	1112.00±1013.25	10.33±1.40	264.20 ±189.52	4.67±5.65
S1	1400.00±116044	13.03±8.7	370.00 ±132.06	4.63±3.01
S2	3200.00±2280.77	361.66±295	1480.00±1200.00	138.33±12.76
S3	4580.00±2902.06	583.33±45.75	1540.00 ±1420.56	250±23.5

S0: control, S1: before discharge point, S2: discharge point, S3: after discharge point
**The number of germs for lettuce is expressed in CFU/100 g of leaves*

Fig. 1. Partial view of two types of faecal sludge disposal systems at Nomayos, Yaounde. (a) discharge into a well, (b) direct discharge on the ground

excreta in the environment does not respect the principles of ecological and health care especially in developing countries. In addition, these authors described disease (amibiasis, ankyloctomiasis, cholera, and typhoid fever) transmission routes: direct contact with faeces during swimming and the consumption of food, especially salads prepared with contaminated water. Waterborne diseases like typhoid fever and diarrhea reportedly affects 32 and 43% respectively of the Cameroonian population [20].

Wastewater contaminated with excreta can be used in agriculture if such water is treated to avoid contamination since municipal sludge and wastewaters reportedly contains pathogens like bacteria, protozoa, viruses, and helminth eggs [17]. The contamination of lettuce reported in this study is favoured by the irrigation method, mainly sprinkling. However, most farmers use wastewater to irrigate their crops by spraying with watering cans. Several authors have studied the health risk associated with consuming wastewater-irrigated food crops and vegetables [14,17,18,19]. Consumption of such wastewater-irrigated food without respect of proper hygiene can result in waterborne diseases.

3.2 Parasitological Characteristics

Mean numbers of helminth eggs in raw faecal sludge were estimated at 3601 eggs/L. In the control site (S0) river Avo'o and the leaves of *Lactuca sativa* irrigated with, the mean number of parasites were 285 eggs/L and 43 eggs/100 g, respectively (Table 2). These parasite concentrations show that faecal sludge is highly polluted and influences the parasitological characteristics of water from river Avo'o and consequently lettuce irrigated with such water. Based on the concentrations of helminth eggs we obtained, the sludge analyzed can be ranked as category a sludge type following the classification of [21]. The high levels of pathogens in irrigation water obtained in this study indicates a possible public health risk that should be prevented. The reuse of wastewater for irrigation in different agricultural systems is associated with different risk factors. Some risk factors are short term and vary in severity depending on the potential for human, animal or microbial pathogens to have contact with the environment [22,9]. It is worth mentioning that most common human microbial pathogens found in wastewater are of enteric origin. Uncontrolled use of wastewater in agriculture has significant health implications for farmers and their families, vendors and producer communities as well as consumers in wastewater irrigated areas [23].

Ascaris lumbricoides, *Enterobius* sp., *Trichuris trichura*, *Tenia* sp., and *Schistosoma* sp. Were the different species of parasites identified? observed at different sampling points (Table 2). Heterogeneity was observed in the type and number of parasite species recorded at the sampling sites. For example, S0 sample site registered only 02 parasite species: *Ascaris lumbricoides (*28.33 eggs/L) and *Trichuris trichura* (3.33 eggs /L). *Ascaris lumbricoides* was the most represented species with a frequency of 84.16%. Sites S0 and samples of wastewater from river Avo'o revealed the presence of 05 species of parasites: *Ascaris lumbricoides*, *Enterobius* sp., *Hookworm* sp., *Trichuris trichura, and Schistosoma* sp. Average numbers of eggs per liter of samples were 540, 123.33, 141.66, 62.25 and 13.26 respectively for each parasite. The presence of parasites in the water samples collected from river Avo'o can be explained by its proximity to the indiscriminate sludge discharge site. This has a direct influence on the quality of water from the site.

3.3 Evolution of Growth Parameters

Plant growth is expressed in terms of number of fresh leaves, height of plant and increase in leaf area with time irrespective of the different treatments applied (Fig. 2). There was a significant difference (P=.001) with Student test in the number of fresh plant leaves, height of plant, length and width of leaves between treatments. After nine weeks of treatment, an increase in the growth parameters studied was recorded. Such increases were proportional to the treatment administered from S0 to S3. The use of wastewater resulted in greater number of plant leaves, plant height and leaf area. Treatment 3 (S3) located after the faecal sludge discharge site showed the largest number of leaves with an average of 34.63 leaves for site 1 and 40.33 leaves for site 2 in the ninth week. A gradual increase in plant height was also observed following an increase in time and the different treatments for all investigation sites. The mean values obtained in the ninth week were 11.5 cm^2 and 13.99 cm^2 for treatment S3 on sites 1 and 2 respectively. Site 2 yielded plants with the largest leaf surfaces. From the first to the fifth week, growth was similarly irrespective of site or treatment. The rapid increase in these parameters could be explained by the presence of excess nutrients found in faecal sludge, all of which are mineralized and rendered bio-available by bacteria present in the soil [8,24,25]. These results corroborate with those obtained by [26,27] which showed that there exist a relationship between an increase in nutrient concentration in wastewater and absorption by plants for growth. Similar observations were made by [14,28,29] while working on the assessment of sewage effluent quality from a sewage treatment plant. The death of plants observed in sites S2 and S3 were about the same, but significantly higher ($P< .001$) than those for S0 and S1 (Fig. 2d). These results could be explained by the fact that S0 is completely located out of the discharge zone while S1 is found upstream and receive very limited amounts of the pollutants poured into river Avo'o. On the other hand, S2 and S3 are found downstream. A previous study by [20] suggested that higher amounts of pollutants in waste water could accelerate the death of plants.

Table 2. Mean number of helminth eggs in different samples n= 33 for of each sample

Parasites	Control (eggs/L)	River Avo'o (eggs/L)	***Lactuca sativa* L. (eggs/100 g)**
Ascaris L.	28.33±10.30	540±221	33±5.61
Enterobius sp.	0	123.33±75.5 9	3.7±1.07
Hookworm sp.	0	141.66±86	4.3±2.02
Trichuris trichura	3.33±11.5	62.25±25	2±0.75
Schistosoma sp.	0	13.26±420	0
Mean number	6.32±4.36	178.10±82.36	8.20±1.89

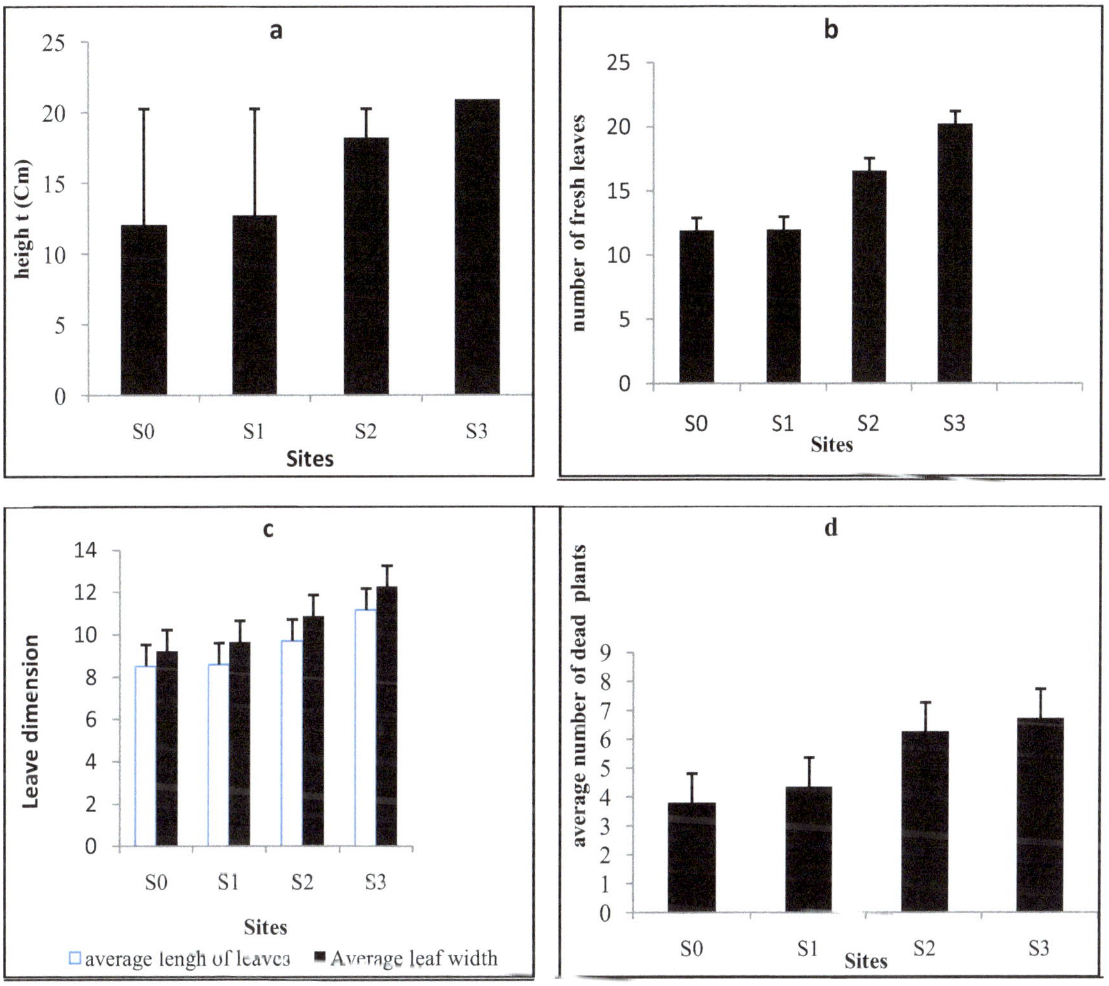

Fig. 2. Evolution of growth parameters of *Lactuca sativa* L. (a: height of plant; b: number of fresh leaves; c: average length and average width of leaves, d: average number of dead plants)

4. CONCLUSION

Most pathogens that affect humans are derived from faeces and transmitted by direct contact during swimming for example or orally through the food we eat and the water we drink. The reuse of wastewater contaminated with faecal sludge for irrigation of crops led to higher crop (*Lactuca sativa* L.) growth rates but was unfortunately associated with high concentrations of faecal coliforms, faecal streptococci and parasites on their leaves. Such high levels of pathogenic micro-organisms highlight an urgent need to better manage faecal sludge in this city. This is important in reducing the associated environmental and health hazards. Based on our results, it is recommended that the government reinforces already existing laws on the treatment of faecal sludge before discharge.

COMPETING INTERESTS

Authors have declared that no competing interests exist.

REFERENCES

1. Niyitegeka D. Bioindicators and bacterial pathogens in Mingoa water and the Yaounde municipal lake: Site conditions, population of pathogens, spatial repartition and temporal fluctuations. PhD Thesis, University of Yaounde I, Cameroon; 2001.
2. Mara D, Sleigh PA, Blumenthal UJ, Carr RM. Health risks in wastewater irrigation: Comparing estimates from quantitative epidemiological studies. J. Water Health. 2007;05(1):39-50.

3. Shuval H, Lampert Y, Fattal B. Development of a risk assessment approach for evaluating wastewater reuses standards for agriculture. Water Sci. Technol. 1997;35(11-12):15-20.
4. Djuikom E, Jugnia LB, Nola M, Foto S, Sikati V. Physicochemical water quality of the Mfoundi River watershed at Yaounde, Cameroon, and its relevance to the distribution of bacterial indicators of faecal contamination. Wat. Sci. Tech. 2009;60:11-18.
5. Ingallinella AM, Sanguinetti G, Koottatep T, Montangero A, Strauss M. The challenge of faecal sludge managem urban areas-strategies regulations and treatment options. Water science and Technology. 2002;46(1):285-294.
6. Mbaye A, Mouster P. Market-Oriented Urban Agricultural Production in Dakar. In: Bukker N. editor. Growing Cities, Growing Food. DSE Feldafing Publishing, Germany; 2000.
7. Strauss M, Larmie SA, Heinss U. Treatment of sludge from onsite sanitation: low-cost options. Water Science and Technology. 1997;35(6):129-137.
8. Silvana AC, Rope PP, Jéferson GS, Célia RM, Gilberto S, Fabio C, Tamara MG, Adolpho JM. Parasitological risk assessment from wastewater reuse for disposal in soil in developing countries. Water Science and Technology. 2012;65(8):6-12.
9. Fouepe Takoundjou A, Kuitcha D, Sigha Nkamdjou L, Ntonga JC. Impact of the vulnerability of surface water on the access to potable water to Yaounde residents: Strategies for adaptation. Presentation at the 6th edition of enterprise day, 2IE, Ouagadougou; 2011.
10. Muhammed I, Zobia A, Faridullah, Arshid P, Amir W. Water use efficiency and accumulation of trace elements in spinach (*Spinacia oleraxea* L.) irrigated with municipal waste water. Journal of Medicinal Plant Research. 2012;6(22):3882-3888.
11. Zahid U, Hizibullah K, Amir W, Qaisar M, Umar F. Water quality assessment of river Kabul at Peshawar, Pakistan: Industrial and urban wastewater impacts. Journal of Water Chemistry and Technology. 2013;35(4):170-176.
12. Nawaz U, Qaisar M, Amir W, Muhammad I, Faridullah Arshad. Assessment of heavy metals in wheat plants irrigated with contaminated wastewater. Polish Journal of Environmental Studies. 2013;23(1):115–123.
13. Assongmo T. Occupation and improvement of wetlands in Yaounde: the case of the Olezoa *pond* (Yaounde). MSc. Thesis University of Yaounde 1; 1990.
14. Kalavrouziotis IK, Koukoulakis PH, Sakelariou-Makrantonaki M, Papanikolaou C. Effects of treated municipal wastewater on the essential nutrient interactions in the plant of *Brassica oleracea* var. Italica. Desalination. 2009;(242):297-312.
15. Seidu R, Heistad A, Amoah P, Drechsel P, Jensen PD, Stenström TA. Quantification of the health risk associated with wastewater reuse in Accra, Ghana: A contribution toward local guidelines. Journal of Water and Health. 2006;6(4):461-471.
16. WHO 2004. World health report 2004: changing history. Burden of disease in dalys by cause, sex and mortality stratum in who regions, estimates for 2002. Genava: world health organization. Accessed date 27 august 2011. Available:http://www.who.int/whr/2004/en/(
17. Kengne IM. Potentials of sludge drying beds vegetated with *Cyperus papyrus* L. and *Echinochloa pyramidalis* (Lam.) Hitchc & Chase for faecal sludge treatment in tropical region. Ph.D Thesis, Univ. Yaounde 1 ; 2008.
18. Petterson SR, Ashbolt N, Sharma A. Microbial risks from wastewater irrigation of salad crops: A screening-level risk assessment". Water Environ. Res. 2001;72:667-672.
19. Hass CN, Rose JB, Gerba CP. Quantitative Microbial Risk Assessment. John Wiley and Sons, New York; 1999.
20. Vymazal J. Types of constructed wetlands for wastewater treatment: Their potential for nutrient removal. In: Backuys Publishers. Transformation of Nutrients in Natural and Constructed Wetlands. Leiden, The Netherlands; 2001.
21. Kuitcha D, Ndjama J, Tita Awah M, Lienou G. Kamgang V, Kabeyene Beyala H, Ateba B, Ekodeck GE. Bacterial contamination of water points of the upper Mfoundi watershed, Yaoundé, Cameroon. Afr. J. Microbiol Res. 2010;4(7):568-574.
22. Delville PL, Boucher L. Wetlands in humid regions of tropical Africa. A guide for diagnostic and intervention. Ed. Ministry of cooperation-CTA. Paris; 1996.

23. Mara D, Cairncross S. Guide lines for the safe use of wastewater and excreta in agriculture and aquaculture. WHO, Switzerland; 1989.
24. Rodier J. Water analysis; natural water, residual water, sea water. 9th edition. Dunod technique, Paris; 2009.
25. Kramkimel JD, Grifono U, Kabeya Mukenyi R. Environmental profile of Cameroon. Provisional report. Ed. Ministry of Environment, Yaounde; 2004.
26. Tanaka H, Asono T, Schroeder ED, Tchobanoglous G. Estimating the safety of wastewater reclamation and reuse using enteric virus monitoring data. Water Environ. Res. 1998;(70):39-51.
27. Tripathi BD, Ssrivastava J, Misra K. Nitrogen and phosphorus removal capacity of four chosen aquatic macrophytes in tropical freshwater ponds. Environ. Conserv. 1991;18:143-147.
28. Jimenez B, Takasshi A. Water reclamation and reuse around the world. In: Jimenez B, Takashi A, editors. Water Reuse: an International Survey of current practice issues and needs. IWA Publishing; 2008.
29. Carr R. Who guidelines for safe wastewater use-more than just numbers". Irrig Drain. 2005;54:103-111.

5

Genetic Characterization of Alupe Napier Grass Accessions Based on Simple Sequence Repeat Markers

Geofrey Kawube[1,2*], Jolly Kabirizi[3] and Clementine Namazzi[3]

[1]National Crops Resources Research Institute, National Agriculture Research Organisation, P.O.Box 7084, Kampala, Uganda.
[2]Faculty of Agriculture and Environment, Gulu University, P.O.Box, 166, Gulu, Uganda.
[3]National Livestock Resources Research Institute, National Agriculture Research Organisation, P.O.Box 96, Tororo, Uganda.

Authors' contributions

This work was carried out in collaboration between all authors. Author GK designed field study, conducted laboratory and data analysis, participated in the writing and revision of the manuscript. Author JK sourced the Napier grass clones, participated in design and planting field experiment and revision of the manuscript. Author CN participated in planting field experiment, collected leaf samples and participated in revision of the manuscript. All authors read and approved the final manuscript.

Editor(s):
(1) Lixiang Cao, Department of Biotechnology, Sun Yat-sen University, Guangzhou, P. R. China.
Reviewers:
(1) Anonymous, India.
(2) Anonymous, Tunisia.

ABSTRACT

Genetic variability of 23 Napier grass clones, 22 of which were collected from Kenya Agricultural Research Institute – at Alupe in Kenya was determined using 17 simple sequence repeat (SSR) markers. All markers were polymorphic with the most discriminative being PSMP2267. Average gene diversity among the Napier grass clones was 0.54 with an average heterozygosity of 0.63. Total number of alleles across all loci was 90 with mean number of alleles per locus of 5.29 and a mean polymorphic information content of 0.50. Results also indicate a high genetic distance among the Napier grass clones with the most distant clone being 16814 while the closest was between kakamega1 and kakamega2 and kakamega2 and kakamega3. Principal Co-ordinates Analysis did not group the clones in a definitive structure, with most clones scattered. However, the SSR

Corresponding author: E-mail: kawgeoff@gmail.com

markers grouped the clones into two clusters derived from a dendogram generated by Neighbour joining analysis with clone 16814 being alone in a cluster. Clone Kakamega1, Kakamega2 and Kakamega3 as very closely related. These findings are key in guiding selection of clones for inclusion in breeding programs and conservation of Napier grass.

Keywords: Genetic diversity; genetic relationships; Pennisetum pupureum; population structure-simple sequence repeats transferability.

1. INTRODUCTION

Napier grass (*Pennisetum purpureum*), also known as elephant grass is a robust perennial forage indigenous to sub-Saharan Africa [1]. The grass is dominant in the fertile crescent along north of Lake Victoria and the western Rift Valley in Uganda [2]. Currently, Napier grass is the principal fodder crop in smallholder intensive and semi intensive livestock production systems in East Africa [3], constituting 40 – 80% of forages used to meet the increasing demand for milk. The demand for Napier grass is growing, mostly among poor households in densely populated areas due to its desirable traits such as tolerance to drought, ability to grow in a wide range of soil conditions, high photosynthetic and water-use efficiency [4]. The grass can also withstand repeated cutting with rapid regeneration, producing a high yield that is very palatable to cattle in the leafy stage [1].

Napier grass productivity in the East African region is limited by several factors especially the emerging new diseases like Napier Grass Stunt Disease and Napier Grass Head Smut Disease, thus constraining the growth of smallholder dairy industry [5]. Therefore, continued utilization of Napier grass as a fodder will depend on exploitation of the genetic variability within and among its populations [6] in search for resistance to these production constraints. This requires a well characterized and inventoried germplasm; which is lacking in the case of Napier grass in East African countries including Uganda [7]. In East Africa Kenya Agricultural Research Institute – Dairy Research Centre at Alupe maintains a collection of *Pennisetum purpureum*; obtained from within Kenya and the International Livestock Research Institute, in Ethiopia whose genetic diversity is not known.

Various methods for estimating diversity in a plant population exist and use of simple sequence repeats (SSRs) has become the method of choice because of the markers multi-allelism, genome specificity, even distribution and high polymorphism. However, the genome of Napier grass has not been sequenced, therefore, Napier grass SSR markers are not known. Besides, Napier grass is a tetraploid (2n = 4x = 28) with triploid and hexaploid hybrids occurring between it and pearl millet [8]. This makes establishing microsatellites that adequately discriminate the different ploidy levels difficult. The available option is through cross-amplification using SSR markers of closely related species [9]. This study, therefore, determined the genetic variability among Napier grass clones maintained at Kenya Agricultural Research Institute – Alupe at station through cross- amplification using SSR markers of closely related organisms.

2. MATERIALS AND METHODS

2.1 Sample Collection and Analysis

Twenty two Napier grass clones (19, 41, 75, 76, 79, 97, 103, 104, 105, 112, 117, 16702, 16789, 16805, 16814, 16815, 79SN, ANF, Kakamega1, Kakamega2, kakamega3 and RBN) obtained from Kenya Agricultural Research Institute – Dairy research Center Alupe and one clone - 16785 obtained from International Livestock Research Institute – Ethiopia were planted in the field at National Crops Resources Research Institute at Namulonge in Uganda in 6 replicates. Two months after planting, samples were collected from the inner most unfolded leaf on one tiller of each plant, placed in a paper bag and silica gels and packed in a box. These were transferred to Bioscience Eastern and Central Africa at International Livestock Research Institute (BecA-ILRI) - Nairobi for genotyping.

In the laboratory, 1.5 g of a leaf was extracted from each leaf sample and ground in mortar in liquid Nitrogen. Total plant DNA was extracted using cetyltrimethylammonium bromide (CTAB) method [10] and diluted to 100 µ using double distilled water. The DNA concentration was determined using Nanodrop UV spectrometry at A260 and A280 while the integrity of DNA was tested on 1.2% agarose gel electrophoresis in TBE buffer stained with gel red. From these, template DNA was made from an aliquot in a 1.5 ml tube and diluted to 50 ng/µ.

2.2 DNA Amplification with Microsatellite Markers

A total of 17 simple sequence repeat microsatellite primer pairs, originally identified in maize, pearl millet and sorghum were conjugated with different dyes (VIC, NED, PET and 6-FAM). These were used in the PCR amplification in 20 µl *AccuPower® Taq* Premix (Bioneer) to which 17 µl of water and 0.5 µl of 5 picomoles of each of the primer pair and 2 µl of template DNA were added. The reaction mixture was subjected to the following PCR conditions: An initial denaturation of 94ºC (3 min) followed by 35 cycles of 94ºC (30 sec); specific primer annealing temperature (1 min) (Table 1); extension at 72ºC (2 min), final extension at 72ºC (10 min) and final hold at 4ºC. The PCR products were run on 1.2% agarose gel electrophoresis stained with gel red in 0.5X TBE buffer at 80 V for 50 minutes and visualized on trans UV and photographed in UVP DIGIDOC – IT system (UVP Biolmaging systems, USA). The PCR products with clear single band amplification on the agarose gel were subjected to capillary electrophoresis with ABI3730 DNA genetic analyser for fragment analysis and allele calls were made using GENEMAPPER software v.3.7 (Applied Biosystems). Primers whose PCR products generated high quality electropherogram peaks of fluorescent intensity above 50 at differing positions in the samples were selected (Table 1) and used for amplification of all the samples.

2.3 Data Analysis

Microsatellite allele distribution data obtained from Genscan®software Version 4.1 were converted into suitable formats for statistical analysis. Allelic size data for each SSR locus was used to estimate percentage of polymorphic loci, Shannon's information index (I), Nei's gene diversity, observed (H_o) and expected (H_e) heterozygosities using Power Marker version 3.25 [11]. Cluster analysis was performed based on Nei's distance matrix using GenALEX6.2 [12].

3. RESULTS

The SSR markers used in this study generated 90 alleles in 23 Napier grass accessions. The number of alleles detected for each primer pair ranged from 2 (CTM10, CTM59, PSMP2235, PGIRD25) to 13 (CTM8) with average of 5.29 (Table 2). The frequency of the major alleles in each marker locus ranged from 0.23 (PSMP2267) to 0.98 (PGIRD25). Polymorphic Information Content for the assayed marker loci ranged from 0.04 (PGIRD25) to 0.85 (PSMP2267) with average of 0.5. The observed and expected heterozygosity generated by the markers was moderate. Marker PSMP2267 had the highest expected and observed heterozygosity of 0.86 and 0.95, respectively while PGIRD25 exhibited the least expected and observed heterozygosity both at 0.04 (Table 2).

The proportion of rare alleles (private alleles) within the Napier grass accessions was very low with seven clones ranged from 0.00 in clones 41, 75, 76, 79, ANF, kakamega1 and RBN while the highest number was recorded in clone 16814. The highest Shannon information index was recorded in clone 16785 while the least was recorded in clone 16814 and 105. In relation, the highest expected heterozygosity was recorded in clone 16785 while the least was recorded in clone 16814 and 105. Napier grass accession 16785 showed the highest number of effective alleles while accession 105 had the least. Similarly, percentage polymorphic loci ranged from 27.8% in clone 16814 and 105 to 77.8% in clone 16785 (Table 3).

Pairwise comparison of genetic distance revealed big difference among the Napier grass clones ranging from 0.11 (between kakamega1 and kakamega2 and kakamega2 and kakamega 3) to 1.00 between clone 16814 and 105 (Table 4). Principal Coordinate Analysis was calculated from dissimilarity coefficients for two first axes coordinates with positive eigen values. The axes accounted for 52.8% total variation with first axis accounting for 29.8% while the second axis accounted for 23.0%. Principal Coordinate Analysis did not group the clones into clear structures. However, clones kakamega1, kakamega2, kakamega3 and 16805 and clone 112 and ANF grouped together, respectively (Fig. 1). The UPGMA dendogram based on pairwise Nei's genetic distance showed two major clusters; one consisting of only clone 16814 and the other consisting of the rest of the clones (Fig. 2).

Table 1. SSR primers used to assess genetic diversity in 23 Napier clones

Primer name	Sequence left primer (forward 5' – 3')	Right primer (reverse 5' – 3')	Annealing temperature (°C)
CTM-10	GAGGCAAAAGTGGAAGACAG	TTGATTCCCGGTTCTATCGA	52
CTM-27	GTTGCAAGCAGGAGTAGATCGA	CGCTCTGTAGGTTGAACTCCTT	52
CTM-59	TCCTCGACATCCTCCA	GACACCTCGTAGCACTCC	54
CTM-8	GCTGCATCGGAGATAGGGAA	CTCAGCAAGCACGCTGCTCT	52
PGIRD21	GCTATTGCCACTGCTTCACA	CCACCATGCAACAGCAATAA	54
PGIRD25	CGGAGCTCCTATCATTCCAA	GCAAGCCACAAGCCTATCTC	58
PGIRD57	GGCCCCAAGTAACTTCCCTA	TCAAGCTAGGGCCAATGTCT	56
PSMP2235	GCTTTTCTGCTTCTCCGTAGAC	CCCAACAATAGCCACCAATAAAGA	54
PSMP2248	TCTGTTTGTTTGGGTCAGGTCCTTC	CGAATACGTATGGAGAACTGCGCATC	58
PSMP2255	CATCTAAACACAACCAATCTTGAAC	TGGCACTCTTAAATTGACGCAT	54
PSMP2266	CAAGGATGGCTGAAGGGCTATG	TTTCCAGCCCACACCAGTAATC	58
PSMP2267	GGAAGGCGTAGGGATCAATCTCAC	ATCCACCCGACGAAGGAAACGA	60
Xipes0093	GGATCTGCAGGTTTGGACAT	CCAAGCACTGAAACATGCAC	57
Phil227562	TGATAAAGCTCAGCCACAAGG	ATCTCGGCTACGGCCAGA	56
Xcup14	TACATCACAGCAGGGACAGG	CTGGAAAGCCGAGCAGTATG	53
Xcup63	GTAAAGGGCAAGGCAACAAG	GCCCTACAAAATCTGCAAGC	53
XTXP278	GGG TTT CAA CTC TAG CCT ACC GAA CTT CCT	ATG CCT CAT CAT GGT TCG TTT TGC TT	50

Table 2. Genetic diversity parameters averaged across all groups and loci for 23 Napier grass clones

Primer	Major allele frequency	Number of alleles	Gene diversity/ expected heterozygosity (He)	Observed heterozygosity (Ho)	Polymorphic information content (PIC)
CTM10	0.52	2	0.49	0.87	0.37
CTM59	0.96	2	0.08	0.09	0.08
CTM8	0.29	13	0.82	0.89	0.80
CTM27	0.43	4	0.64	0.96	0.57
PGIRD21	0.37	9	0.80	0.52	0.78
PGIRD57	0.76	5	0.40	0.04	0.38
PSMP2248	0.63	4	0.53	0.22	0.47
Xipes0093	0.60	5	0.58	0.80	0.53
Phil227562	0.54	4	0.58	0.93	0.50
Xcup14	0.79	4	0.35	0.36	0.38
PSMP2266	0.47	5	0.66	1	0.61
PSMP2235	0.59	2	0.48	0.65	0.37
PGIRD25	0.98	2	0.04	0.04	0.04
PSMP2267	0.23	11	0.86	0.95	0.85
PSMP2255	0.33	7	0.76	0.78	0.72
XTXP278	0.64	6	0.53	0.72	0.47
Xcup63	0.48	5	0.66	0.96	0.61
Mean	0.57	5.29	0.54	0.63	0.50

Table 3. Mean number of effective loci (*ne*), shannon index (*i*), proportion of private alleles, expected heterozygosity (*he*) and percentage polymorphism across the 23 Napier grass clones

Population	Ne	I	Proportion of private alleles	He	% Polymorphism
19	1.500	0.385	0.222	0.278	55.6
41	1.667	0.462	0.000	0.333	66.7
75	1.444	0.347	0.000	0.250	50.0
76	1.667	0.462	0.000	0.333	66.7
79	1.556	0.385	0.000	0.278	55.6
97	1.389	0.347	0.111	0.250	50.0
103	1.556	0.424	0.056	0.306	61.1
104	1.611	0.424	0.111	0.306	61.1
105	1.000	0.193	0.056	0.139	27.8
112	1.278	0.308	0.111	0.222	44.4
117	1.556	0.462	0.056	0.333	66.7
16702	1.222	0.308	0.056	0.222	44.4
16785	1.778	0.539	0.111	0.389	77.8
16789	1.500	0.385	0.056	0.278	55.6
16805	1.556	0.385	0.111	0.278	55.6
16814	1.222	0.193	0.278	0.139	27.8
16815	1.556	0.424	0.056	0.306	61.1
79SN	1.444	0.347	0.056	0.250	50.0
ANF	1.333	0.308	0.000	0.222	44.4
kakamega1	1.500	0.385	0.000	0.278	55.6
kakamega2	1.556	0.385	0.056	0.278	55.6
kakamega3	1.556	0.385	0.056	0.278	55.6
RBN	1.222	0.308	0.000	0.222	44.4

Table 4. Nei's unbiased genetic distance of the 23 Napier grass clones based on SSR analysis

19	41	75	76	79	97	103	104	105	112	117	16702	16785	16789	16805	16814	16815	79SN	ANF	kaka1	kaka2	kaka3	RBN	
0																							19
0.263	0																						41
0.312	0.259	0																					75
0.208	0.239	0.179	0																				76
0.345	0.105	0.207	0.181	0																			79
0.6	0.288	0.519	0.442	0.207	0																		97
0.412	0.155	0.315	0.238	0.262	0.373	0																	103
0.587	0.426	0.612	0.426	0.38	0.503	0.562	0																104
0.938	0.839	0.872	0.795	0.725	0.749	0.913	0.699	0															105
0.974	0.57	0.852	0.822	0.552	0.576	0.635	0.674	0.698	0														112
0.683	0.365	0.511	0.434	0.292	0.377	0.359	0.535	0.634	0.57	0													117
0.741	0.608	0.647	0.775	0.587	0.685	0.897	0.597	0.852	0.922	0.87	0												16702
0.464	0.123	0.35	0.273	0.149	0.318	0.179	0.367	0.693	0.342	0.241	0.588	0											16785
0.652	0.292	0.563	0.486	0.26	0.207	0.444	0.38	0.687	0.422	0.353	0.7	0.233	0										16789
0.652	0.235	0.462	0.451	0.158	0.285	0.38	0.55	0.805	0.453	0.292	0.783	0.176	0.234	0									16805
0.615	0.563	0.424	0.496	0.427	0.541	0.738	0.913	1.000	0.939	0.672	0.939	0.693	0.687	0.687	0								16814
0.55	0.209	0.469	0.359	0.157	0.315	0.294	0.419	0.822	0.635	0.327	0.758	0.179	0.319	0.29	0.699	0							16815
0.637	0.288	0.423	0.377	0.258	0.336	0.286	0.538	0.788	0.447	0.377	0.808	0.201	0.232	0.285	0.64	0.344	0						79SN
0.828	0.465	0.685	0.731	0.422	0.417	0.635	0.459	0.698	0.47	0.534	0.599	0.342	0.23	0.392	0.735	0.526	0.388	0					ANF
0.504	0.13	0.462	0.451	0.158	0.258	0.349	0.478	0.725	0.453	0.385	0.661	0.149	0.182	0.158	0.582	0.235	0.232	0.335	0				kaka1
0.47	0.235	0.399	0.353	0.234	0.285	0.262	0.444	0.764	0.453	0.385	0.741	0.122	0.158	0.234	0.651	0.29	0.232	0.335	0.11	0			kaka2
0.575	0.235	0.43	0.417	0.158	0.258	0.319	0.319	0.725	0.453	0.353	0.783	0.176	0.158	0.208	0.582	0.182	0.232	0.281	0.134	0.11	0		kaka3
0.7	0.465	0.611	0.534	0.364	0.332	0.597	0.561	0.629	0.708	0.465	0.634	0.475	0.308	0.422	0.698	0.459	0.509	0.533	0.392	0.422	0.422	0	RBN

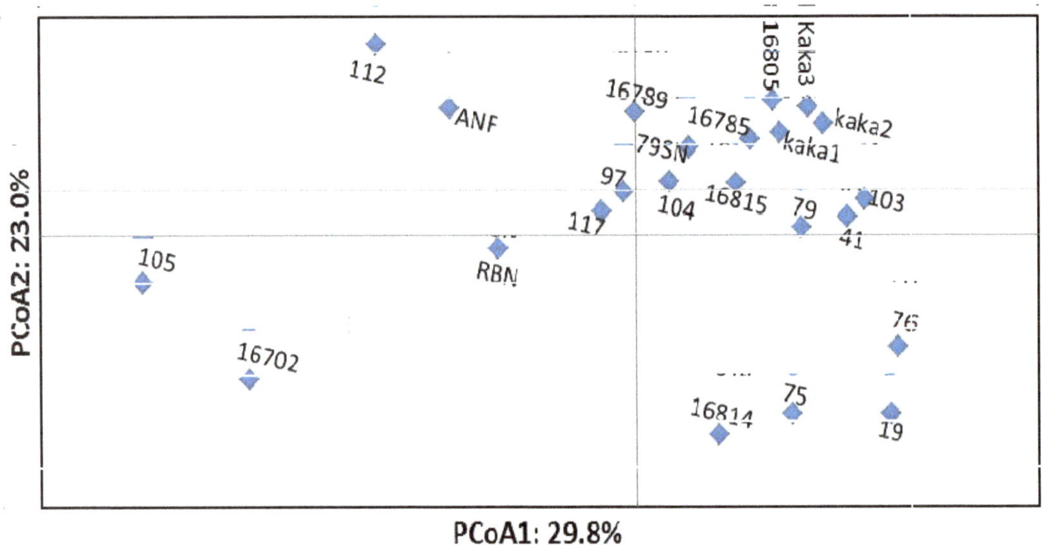

Fig. 1. PCoA scatter plot showing the clustering of the 23 Napier grass clones
Kaka = Kakamega

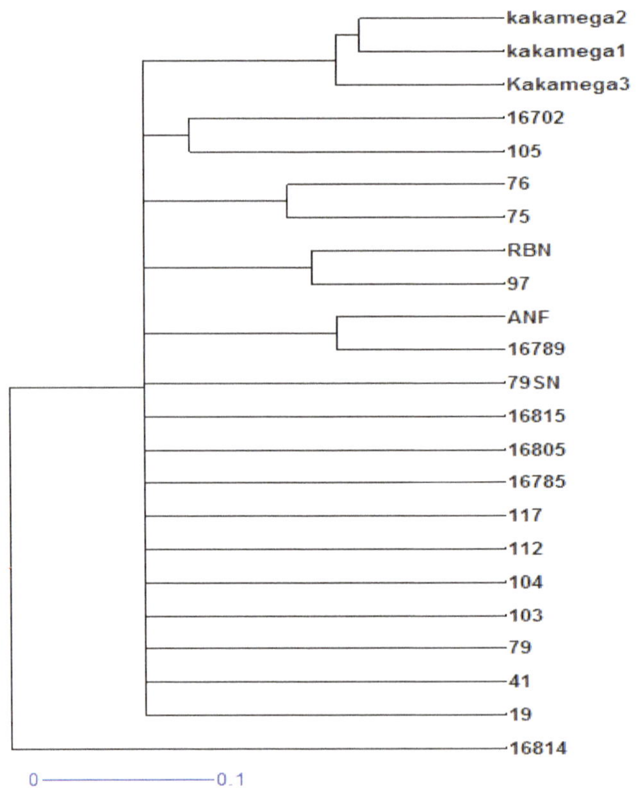

Fig. 2. UPGMA neighbour joining dendogram of 23 napier grass clones computed from 17 ssr markers using darwin hierachial clustering

4. DISCUSSION

Genetic characterization of cultivars is an important step in any breeding programs for selection of appropriate parental lines [13]. Several DNA marker systems for germplasm genetic characterization are available and SSRs have been found most adequate in detecting relationships among closely related materials as well as obtaining specific genetic fingerprints [14]. In this study, 17 SSR markers used produced high mean polymorphic information content suggesting that they are highly informative and able to discriminate among the different clones. According to Elibariki et al. [15] the ability to discriminate, however, varies from one marker to another, thus the most polymorphic marker was CTM8 while the least polymorphic was PGIRD25.

Both gene diversity and observed heterozygosity averaged across all loci was moderate. This result is in agreement with the findings of Wanjala et al. [16] who while working on Napier grass from east Africa region using AFLPs found moderate diversity among accessions. According to Bhandari et al. [17], Napier grass is of free pollination and high genetic diversity is expected from its natural crossings. The moderate genetic diversity revealed in this study is due to the fact that Napier grass grown onfarm is predominantly propagated by cuttings and subjected to high selection intensity by farmers. The markers revealed high number of private alleles in majority of the Napier grass clones. These, if included in breeding programs increase the chances of getting clones with farmer preferred traits.

The genetic distance revealed between the clones was generally high, with the highest distance being between clones 16814 and 105. This was further supported by the dendogram in which clone 16814 clustered different from the rest. This provides a basis for developing heterotic pool [18] from which crosses between genetically diverse parents can be made to produce progenies with higher genetic variation than those produced by closely related parents. The grouping of clone 16814 different from clone 16815 and 16805, yet all originate from United States of America [19] shows that the clustering was not based on the origin of the clones. This view contradicts the findings of Lowe et al. [1] who while using RAPDS reported that Napier grass accessions cluster corresponding to geographical location. However, it is in agreement with Wanjala et al. [16] who while using AFLPs reported that Napier grass did not cluster depending on their origin. The clustering together of the other clones most of which originate in Africa is a proof that Africa is the center of diversity [9], as such it houses majority of the *pennisetum* gene pools [8]. The loose clustering of accessions as revealed by PCA is possibly due to absence or low gene flow since Napier grass is clonally propagated. The genetic closeness of Kakamega 1, Kakamega 2 and Kakamega 3 indicates that they share most alleles and were collected from the same area known as Kakamega in Kenya.

5. CONCLUSION

Based on the foregoing, clones evaluated in this study are diverse with multitudes of private alleles which if found useful can be exploited in breeding to improve Napier grass. As such, Clone 16814, which is the most distant to all, is better suited for improvement of the rest of the clones if its attributes are found superior to those in others. Clones Kakamega1, Kakamega2, and Kakamega3 are more less the same, hence if any genetic improvement is to be carried out; it has to be with other distant clones.

ACKNOWLEDGEMENTS

This study was funded by the World Bank under Eastern Africa Agricultural Productivity Project (EAAPP) in improving smallholder dairy cattle production in Uganda.

COMPETING INTERESTS

Authors have declared that no competing interests exist.

REFERENCES

1. Lowe AJ, Thorpe W, Teale A, Hanson J. Characterisation of germplasm accessions of Napier grass (*Pennisetum purpureum* and *Pennisetum purpureum* x *Purpureum glaucum* Hybrids) and comparison with farm clones using RADP. Genetic resources and Crop Evolution. Kluwer Academic Publishers. Printed in the Netherlands. 2003;50:121-132.
2. Farrell G, Simons SA, Hillocks RJ. Pests, diseases and weeds of Napier grass, *Pennisetum purpureum*: A review. International Journal of Pest Management. 2002;48(1):39-48.
3. Staal SJ, Kruska R, Balteweck J, Kenyanjui M, Wokobi A, Njubi D, Thornton P, Thorpe W. Combined household and

GIS analysis of smallholder production systems. An application to intensifying smallholder dairy systems approaches for Agricultural Development (SAAD-III) held in Lima, Peru. National Agrarian University, La Molina, Lima, Peru; 1999.

4. Anderson WF, Dien BS, Brandon SK, Peterson JD. Assessment of bermuda grass and bunch grasses as feed stocks for conversion to ethanol. Applied Biochemistry and Biotechnology. 2008;145:13–21.
5. New Agriculturists. A healthy aversion to smut. Available:http://www.new-ag.info/98-2/newsbr.html (1November 2009).
6. Faisal M, Ahmed SN, Anis M. An efficient micro propagation system for *Tylophora indica*: An endangered, medicinally important plant. Plant Biotechnology Reports. 2007;1:155-16.
7. Kawube G, Alicai T, Otim O, Mukwaya A, Kabirizi J, Talwana H. Resistance of Napier grass clones to Napier grass stunt disease. African Crop Science Journal. 2014;22(3):109-115.
8. Techio VH, Davide LC, Cagliari A, Barbosa S, Pereira AV. Karyotipic asymmetry of both wild and cultivated species of *Pennisetum.* Bragantia Campinas. 2010;69:273–279.
9. Azevedo ALS, Costa PP, Machado JC, Machando MA, Pereira AV, Ledo FJS. Cross species amplification of *Pennisetum glaucum* microsatellite markers in *Pennisetum purpureum* and genetic diversity of Napier grass accessions. Crop Science. 2012;4:1776–1785.
10. Doyle, JJ and Doyle JL. Isolation of plant DNA from fresh tissue. Focus. 1990;12:13-15.
11. Liu K, Muse SV. Power marker: Integrated analysis environment for genetic marker data. Bioinformatics. 2005;21:2128-2129.
12. Peakall R, Smouse PE. GENALEX 6. Genetic analysis in Excel. Population genetic software for teaching and Research. Molecular Ecology Notes. 2006;6:288-295.
13. Xie XM, Zhou F, Zhang XQ, Zhang JM. Genetic variability and relationships between MT-1 elephant grass and closely related cultivars assessed by SRAP markers. Journal of Genetics. 2009;88:281–290.
14. Munoz-Falcon J, Prohens J, Vilanova S, Nuez F. Diversity in commercial varieties and landraces of black eggplants and implications for broadening the breeders' genepool. Annals of Applied Biology. 2009;154:453 – 465.
15. Elibariki Gladness, Njahira Moses, Wanjala Bramwel, Hosea Kenneth, Ndunguru Joseph. Genetic diversity and identification of duplicates in selected Tanzanian farmer preferred cassava landraces using simple sequence repeat (SSR) markers. International Journal of Research in plant science. 2013;3(4):81-87.
16. Wanjala BW, Obonyo M, Wachira FN, Muchugi A, Mulaa M, Harvey J, Skilton RA, Proud J, Hanson J. Genetic diversity in Napier grass (*Pennisetum purpureum*) cultivars: Implications for breeding and conservation. AoB PLANTS. 2013;5:plt022. DOI:10.1093/ aobpla/plt022.
17. Bhandari AP, Sukanya DH, Ramesh CR. Application of isozyme data in fingerprinting Napier grass (*Pennisetum purpureum* Schum.) for germplasm management. Genetic Resources and Crop Evolution. 2006;53:253–264.
18. Fregene MA, Suarez M, Mkumbira J, Kulembeka H, Ndedya A, Kulaya A, Mitchel S, Gullberg U, Rosling H, Dixon AGO, Dean R, Kresovich S. Simple sequence repeat marker diversity in cassava landraces: Genetic diversity and differentiation in an asexually propagated crop. Theoretical and Applied Genetics. 2003;107:1083-1093.
19. Mark Van De Wouw, Jean Hanson, Samuel Luethi. Morphological and agronomic characterisation of a collection of Napier grass (*Pennisetum purpureum*) and *P. purpureum X P. glaucum. Tropical Grasslands.* 1999;33:150–158.

6

Plant Regeneration from Shoot Tips-derived Callus of Ginger (*Zingiber officinale* Rosc.)

Diaa A. Ibrahim[1*], Gharbia H. Danial[1], Vian M. Mosa[1] and Belan M. Khalil[1]

[1]Department of Scientific Research Center, Faculty of Science, University of Duhok, Iraq.

Authors' contributions

This work was carried out in collaboration between all authors. Authors DAI and GHD designed the study, wrote the protocol and wrote the first draft of the manuscript and reviewed the experimental design and all drafts of the manuscript. Author GHD performed the statistical analysis. Authors VMM and BMK managed the analyses of the study and identified the plants. All authors read and approved the final manuscript.

Editor(s):
(1) Lixiang Cao, Department of Biotechnology, Sun Yat-sen University, P. R. China.
(2) Anonymous.
Reviewers:
(1) Arvinder Kapur, Department of Obstetrics and Gynaecology, University of Wisconsin, Madison, USA.
(2) Anonymous, India.
(3) Anonymous, Malaysia.
(4) Anonymous, Malaysia.

ABSTRACT

A protocol was described to regenerate plants from callus culture of ginger (*Zingiber officinale* Rosc.). Callus cultures were induced from ginger shoot tips inverted cultured on semi-solid MS medium supplemented with 2, 4-D (0.5-1.0 mg L^{-1}) and BA (0.5- 1.0 mg L^{-1}). Maximum callus induction was obtained on MS medium supplemented with BA (0.5 mg L^{-1}) either alone or with 2,4-D (0.5 – 1.0 mg L^{-1}), while the highest callus weight was on MS medium supplemented with BA (0.5 mg L^{-1}) and 2,4-D (0.5 mg L^{-1}). Induced callus was tested for regeneration on MS medium supplemented with different concentrations of 2,4-D and BA and the best regeneration was observed on a medium containing 2.5 mg L^{-1} BA. The regenerated plants were rooted and successfully established in the field after few days of acclimatization.

Keywords: Ginger; callus induction; plant regeneration; shoot tips.

**Corresponding author: E-mail: gha.2008@yahoo.com, Diaa65@yahoo.com;*

1. INTRODUCTION

Zingiber officinale Rosc. (Ginger) of the family *Zingiberaceae* is an important tropical horticultural plant, values all over the world as an important spices for its medicinal properties. The *Zingiberaceae* is an herbaceous moderate sized family of relatively advanced monocotyledonous plant of the order *Zingiberales*. Zingiberaceous plants are rhizomatous, perennial and aromatic herbs often of large size, bearing flowers either terminally on aerial leaf shoots or from ground level [1]. Ginger's pungent aromatic rhizome is consumed all over the world as a spice, culinary herb, condiment, home remedy, and medicinal agent [2]. It is vegetatively propagated through underground rhizomes, with a very low multiplication rate [3]. The common problems in increasing ginger production are soil-borne disease infection and degeneration of rhizomes [2], and the breeding of ginger has been severely hampered by poor flowering and seed set [4], and even the rhizome cannot be stored for long time as it is susceptible to fungal diseases, which affect the quality of tubers [5]. *In vitro* culture techniques, especially meristem culture, provide an alternative way of plant propagation and maintenance of disease free germplasm banks [6]. There are many reports on ginger tissue culture. [7] Reported that maximum six shoots per bud from *in vitro* culture, with low survival under field conditions. Organ culture [3], somatic embryogenesis [4] and [8], organogenesis [9], protoplast culture [10], microrhizome production [11], shoot tips culture [12], rhizome buds cultured at different size [13] and germplasm preservation [14] have been conducted. Ginger seed production also was suggested by [15] by *in vitro* propagation methods, however these methods are not efficient, expensive and do not meet ginger germplasm conservation requirements.

The present work aimed to describe an efficient protocol for; calli production; shoot regeneration, keeping in view the potential action against various ailments and its regeneration problems in natural environment. Micropropagation of this medicinal important plant was accomplished to provide stable supply of the *in vitro* raised plantlets for the mass production for this important medicinal plant in Iraqi Kurdistan Region and to become a powerful tool for future studies.

2. MATERIALS AND METHODS

2.1 Callus Induction

This investigation was conducted at Plant Tissue Culture Lab. at the Scientific Research Center, Faculty of Science at the University of Duhok. Shoot tips of *in vitro* cultured ginger on MS medium [16] supplemented with 1.0 mg/l BA were cut and placed in inverted form on the culture medium. Various concentrations of 2,4-dichloro-phenoxyacetic acid (2,4-D) and benzylaminopurine (BA) (0.0, 0.5, 1.0 mg/l) in MS medium containing 3% sucrose, 100 mg/l inositol were evaluated for inducing callus. The pH of the medium was adjusted to 5.7±0.1 with 1N NaOH or HCl, prior to the addition of agar (7. 0) g/l. The medium was brought up to the final volume, subsequently, 25 ml of medium was dispensed into 250 ml Mason jars and capped with colorless PVP covers and fitted with rubber bunds. The medium was sterilized by autoclaving for 15 min at 121°C temperature under 1.04 kg/cm² and allowed to solidify under room temperature. Three explants were cultured in each vessel with five replications. The cultures were incubated in the culture room under 25± 1°C temperature and 16 hours daily exposure to 1000 Lux cool white light, followed by 8 hours of darkness. The results were recorded after 8 weeks of culture.

2.2 Plant Regeneration

Two months old callus cultures were transferred to MS medium supplemented with BA (0.0, 2.5, and 5 mg/l) and 2, 4-D (0.0, 0.2 and 0.4 mg/l) either alone or in combination. After eight weeks the shoots length, number of shoots and leaves/ explant were recorded. The plantlets were suppurated from the callus and cultured on MS medium was supplemented with 2.5 mg/L kinetin for both multiplication and rooting stages [17].

Eventually, for hardening stage, a number of well and successfully rooted plantlets were removed from culture vessels and their roots were washed with distilled water and immersed in Benlate fungicide (0.1% for 10 minutes). Finally, they were planted in pots containing steam sterilized soil mixture of peat moss: loam: stryrofoam (1:1:0.5) under tightly controlled atmosphere of the greenhouse.

2.3 Data Analysis

All the comparison between means was carried out according to Duncan's multiple range test ($P < 0.05$) using a computerized program of SAS [18].

3. RESULTS AND DISCUSSION

After 10 days of culture, there was a notable enlargement of inverted shoot tips with a whitish to brown colour and fragile texture (Fig. 1A). The callus growth was slow and highly sensitive to mechanical force or fragile, the surface was covered with small hairy structure.

The caulogenic response of the shoot tips to various concentrations and combinations of cytokinins and auxins is shown in (Fig. 2) For callus induction the highest and best callus induction (100%) was noticed on MS medium containing BA either alone or in combination with 2,4-D (Fig. 1, B), while those devoid of BA showed lowest response. Similar results were obtained by [19] and [9] where they reported that 2, 4-D is the most effective auxin for callus induction in ginger and turmeric [20].

The highest callus weight (1.43 g) showed in (Fig. 3) observed on MS medium supplemented with 0.5 mg/l from BA and 2,4-D (Fig. 1, C) followed by MS medium containing only BA at concentration 0.5 mg/l (1.34 g) , no significance different between the treatments. While the least weight (0.17 g) was observed on MS medium devoid of growth regulators.

The well developed, white and fragile callus was transferred to regeneration medium (Fig. 4). The highest mean number of regenerated shoots (6.0 shoots/explants) was observed on MS media containing BA 2.5 mg/l (Fig. 1, D) followed by MS medium supplemented with BA 5.0 mg/l which was 5.0 shoots/explant, but did not different significantly. Similar results were recorded by [20] they observed that 5.0 mg/l BA produced the highest number of shoots form callus in *Curcuma attenuate*, [21] recorded that BA combination with NAA or TDZ was effective for shoot regeneration from anther-derived -callus of *Curcuma attenuate*. The MS medium fortified with BA and 2, 4-D showed the lowest mean numbers of shoots. Similar results were recorded by [19], where 1 to 3 mg/l of BA was the best concentrations for shoot regeneration from callus tissues. [9] State that 2, 4-D at lower concentration would enhances organogenesis. Since there is no significant difference between 2.5 and 5.0 mg/l of BA, it is advisable to use the lowest concentration to reduce the mass production cost.

Fig. 1A. Shoot tip culture in inverted in MS medium, B. Callus induction in MS medium supplemented with BA and 2, 4-D after 8 weeks, C. High callus weight in MS medium supplemented with 0.5 mg/l BA and 2, 4-D, D. plantlets regenerated from the callus, E Acclimatization of plantlets in the soil, F. Hardened plant ready for transfer to field with formed rhizome

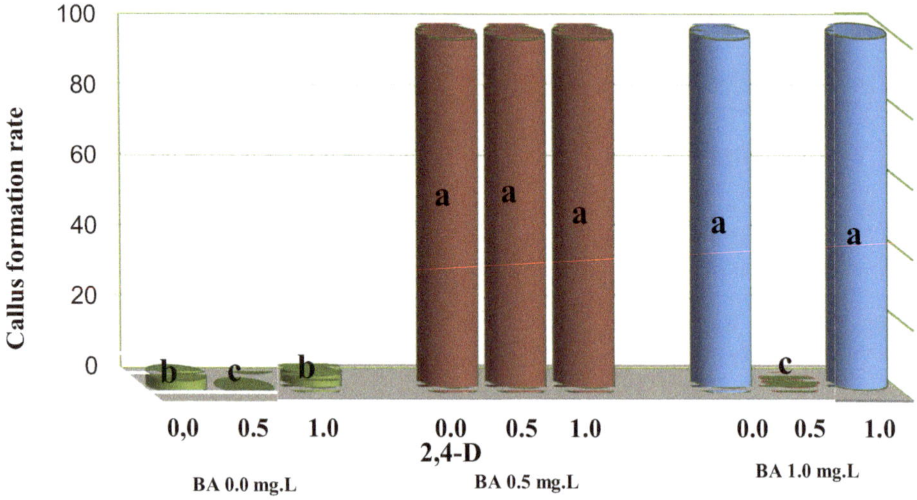

Fig. 2. Effects of different concentrations of BA and 2, 4-D on callus formation from shoot tips of *Zingiber officinale*

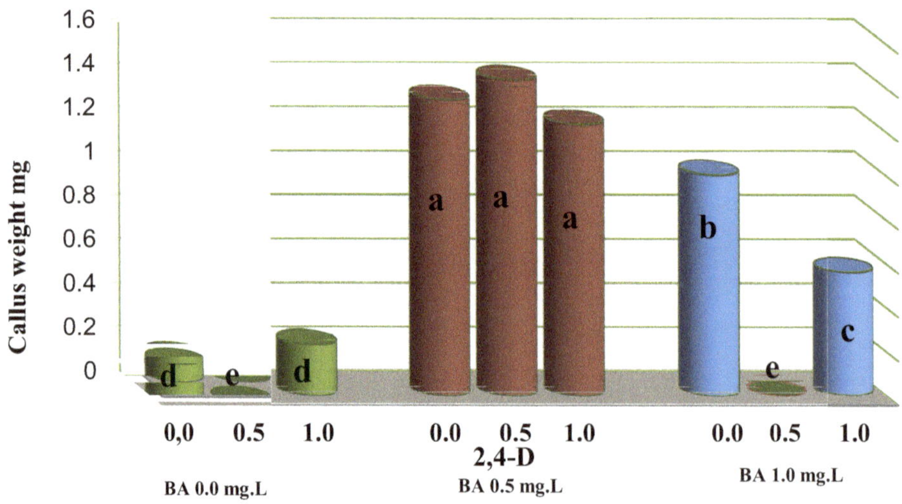

Fig. 3. Effects of different concentrations of BA and 2, 4-D on callus weight from shoot tips of *Zingiber officinale*

These results emphasize the importance of cytokinins in shoot regeneration in *in vitro* plant propagation. Cytokinins activate RNA synthesis; stimulate protein synthesis and the activities of some enzymes. Generally, herbaceous plants are highly responsive to BA treatments and most herbaceous species produce well formed shoots suitable for further shoot proliferation [22].

On the other hand, the highest mean of shoot length (6.00 cm) (Fig. 5) and highest mean of leaves number (5.0 leaves/explant) (Fig. 6) was found in a medium containing 0.2 mg/l 2, 4-D only without adding any cytokinins and this was highly significant than all other means.

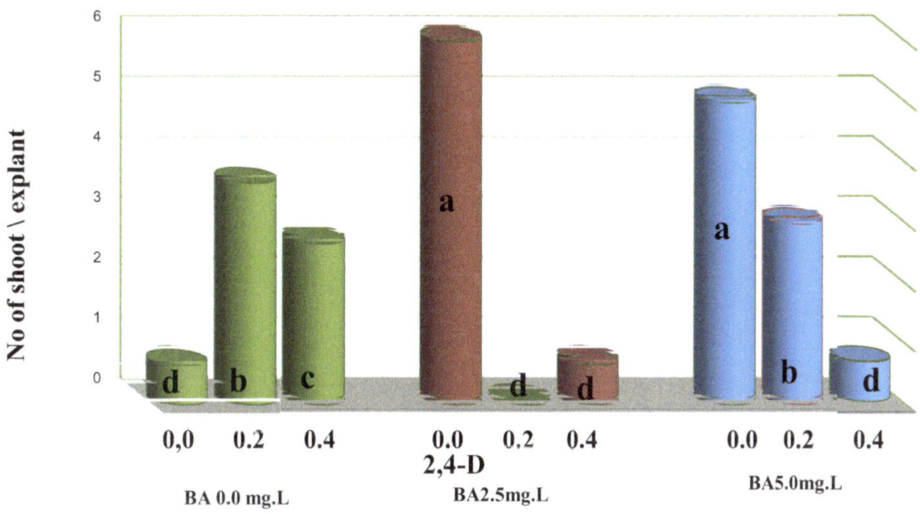

Fig. 4. Effects of different concentrations of BA and 2, 4-D in No. of shoot\explant from callus culture of *Zingiber officinale*

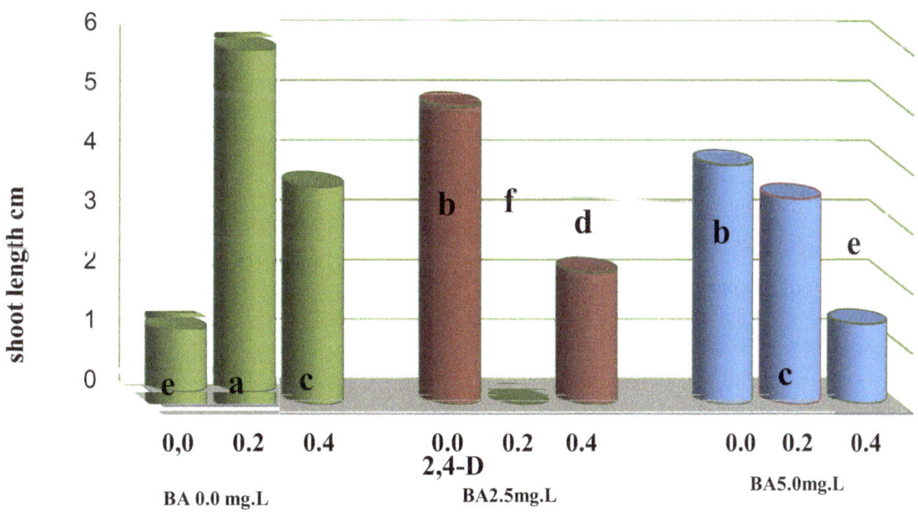

Fig. 5. Effects of different concentrations of BA and 2, 4-D in shoot length (cm) from callus culture of *Zingiber officinale*

The shoots regenerated from calli were transfer and successfully multiplied in 2.5 mg/l kinetin. Roots were also formed in the same medium or in phytohormone-free medium. These records are in agreement with those found by [17] and [23] which indicates that the endogenous auxin level in the plant itself is so sufficient that can form roots without applying any exogenous auxins. Plantlets were successfully transplanted into pots after sufficient development of roots to establish the field condition and rhizome formation in the soil (Fig. 1, E, F).

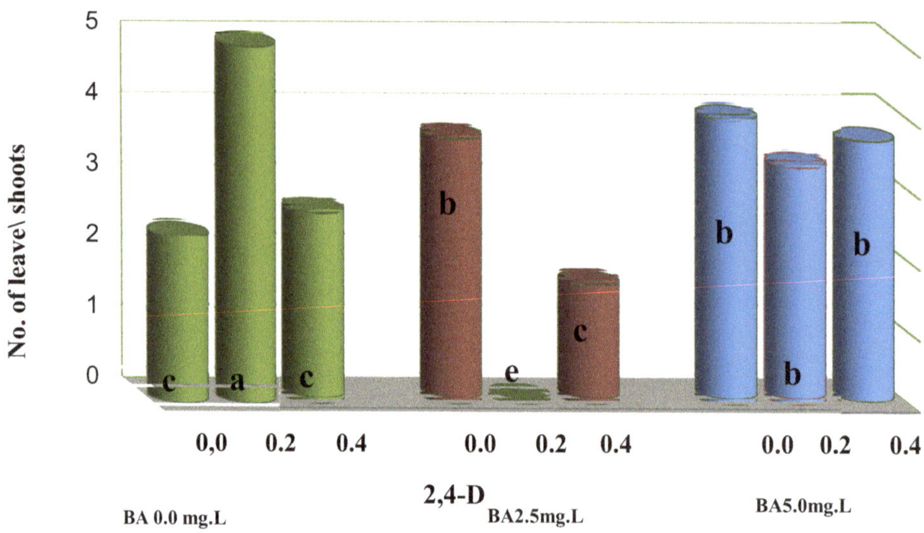

Fig. 6. Effects of different concentrations of BA and 2, 4-D in No. of leaves\ shoots from callus culture of *Zingiber officinale*

4. CONCLUSION

It was concluded that it is possible to regenerate plants from callus culture of ginger (*Zingiber officinale* Rosc.), which is induced from ginger shoot tips explants and are suitable for propagation. Plant regeneration was successfully developed for shoot tips-derived callus of ginger and these explants may be used according to their higher rate of shoot multiplication. The plant regeneration is varies dependent on phytohormone concentrations. For ginger shoot tip explants, the developed protocol is suitable for micpropagation for future work.

ACKNOWLEDGMENT

We would like to express our thanks to the administration of Scientific Research Center/University of Duhok for offering the facilities to conduct this research.

COMPETING INTERESTS

Authors have declared that no competing interests exist.

REFERENCES

1. Kambaska KB, Santilata S. Effect of plant growth regulator on micropropagtion of ginger (*Zingiber officinale* Rosc.) cv-Suprava and Suruchi. Journal of Agricultural Technology. 2009;5(2):271-280.
2. Guan Q, Guo Y, Wei Y, Meng F, Zhang Z. Regeneration of somatic hybrids of ginger via chemical protoplast fusion. Plant Cell Tiss. Organ Cult. 2010;102:279–284.
3. Sharma TR, Singh BM. High-frequency in vitro multiplication of disease-free *Zingiber officinale* Rosc. Plant Cell Reports. 1997;17:68–72.
4. Kackar A, Bhat SR, Chandel, KPS, Malik SK. Plant regeneration via somatic embryogenesis in ginger. Plant Cell, Tissue and Organ Culture. 1993;32:289-292.
5. Jamil M, Kim JK, Akram Z, Ajmal SU, Rha ES. Regeneration of ginger plant form callus culture through organogenesis and effect of CO2 enrichment on the differentiation of regenerated plant. Biotechnology. 2007;6(1):101-104.
6. Routs GR, Palai SK, Samantaray S, Das P. Effect of growth regulator and culture conditions on shoot multiplication and rhizome formation in ginger (*Zingiber offincinale Rosc.*) in vitro. In Vitro Cell. Dev. Biol. Plant. 2001;37:814-819.
7. Hosoki T, Sagawa Y. Clonal propagation of ginger (*Zingiber officinale* Rosc.) through tissue culture. Hort Science. 1977;12:451–452.

8. Lincy AK, Remashree AB, Sasikumar B. Indirect and direct somatic embryogenesis from aerial stem explants of ginger (*Zingiber officinale* Rosc.). Acta Botania Croatica. 2009;68(1):93-103.
9. Babu KN, Samsudeen K, Ravindran PN. Direct regeneration of plantlets from immature inflorescences of ginger (*Zingiber officinale* Rosc.) by tissue culture. J. Spices Aromat. Crops. 1992;1:43–48.
10. Geetha SP, Babu KN, Rema J. Isolation of protoplasts from cardamom (*Elettaria cardamomum Maton.*) and ginger (*Zingiber officinale* Rosc.). J. Spices Aromat. Crops. 2000;9:23–30.
11. Zheng Y, Liu Y, Ma M, Xu K. Increasing in vitro microrhizome production of ginger (*Zingiber officinale* Roscoe). Acta Physiol Plant. 2008;30:513–519.
12. Ayenew B, Tefera W, Kassahu B. In vitro propagation of Ethiopian Ginger (*Zingiber officinale Rosc.*) cultivars: Evaluation of explants types and hormone combinations. Afr. J. Biotechnal. 2012;11(16):3911-3918.
13. Sathyagowri S, Thayamini H. In vitro plant regeneration of ginger (*Zingiber officinale* Rosci.) with emphasis of initial culture establishment. Int. J. Med. Arom. Plants. 2011;1(3):195-202.
14. Dekker AJ, Rao AN, Gob CJ. In vitro storage of multiple shoot cultures of ginger at ambient temperatures of 24–29C°. Sci. Hort. 1991;47:157–167.
15. Uozumi M, Asamo Y, Kobayashi TK. Micropropagation of horseradish hairy root by means of adventitious shoot primordia. Plant Cell Tissue Org. Cult. 1994;36:183-190.
16. Murashige T, Skoog F. A revised medium for rapid growth and bioassay with tobacco tissue cultures. Physiol. Plant. 1962;15:473-497.
17. Toma SR, Danial GH, Rashad RH. In vitro propagation of ginger (*Zingiber officinale* Rosc.) from a single bud explant as affected by kinetin and benzyl adenine. Journal of Duhok University. 2012;15(1): 423-427.
18. SAS, SAS/STAT User's Guide for Personal Computers. Release 6.12. SAS Institute Inc. Cary, NC: USA; 200.
19. Malamug JF, Inden A, Asahira T. Plantlet regeneration and propagation from ginger callus. Sci. Hortic. 1991;48:89 - 97.
20. Jala A. The effect of the 2,4-Dichlorophenoxy acetic acid, benzyl adenine and paclobutrazol, on vegetative tissue derived somatic embryogenesis in turmeric (*Curcuma* Var Chattip). Int. Trans. J. Eng. Manag. Sci. Tech. 2013;4(2):105-110.
21. Yaping K, Guohua M, Jaime A, Teixeira S. Callus induction and shoot organogenesis from anther cultures of Curcuma attenuate Wall. Plant Cell Tiss. Organ Cult. 2013; 112:1-7.
22. Debergh PC, Zimmerman RH. Micropropagation technology and application: Kluwer Academic Publishers. Dordecht the Netherlands; 1991.
23. Kavyashree R. An efficient in vitro protocol for clonal multiplication of ginger - var. Varada. Indian Journal Biotechnology. 2009;8:328-331.

7

Partitioning of Rice (*Oryza sativa* L.) Genotypes Based on Morphometric Diversity

Ahasanul Hoque[1*], S. N. Begum[2], A. H. K. Robin[1] and Lutful Hassan[1]

[1]*Department of Genetics and Plant Breeding, Bangladesh Agricultural University, Mymensingh-2202, Bangladesh.*
[2]*Plant Breeding Division, Bangladesh Institute of Nuclear Agriculture, Mymensingh-2202, Bangladesh.*

Authors' contributions

This work was carried out in collaboration between all authors. Author AH wrote the protocol, designed the study, executed the field experiment, performed the statistical analyses, and wrote the first draft of the manuscript. Authors SNB and AHKR revised the protocol, supervised the experiment, and helped to manage the analyses of the study. Author LH managed the literature searches and revised the manuscript. All authors read and approved the final manuscript.

Editor(s):
(1) Lanzhuang Chen, Laboratory of Plant Biotechnology, Faculty of Environment and Horticulture, Minami Kyushu University, Miyazaki, Japan.
Reviewers:
(1) Anonymous, Nigeria.
(2) Gulzar Singh Sanghera, Plant Breeding and Genetics, Punjab Agricultural University, Regional Research Station, Kapurthala, India.

ABSTRACT

Aims: The Objectives of this study were to partition the rice genotypes into different clusters, identification of heterotic groups, and most important traits contributing to divergence to utilize them for specific objective-oriented breeding programs in future.
Study Design: The experiment was set out in randomized complete block design with three replications.
Place and Duration of Study: The experiment was carried out at experimental farm of Department of Genetics and Plant Breeding, Bangladesh Agricultural University, Bangladesh during *Aus* (Kharif) season of 2012.
Methodology: The diversity among sixty rice genotypes and contribution of thirteen traits towards diversity were analyzed using Mahalanobis's D^2 statistics and principal component analysis.
Results: Analysis of variance showed highly significant variation among the genotypes for all the traits. Cluster analysis based on D^2 values exhibited seven distinct clusters. The highest intra-

Corresponding author: E-mail: ahasanul102@bau.edu.bd;

cluster distance (21.95) was observed in cluster II whereas that was lowest (7.62) for cluster VI. Maximum inter-cluster distance was observed between cluster III & VII (46.75) followed by cluster II & VII (42.91), cluster V & VII (38.48), and cluster III & VI (30.87). In all cases inter-cluster distance was higher than the intra-cluster distance suggesting wider diversity among the genotypes. All the short duration genotypes with high yield, high tiller number per hill and more filled grain per panicle were grouped in cluster VII whereas tall, long duration genotypes with low yield, wider flag leaf area, long panicle and more unfilled grain per panicle were grouped in cluster II. Cluster III composed of long duration & moderate yielded genotypes, but cluster V composed of genotypes with long duration and high yield. First three principal components explained about 81% of the total variation. Results of PCA suggested that traits such as number of filled grains per panicle, number of unfilled grains per panicle, flag leaf area, plant height and days to maturity were the principal discriminatory characteristics.

Conclusion: The studied rice genotypes showed considerable divergence for most of the traits. These results can now be used by the breeders to develop rice varieties having desirable characteristics and new breeding strategies for rice improvement.

Keywords: Rice; D^2 statistics; cluster analysis; principal component analysis.

1. INTRODUCTION

Rice (*Oryza sativa* L.), belonging to the family gramineae, is the main source of nutrition for 50% of world population [1] and supplies staple food for one third of the global population [2]. Among the most cultivated cereals in the world, it ranks second next to wheat [3]. Bangladesh being an agro-based and over populated country where peoples mainly depend on rice for their livelihood. The demand for rice is increasing day by day as nearly three million people being added each year to the total population [4]. Therefore, to meet the food demand of the growing population and to achieve food security, the present production level need to bc increased, which is possible through heterosis breeding and other innovative breeding approaches.

Magnitude of heterosis depends on the choice of appropriate parental lines [5]. Chances of attaining higher heterosis and wide range of variability in segregating generations can be increased by the inclusion of more diverse parents in hybridization program [6,7]. However, in some cases crossing between moderately diverse parents also have displayed maximum heterosis [8,9]. Precise assessment of the levels and patterns of genetic diversity facilitates the exploration of genetic variability in germplasm [10], identification of diverse parental combinations to create segregating progenies with maximum genetic variability for further selection [11] and introgression of desirable genes from wild to adapted high yielding germplasm [12]. Thus information about genetic diversity simplifies the selection of parental genotypes from random populations before attempting crosses and hence saving time and resources [13] and also helps to make decision on management procedures of the germplasm.

With the advancement of multivariate statistics, the assessment of degree of divergence among biological populations and relative contribution of different traits to total divergence have now become possible. In the bid to generate genotypes having desirable attributes, the breeder would like to choose genetically distant parents for hybridization. The process of classification of germplasm using Mahalanobis's D^2 statistics can help to practical reach [14]. In this regard, different methods were used for classification [15], but a procedure combining principal component and D^2 analysis was found expedient [16]. From this point of view, the present investigation was done using sixty rice genotypes to assess the genetic diversity for determining the heterotic groups and the relative contribution of different traits towards the total divergence for further management and utilization of those lines in hybridization program using Mahalanobis's D^2 statistics and Principal Component Analysis (PCA).

2. MATERIALS AND METHODS

A field experiment was conducted to study the extent of genetic diversity in rice for growth parameters and yield contributing traits. Sixty genotypes of rice were evaluated under field condition during the period from March to December, 2012, at the experimental farm of the Department of Genetics and Plant Breeding, Bangladesh Agricultural University, Mymensingh in Randomized Complete Block Design (RCBD) with three replications. Twenty six days old seedlings were transplanted in 2 m^2 plot with inter-row and intra-row spacing of 20 × 15 cm. All recommended practices were followed to maintain uniformity in plant population and to ensure good crop growth. Five representative plants for each treatment in each replication were randomly selected to record days to panicle initiation, plant height (cm), flag leaf area (cm^2), number of effective tillers per hill, panicle length (cm), number of filled grains per panicle and number of unfilled grains per panicle. Traits related to growth duration such as days to 50% flowering, days to flowering completion, days to milking stage, days to dough stage, days to maturity and yield per plot were computed on plot basis. For calculation of days to panicle initiation five main (primary) tillers from five randomly selected hills of each plot for each replication were collected at maximum tillering stage. The collected tillers were then exposed and the growing zone of first node was examined carefully under stereomicroscope and the length of the newly developed panicle was measured using a millimeter graph paper. Then the stage of panicle development was determined by the guidelines described in Manual for Hybrid Rice Seed Production [17]. The days to panicle initiation was recorded as days from sowing to first stage of panicle development using the following formula:

$$DPI = T_n - T_1$$

Here,

DPI = Days to panicle initiation
T_n = Days from sowing to observed stage of panicle development
T_1 = Time gap between 1st and observed stage of panicle development according to manual

For calculating flag leaf area, twenty flag leaves were clipped from randomly selected plants from different plots and their areas were measured by leaf area meter. The length and breadth of those flag leaves were measured by using meter scale. Then K-value was measured by using the following formula:

K-value = Leaf area measured by leaf area meter / (leaf length x leaf breadth)

Finally the flag leaf area of five tillers from five randomly selected hills of each plot was measured by using the following formula and average was taken

Flag leaf area = Length× Breadth× K-value

To test the significance of variation among the genotypes for each trait, the analysis of variance was performed using the plant breeding statistical program PLABSTAT, version 2N [18]. The Mahalanobis's generalized distance (D^2) and principal component analysis were done using statistical software R, version 3.1.1 [19]. The clustering of the lines was done following the Tocher's method [20].

3. RESULTS AND DISCUSSION

The analysis of variance (Table 1) revealed that the genotypes differed significantly (*P*=.01) for all the traits studied, indicating the presence of remarkable genetic variability among them. This implied that it would be sensible to classify the lines on the basis of degree of divergence considering all the thirteen studied traits.

Depending upon the range of diversity, 60 genotypes were grouped into seven clusters (Table 2). The distribution pattern revealed maximum number of genotypes (16 genotypes) in cluster I whilst cluster V included minimum number of genotypes (3 genotypes). Cluster II, III, IV, VI and VII comprised of 4, 10, 9, 11 and 7 genotypes, respectively. The less number of genotypes in the clusters V and II was probably due to small number of traits as well as duplication effect of the traits included in the study. In this connection, in case of hierarchical and dynamic clustering, the frequency of genotypes in a given cluster is increased by increasing the number of traits under study [21]. In all the cases the inter-cluster distances were greater than the intra-cluster distances (Table 3) suggesting wider diversity among the genotypes of the distant groups. The similar results regarding inter and intra-cluster distances have been described in different crops by various researchers [22-26].

The intra-cluster divergence was maximum in cluster II (21.95) and minimum in cluster VI (7.62), indicating that the genotypes in cluster II were more heterogeneous than those in cluster VI. The intra-cluster distances were comparatively high for all the seven clusters with the range of 7.62 to 21.95, indicating heterogeneous nature of the genotypes within each cluster. In previous experiment, homogeneous nature of the genotypes within the clusters for irrigated traditional and modern rice germplasm was reported [27], which is contrarily to the present findings. The possible reason could be variation in number of genotypes and traits studied in present experiment with that of the previous one.

Regarding the inter-cluster distance, cluster III showed maximum distance from cluster VII (46.75) followed by the distance between cluster II and VII (42.91), cluster V and VII (38.48) and cluster III and VI (30.87) which reflected wider diversity among these clusters. Minimum distance was found between the genotypes of the cluster I and VI (12.40) followed by the genetic distance between clusters IV and VI (13.56) (Table 3).

Table 1. Analysis of variance (ANOVA) for 13 traits of 60 rice genotypes

Traits	Sources of variation (Mean of square)		
	Replication (d.f.= 2)	Genotype (d.f.=59)	Error (d.f.=118)
DPI	38.046	249.725**	2.411
DF	1.250	288.547**	0.900
DFC	1.572	292.202**	0.730
DMI	2.067	288.766**	1.010
DD	11.872	290.943**	1.629
FLA	371.772	355.217**	24.094
DM	32.706	332.058**	3.299
PH	156.606	408.012**	14.069
ETN	0.617	5.702**	2.289
PL	3.517	12.649**	1.810
FG	385.089	1355.493**	49.761
UFG	192.772	356.943**	17.558
YPP	0.012	0.202**	0.003

DPI: days to panicle initiation, DF: days to 50% flowering, DFC: days to flowering completion, DMI: days to milking stage, DD: days to dough stage, FLA: flag leaf area (cm^2), DM: days to maturity, PH: plant height (cm), ETN: number of effective tillers per hill, PL: panicle length (cm), FG: number of filled grains per panicle, UFG: number of unfilled grains per panicle, YPP: yield per plot (kg), d.f.: degrees of freedom
*** Significant at 1% level*

Table 2. Clustering of 60 rice genotypes based on D^2 statistics

Cluster no.	No. of genotypes	Name of genotypes
I	16	G1, G9, G10,G15, G20, G21, G22, G35, G36,G39,G41, G45, G46, G52, G53, G59
II	4	G2, G3, G5, G19
III	10	G4,G7,G8, G11,G12,G14,G43,G44,G58, G60
IV	9	G6, G17, G26, G30, G31, G40, G47, G51, G57
V	3	G13, G16, G32
VI	11	G18, G23, G24, G25, G27, G29, G33, G37, G38, G42,G48
VII	7	G28, G34, G49, G50, G54, G55, G56

Table 3. Intra and inter cluster average distances among 60 rice genotypes

Cluster	I	II	III	IV	V	VI	VII
I	**11.81**	27.01	25.92	13.81	20.21	12.40	24.77
II		**21.95**	18.81	19.77	20.37	27.50	42.91
III			**11.08**	22.92	16.86	30.87	46.75
IV				**9.62**	19.95	13.56	29.81
V					**13.83**	25.01	38.48
VI						**7.62**	20.14
VII							**11.30**

Values in bold illustrate the intra cluster distance and others show inter cluster distance

Mean performance of different clusters for studied traits (Table 4) reflected that all the short duration genotypes were grouped into cluster VII whereas cluster III included long duration genotypes indicating maximum contribution of growth duration towards the divergence between cluster III and VII. Again all the high yielding genotypes with high number of filled grains per panicle were grouped into cluster VII whereas cluster II included low yielding genotypes with less number of filled grains per panicle indicating maximum contribution of these traits towards the divergence between cluster II and VII. The cluster V was divergent from cluster VII mainly due to days to panicle initiation, plant height, flag leaf area and number of effective tillers per hill indicating maximum contribution of these traits towards the divergence. The cluster III was divergent from cluster VI mainly due to days to 50% flowering, days to maturity and flag leaf area. In this context, traits contributing maximum towards the divergence should be given greater emphasis for deciding the type of cluster for the purpose of further selection and the choice of parents for hybridization [28].

The contribution of different traits towards the divergence was measured through principal component analysis (Table 5). 'Proper values' measure the importance and contribution of each component to total variance, whereas each coefficient of proper vectors indicates the degree of contribution of every original variable with which each principal component is associated. The higher the coefficients, regardless of the direction (positive or negative), the more effective they will be in discriminating between genotypes. For determining the number of factors to extract, scree test was done. According to scree test (Fig. 1), first three components were found most effective which accounted for more than 80% of total variation (Table 5), giving a clear idea of the structure underlying the variables analyzed. However, to determine the cutoff limit for the co-efficients of the proper vectors, the criteria that treated co-efficients greater than 0.3 as having a large effect, while traits having a coefficient less than 0.3 were considered not to have important effects on the overall variation observed in the present study.

The first principal component accounted for more than 56% of total variance, whereby days to panicle initiation, days to 50% flowering, days to flowering completion, days to milking stage, days to dough stage and days to maturity were the variables that contributed most negatively (Table 5). As a result, the first component differentiated those genotypes that flower and mature earlier in the season. The first component identified mainly phenological variables presenting negative contributions. Similar results for days to 50% flowering and days to maturity in rice were also observed by the previous researchers [29,30]. The second principal component accounted for more than 16% of total variance. Variables highly and negatively correlated were number of filled grain per panicle and yield per plot. In contrast, number of unfilled grains per panicle contributed most positively. Thus, the second principal component differentiated the high yielding genotypes. The third principal component accounted for 8% of total variance and was associated with flag leaf area, plant height, number of effective tillers per hill and panicle length, thus differentiating those genotypes with good architecture. These findings agree with Caldo et al. [31], who reported that maturity, heading, plant height, culm length, leaf length, and tillering ability were the major factors contributing to the variation of parental lines of modern Philippine rice cultivars.

It is expected that maximum amount of heterosis will be manifested in cross combinations involving the parents belonging to the most divergent clusters. There are also reports that selection of parents for hybridization should be from two clusters having wider inter-cluster distance to get maximum variability in the segregating generations and subsequent selection of ideotypes [32,33]. In the present study, relatively high genetic distance existed between cluster III and VII, II and VII followed by V and VII, III and VI. Keeping in view the above observations, crossing between the genotypes from cluster III and VII will give heterosis for growth duration and breeder can choose desirable one (short, intermediate, long) from the segregating generations as these two clusters are diversified from each other mostly for growth duration contributing traits. Again to get heterosis in hybrid and variations in segregating generations for yield per plot & filled grain per panicle; days to panicle initiation, plant height & number of effective tillers per hill; 50% flowering, days to maturity & flag leaf area crossings between the genotypes from cluster II & VII; V & VII; III & VI, respectively are more sensible.

Cluster no.	DPI	DF	DFC	DTM	DTD	FLA	DM	PH	TN	PL	FG	UFG	YPP
I	65.88	95.13	99.73	105.06	112.13	45.40	126.15	101.29	8.75	26.08	106.71	23.40	0.70
II	76.25	106.17	109.92	114.84	122.42	70.33	137.42	120.00	6.75	28.17	59.67	61.00	0.21
III	80.63	111.03	115.33	120.83	128.43	59.73	142.83	108.33	8.40	27.33	102.84	23.63	0.60
IV	67.59	97.67	102.11	107.74	115.15	49.93	128.92	103.07	8.44	26.18	70.00	35.93	0.35
V	72.33	105.56	110.00	114.00	121.66	67.56	137.00	130.00	5.89	25.33	116.78	24.78	0.82
VI	60.67	91.00	95.79	101.33	109 79	39.91	122.94	97.79	8.51	24.70	85.67	26.33	0.50
VII	53.62	80.76	84.29	89.71	96.67	38.67	109.67	99.57	9.52	23.19	115.33	23.90	0.87

DPI: days to panicle initiation, DF: days to 50% flowering, DFC: days to flowering completion, DMI: days to milking stage, DD: days to dough stage, FLA: flag leaf area (cm2), DM: days to maturity, PH: plant height (cm), ETN: number of effective tillers per hill, PL: panicle length (cm), FG: number of filled grains per panicle, UFG: number of unfilled grains per panicle, YPP: yield per plot (kg)

Table 5. Principal components (PCs) for thirteen traits in 60 rice genotypes

	PC1	PC2	PC3
Eigen value	7.85	2.27	1.21
% Variance	56.10	16.20	8.60
Cumulative (%) total variance	56.10	72.30	80.90
Co-efficient vector			
Days to panicle initiation	-0.337	-0.126	-0.116
Days to 50% flowering	-0.349	-0.110	-0.082
Days to flowering completion	-0.349	-0.104	-0.087
Days to milking stage	-0.348	-0.099	-0.108
Days to dough stage	-0.346	-0.093	-0.118
Flag leaf area (cm^2)	-0.221	0.126	0.557
Days to maturity	-0.343	-0.096	-0.120
Plant height (cm)	-0.197	0.079	0.456
Number of Effective tillers per hill	0.098	-0.123	-0.305
Panicle length (cm)	-0.181	0.170	0.339
Number of filled grains per panicle	0.087	-0.570	0.305
Number of unfilled grains per panicle	-0.107	0.515	0.104
Yield per plot (kg)	0.125	-0.514	0.305
% Variation explained	56.10	16.20	8.60

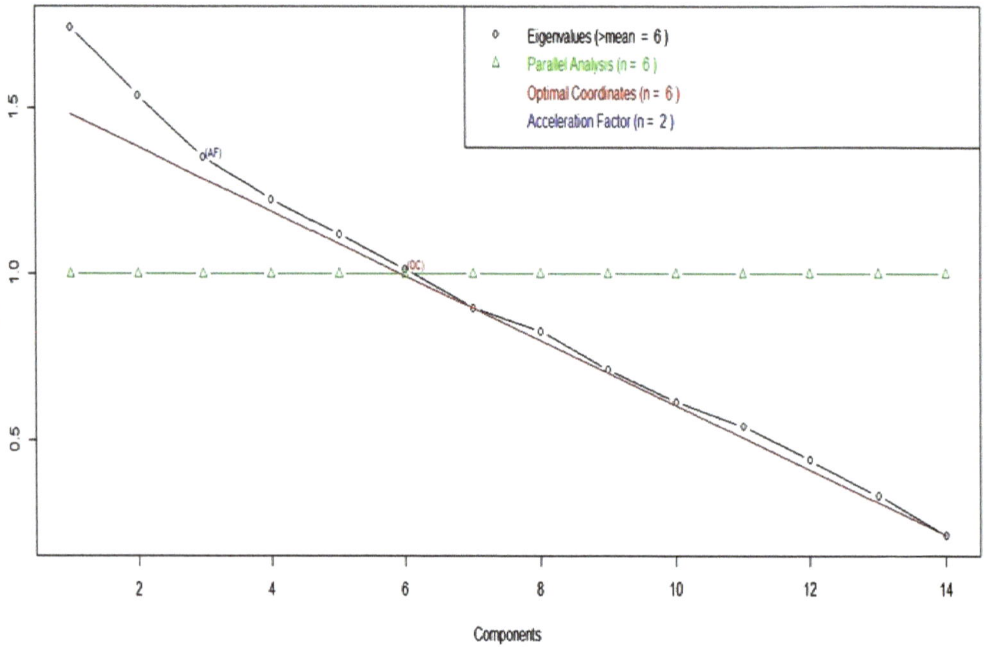

Fig. 1. Scree plot of principal component analysis of 60 genotypes of rice

Several reports showed that the parents separated by the medium magnitude of divergence also show high heterosis [34]. So genotypes from clusters differentiated from each other with medium magnitude of divergence are also important. Breeder can also perform hybridization among the genotypes of the same cluster having high intra-cluster distance to get heterosis and variation. Again for improving several desirable traits, combination breeding could be done among the genotypes from different clusters.

4. CONCLUSION

The multivariate analysis using Mahalanobis's D^2 statistics in association with principal component analysis clearly displayed the existence of wide diversity among the studied rice genotypes. Cluster analysis based on morphological traits suggested that the genotypes could be grouped, such assemblages are valuable for breeders to identify suitable parents for further breeding program to improve the specific traits. As morphological variations are influenced by environmental factors also, molecular markers can be used for further confirmation of such groupings.

ACKNOWLEDGEMENTS

The authors acknowledge the Ministry of Science and Technology of Bangladesh for providing the funds to accomplish the research work under the National Science and Technology (NST) fellowship. The authors are also thankful to the Department of Genetics & Plant Breeding, Bangladesh Agricultural University, Bangladesh for providing seeds of the rice genotypes and field facilities of the experimental farm.

COMPETING INTERESTS

Authors have declared that no competing interests exist.

REFERENCES

1. Sasaki T, Burr B. International rice genome sequencing project: The effort to completely sequence the rice genome. Curr Opin Plant Biol. 2000;3(2):138-142.
2. Chakravarthi BK, Naravaneni R. SSR marker based DNA fingerprinting and diversity study in rice *(Oryza sativa.* L). Afr J Biotechnol. 2006; 5(9):684-688.
3. Abodolereza A, Racionzer P. Food Outlook: Global market analysis. 2009;23-27.
4. Chowdhury MR. Population challenge facing Bangladesh. Public Journal of Social Sciences. 2011;11:2. Available:www.google.com/site/openjournals/journal-of

5. CIMMYT. The genetics and exploitation of heterosis in crops: An International Symposium, 17-22 August 1997, Mexico City, Mexico: Book of Abstracts. CIMMYT; 1997.
6. Joshi AB, Dhawan NL. Genetic improvement in yield with special reference to self- fertilizing crops. Indian J. Genet. 1996;26:101-13.
7. Anand I, Murty B. Genetic divergence and hybrid performance in linseed. Indian J. Genet. Plant Breed. 1968;28(2):178-185.
8. Prasad S, Singh T. Heterosis in relation to genetic divergence in maize *(Zea mays* L.*).* Euphytica. 1986;35(3):919-924.
9. Chauhan V, Singh B. Heterosis and genetic variability in relation to genetic divergence in soybean. Indian J. Genet. Plant Breed. 1982;42(3):324-328.
10. Cox T, Murphy J, Rodgers D. Changes in genetic diversity in the red winter wheat regions of the United States. Proceedings of the National Academy of Sciences. 1986;83(15):5583-5586.
11. Barrett B, Kidwell K, Fox P. Comparison of AFLP and pedigree-based genetic diversity assessment methods using wheat cultivars from the Pacific Northwest. Crop Science. 1998;38(5):1271-1278.
12. Thompson JA, Nelson RL, Vodkin LO. Identification of diverse soybean germplasm using RAPD markers. Crop Science. 1998;38(5):1348-1355.
13. Hallauer AR, Miranda J. Quantitative genetics in maize breeding.2nd ed. Iowa State Univ Press, Ames, IA. 1988;28:869.
14. Arunachalam V. Genetic distance in plant breeding. Indian J. Genet. Plant Breed. 1981;41:226-236.
15. Murty B, Arunachalam V, Saxena M. Cataloguing and classifying a world collection of genetic stocks of sorghum. Indian J. Genet. Plant Breed. 1967;27:1-312.
16. Vairavan S, Siddiq EA, Arunachalam V, Swaminathan MS. A study on the nature of genetic divergence in rice from Assam and North East Himalayas. Theor. Appl. Genet. 1973;43(5):213-221.
17. Virmani SS, Sharma H. Manual for hybrid rice seed production. Int. Rice Res. Inst. 1993;34-36.
18. Utz H. PLABSTAT (Version 2N), A computer program for the computation of variances and covariances. Institut of Plant Breeding, Seed Science and Population Genetics, Universität Hohenheim, Stuttgart; 2007.
19. Team RC. R: A language and environment for statistical computing; 2012.
20. Rao CR. Advanced statistical methods in biometric research. John Wiley & Sons, New York; 1952.
21. Huidong M, Shiliang G. Cluster analysis for agronomic characters of barley varieties in Jiangsu-Zhejiang, Shanghai area. Scientia Agricultura Sinica. 1987;3:004. Chinese.
22. Panigrahi KK, Sarka KK, Baisakh B, Mohanty A. Assessment of genetic divergence in potato (*Solanum Tuberosum* L.) genotypes for yield and yield attributing traits. International Journal of Agriculture, Environment and Biotechnology. 2014; 7(2):247-254.
23. Gohil R, Pandya J. Genetic diversity assessment in physic nut *(Jatropha curcas* L.). International Journal of Plant Production. 2008;2(4):321-326.
24. Gashaw A, Mohammed H, Singh H. Genetic divergence in selected durum wheat genotypes of Ethiopian plasm. Afr Crop Sci J. 2007;15(2):67–72.
25. Yatung T, Dubey RK, Singh V, Upadhyay G. Genetic diversity of chilli (*Capsicum annuum* L.) genotypes of India based on morpho-chemical traits. Aust J Crop Sci. 2014;8(1):97-102.
26. Zia-ul-Qamar, Akhtar J, Ashraf M, Akram M, Hameed A. A multivariate analysis of rice genetic resources. Pak J. Bot. 2012;44(4):1335-1340.
27. Akter K, Bashar M, Iftekharuddaula K, Ahmed M, Rashid E. Genetic diversity among irrigated traditional and modern rice germoplasm. J. Biol. Sci. 2002;2(10):659-661.
28. Latif M, Rahman M, Kabir M, Ali M, Islam M, Rafii M. Genetic diversity analyzed by quantitative traits among rice (*Oryza sativa* L.) genotypes resistant to blast disease. Afr J Microbiol Res. 2011;5(25):4383-4391.
29. Guei R, Sanni K, Abamu F, Fawole I. Genetic diversity of rice (*Oryza sativa* L.). Agronomie Africaine. 2005;5:17-28.
30. Sanni KA, Fawole I, Ogunbayo A, Tia D, Somado EA, Futakuchi K, Sié M, Nwilene FE, Guei RG. Multivariate analysis of diversity of landrace rice germplasm. Crop Science. 2012;52(2):494-504.
31. Caldo R, Sebastian L, Hernandez J. Morphology-based genetic diversity

analysis of ancestral lines of Philippine rice cultivars. Philippine Journal of Crop Science. 1996;21(3):86-92.

32. Rahman M, Acharya B, Sukla S, Pande K. Genetic divergence in low land rice genotypes. Oryza. 1997;34(3):209-212.

33. Bose LK, Pradhan SK. Genetic divergence in deep water rice genotypes. J.C. European Agric. 2005;6:635–640.

34. Mian M, Bahl P. Genetic divergence and hybrid performance in chickpea. Indian J. Genet. Plant Breed. 1989;49(1):119-124.

8

Nutrients Metabolizability of Nectar for Long-Term Maintenance of *Amazilia amazilia* and Effect of a Dietary Fiber Source

Carla Fabro[1*], Stefano Rimoli[2] and Piero A. Susmel[1]

[1]Department of Agricultural and Environmental Science, University of Udine, Italy.
[2]Margherita Hack Hummingbirds Centre, Torreano di Martignacco (UD), Italy.

Authors' contributions

This work was carried out in collaboration between all authors. Author CF has made the contribution to design of experimental work, to acquisition, analysis and interpretation of data. She was also involved in the writing of manuscript. Author SR has made the contribution to formulation of the diets and in the revision of the paper. Author PAS has coordinated all the experimental work and has made contribution in writing manuscript and in the critical revision. Each author should have participated sufficiently in the work to take public responsibility for appropriate portions of the content. All authors read and approved the final manuscript.

Editor(s):
(1) Yeamin Hossain, Department of Fisheries, Faculty of Agriculture, University of Rajshahi, Bangladesh.
(2) Anthony N. Rezitis, Department of Business Administration of Food and Agricultural Enterprises, University of Patras, Greece.
Reviewers:
(1) Anonymous, USA.
(2) Anonymous, France.

ABSTRACT

Aims: To formulate an artificial nectar for long-term maintenance of captive hummingbird, to measure the metabolizability of nutrients in the artificial nectar and to evaluate the effects of dietary fiber (inulin) addition on its digestive utilization.
Study Design: Two groups of seven hummingbirds randomly distributed in two diets with different levels of inulin (0-7% DM).
Methodology: The trial was conducted on two groups of seven non-reproductive *Amazilia amazilia* averagely weighing 4.0 g, giving nectar containing about 179 g/l dry matter (DM). DM content of sucrose and hexoses was 92.3%, (0.75 M), that of crude protein (CP) 4.3% and fat (EE) 2.3%. Inulin was added in the ratio of 7 g/100 g, as fed, of nectar powder. The experiment lasted 24 days; the

Corresponding author: E-mail: piero.susmel@uniud.it;

birds were separately housed in experimental cages.
Results: The daily DM intake of hummingbirds was about 2054 mg/d and upraised by 14%, to 2348 mg/d, when inulin was added to the nectar; the energy intake averaged 34 kJ/d. The average metabolizabilities of nectar were about 94% for DM and energy, 98.4% for sugars, 57% for CP and 44% for EE. The absorption of DM, sugars, energy and protein were statistically reduced by inulin addition, instead that of ash increased. From the balance trial it results that about one third of inulin should undergo a fermentation in the intestine. Adjustment of sugar and protein contents and of protein quality to hummingbirds' species and rearing environment should be useful.

Keywords: Captive hummingbirds; Amazilia amazilia; long-term maintenance; artificial nectar; inulin; hummingbird metabolizability.

1. INTRODUCTION

Hummingbirds, according to their feeding habit, are defined as nectivorous. Around 320 species, weighing from 2 up to 20 g have being classified in various environmental and ecological niches. They collect nectar from many flower species, but also eat pollen, ash, sand, tree sap, spiders, insects and other small arthropods [1,2]. For a very long time in some area of North, Central and South America wild hummingbirds are attracted offering artificial nectars in feeders [3]. Many commercial types of nectar, containing sugars and other few ingredients, are offered to attract free-living hummingbirds, e.g. Roudybush, Wombaroo, Nectar+, Perky-Pet.

More problematic is to formulate soluble complete diets to raise, maintain and breed hummingbirds in captivity for life. From the available information, it appears that hummingbirds have been kept in captivity since the middle of the 19th century, but the formulation of a complete diet, more or less purified, was only occasionally investigated and made known in the past. [4] proposed an artificial nectar containing honey or sugar, meat extract and condensed milk, almond oil and vitamins. [5] fed hummingbirds various diets and different ingredients recording the results of chemical analysis and the animal response, obtaining successful results in long-term maintenance and in reproduction. [6] offered a sucrose solution 0.5 M (171 g/l solution) added with 5.2 g/l of concentrate supplements containing protein, fat and carbohydrates. [7] developed a purified liquid diet containing about 25% of dry matter (DM), 80% sucrose, 3% protein (CP), delivered as commercial isolated soy protein supplemented with DL-methionine, starch (5%), corn oil (3,7%), mineral and vitamins. [8] use a maintenance diet composed by 20% sucrose, 0.5% amino acids (Glicopan P), 3% oil, 0.03% vitamins (Vita Gold) and 1% calcium carbonate. [9] dealt with the subject of nutritional exigencies of captive hummingbirds discussing many issues and suggested the formulation of a diet containing 88.5% of sugars, oil 2%, 6.5% of CP and a defined amount of each of 20 amino acids, supplemented with a complete array of mineral and vitamins. All these formulations show some common issues and give essential indication on how to devise artificial nectars for maintaining different hummingbird species for long time. It also needed to consider the results of scientific research carried out either on free-living birds, focusing on feeding habit and feeds preference and intake, or on captive individuals, for functional physiology and nutritional requirements.

In practice, to successfully maintain and reproduce captive hummingbirds, the artificial nectar must satisfy all nutritional needs, miming the inputs of a natural feeding. Energy and main nutrients requirements for maintenance, flight and reproduction have been differently studied and indicated [9-12]. From on hand knowledge, it appears that hummingbirds do not deliberately forage for fiber, intended as insoluble and indigestible carbohydrates of plants cell wall. Conversely, they prey variable amount and species of arthropods rich of chitin, 5÷20% DM [13], a polysaccharide which is main constituents of the exoskeletons of many arthropods, classified as added fiber having physiological properties similar to dietary fiber [14,15]. The digestion of chitin requires chitinases [16]. Many bacteria and fungi species, as well as certain invertebrates, fishes, birds, terrestrial mammals and humans produce chitinases, which permit the breakdown of polymeric chitin to oligosaccharides and monomeric sugar subunits [13]. Either chitinase production in birds and mammals could occur through a chitinoclastic bacterial gut flora, or by glandular production of chitinases [17] instead, according to [18], mammals lack chitinases activity. The chitin and chitosan digestibility, mainly measured in marine species, but also in mammals and birds, is variable and affect lipids digestion [19-22]. This topic appears not yet investigated in hummingbirds as well as the role of this insoluble dietary fiber on digestion and absorption of other nutrients and its potential beneficial physiologic and healthy effects.

If arthropods are not regularly available, other source of dietary fiber might be given in the artificial nectar. Chitin, chitosan, produced by alkaline deacetylation of chitin obtained mainly from marine crustacean, and other vegetable insoluble fiber sources are impractical, because insoluble deposit sediments or surfaces in the feeders. Inulin, a heterogeneous assortment of fructose polymers found in a number of vegetables, fruit, and grains, was already used as soluble fibrous ingredients in animal nutrition and appears suitable for the scope [23,24]. For this reason, we consider to evaluate the effects of the addition of inulin on digestive utilization of the artificial nectar. Beside the nutritional facts, when hummingbirds are maintained *ex situ,* also health problems might arise and some aspects as intestinal mycoses, liver steatosis should be monitored and prevented.

Available nutritional data are still incomplete and sketchy and some physiological aspects and behaviors of hummingbirds are undisclosed so that some solution must be attained by trial and error.

The objective of our experiments is to measure

i. the metabolizability of the nutrients in a complete diet formulated to maintain in good health captive adult *Amazilia amazilia,* after a critical review of the existing literature, and
ii. the nutritional effect of inulin addition to the nectar.

2. MATERIALS AND METHODS

In Miramare park (Trieste – Italy), at the Centre for the Safeguard of hummingbirds, in fully air conditioned aviaries in a natural mimic environment with tropical plants and where fruit flies (*Drosophila* sp.) were grown in buckets containing fruits, lived about 90 hummingbirds mainly belonging to *Amazilia amazilia* species. *Amazilia* are medium-sized trochilid (length 8-11 cm, weight 4-7 g) that inhabits arid and semiarid habitats of the tropical Pacific lowlands and adjacent subtropical Andean slopes, eg: edges of deserts scrub and dry forests, of Ecuador and Peru [25]. It is a territorial and nectivorous bird, which by individuals or pairs defends its territories against conspecifics or other nectivorous species [26].

The nectar was formulated on the basis of a critical review of results from studies conducted in the wild and in the laboratories, which has been summarized at the beginning of the Discussion (§ 4) and progressively adjusted considering the intake and few vital and health parameters. Finally, we have evaluated the nutrients metabolizability of the complete artificial nectar and the effect of the addition of inulin on intake and metabolizabilities.

The experiment was conducted during the non-reproductive period. Adult *Amazilia amazilia* hummingbirds were randomly assigned to two groups of seven hummingbirds each, 4 males and 3 females, fed artificial nectar (Nt) or artificial nectar added with inulin (Ni). A week before the beginning of the trial, each bird was housed separately in an experimental cage (0.80 m x 0.50 m x 0.90·m) fitted with a removable tray for *excreta* collection. Cages were allocated in a separate experimental room kept at 25°C and provided with constant light artificial photoperiod of 12·h L: 12·h D and with full spectrum lights bulbs for the "sun bath" (3 times/d for 30-40 min). In the experimental environment, fruit flies were not present. During acclimation and experimental periods, of 7 and 17 days, respectively, the nectar was offered *ad libitum* in a 13 ml glass calibrated tube placed through a hole of the cage. The nectar powder (Nt) is prepared once a week using the same ingredients. The nectar contains sucrose 47.2%, glucose 19.2%, fructose 19.2%, hydrolyzed soy protein and collagen (Wellesse–Ferndale USA) 1.5%, a mix of essential amino acids 2.2% (Nutricia –Danone), maize and soybean oils 2.5%, flower's Bio-pollen 4.6% (Fior di Loto – Torino, Italy), silimarin (Legalon – Rottapharm-Madaus) 1.1%, Bio-Mos (Alltech) 1.1%, mineral and vitamins supplement 1.3%. As for sugar concentration, Nt and Ni were equimolar, 0.75 M or 0.5 M sucrose equivalent. Inulin was added to the nectar powder (Ni) in the ratio of 7 g/100 g, as fed. We used inulin 90% (Farmalabor-Italy), extracted from *Dahlia variabilis, Helianthus tuberosus* and other *Compositae*. This inulin is classified as "soluble non digestible" fiber (solubility 120 g/l) and is composed by fructans which, having a ß-configuration of the anomeric C2 in its fructose monomers, resists hydrolysis by intestinal digestive enzymes, which are specific for α-glycosidic bonds [27,28]. To reduce the fractionation, due to precipitation or surfacing, of less soluble or insoluble ingredients and the fermentation, nectar was renewed twice daily, at 7 am and at 3 pm; the volumetric nectar intakes were individually recorded at the same time. Water was offered *ad libitum* in 10· ml water feeders and in birdbaths. Nectars were daily diluting in 200 ml of distiller water 40 g, as fed, of Nt powder, or 43 g (inulin 3 g) to prepare Ni. Ni contains 7.11% of inulin on DM basis. The volume of Nt solution, equals 210 ml and that of Ni 226 ml. *Excreta* were collected from removable trays twice daily using plastic spatulas, put in plastic bottles, immediately weighed and frozen at -20°C. Before the chemical analysis, the two pools of feces were weighed and thoroughly mixed in 250 ml of distilled water. DM was measured on 15 ml of sample by drying at 60°C to constant weight. On the remaining CP (Nx6.25), Ether Extract (EE), ash [29], Neutral Detergent Fiber (NDF) [30], were determined. The gross energy (GE) content (expressed in kJ) was either measured, using an adiabatic calorimeter (IKA C7000), and checked using the average energy values of different nutrients proposed by [31] Water-soluble carbohydrate contents (NFE) of DM was calculated according to the following: NFE= 100% -% CP-% ash -% EE -% NDF. The results of feeds analysis, which are the average of duplicate determinations, are reported in Table 1.

The apparent metabolizability of different nutrients was calculated using the following formula: Metabolizability (%)=[1-Intake (g)/*Excreta* (g)] x 100]. Data were statistically analyzed by ANOVA, and the differences among group means were separated by Dunnett test, setting the significance levels at P<0.01 (SPSS Statistics 17.0). In the tables F is the ratio between variance of the group means / mean of the within group variances; the significance level is obtained from the F distribution with numerator and denominator degrees of freedom (df).

The coefficient of variability (CV), calculated as a percentage of average/standard deviation ratio, was used to describe the range of variation of weights (LW), DM intake and excretion among animals. Body weight was recorded throughout the acclimation and experimental period with an analytical balance (accuracy±0.001·g). Initial LW were 4.08 g and 4.14 g for Nt and Ni groups and at the end of the experimental period 3.98 g and 3.80 g, respectively.

3. RESULTS

The chemical composition of nectars and the nutrients' content of 1 l of solutions are in Table 1. Carbohydrates (NFE) make up the 92.3% DM of Nt, the CP and EE contents are 4.28% and 2.30% of DM, respectively. Sucrose constitutes the 55.1% of added sugars, the balance is equally provided by glucose and fructose, 22.4%; the difference between added sugars and NFE is due to carbohydrates in the additives and other ingredients. The NDF content come from the natural ingredients used to make up the nectar, as pollen and probiotics, and corresponds to about 0.2% on DM.

The addition of inulin – 7.1% of total DM - diluted the nutrients content of Ni. The nectar solutions contained the same amount of DM, 179 g/l, and were equimolar 0.75 M, or 0.5M sucrose equivalent.

The average daily intakes of hummingbirds are reported in Table 2. When hummingbirds received the nectar containing inulin, the DM intake was significantly increased (P<0.01) as those of the remainder nutrients. Inulin is included in NFE and, as expected, both DM and NFE intakes largely differ between the two nectars; though the intakes and the differences among the intakes of other nutrients between Nt and Ni were lesser, resulted statistically highly significant (Table 2). The daily energy intakes of Nt and Ni, 34.16 and 39.12 kJ/d respectively, were also significantly different (P<0.01).

The DM content of droppings was much lower than that of nectars, but its nutrient concentration was noticeably higher. The presence of inulin significantly increased the water content of droppings, which appear more viscous. The bulking effect of dietary fiber was evident: the excretion of DM and nutrients was significantly increased (Table 3). In consequence, the DM composition of droppings notably changed. As expected, a bigger quantity of NFE was excreted, due to the inulin intake (Tables 2, 3). In addition, the contents of CP and EE were significantly increased, while that of ash has been reduced. The amounts of NDF excreted are identical, nearly matching the intakes.

The CV of individual nectar intake within days and animals was similar between the two groups. The CV was of 6.0%, lower than that observed for the excretion (CV=19.1%).

Hummingbirds receiving Nt on average maintain their LW (-11 mg), those of the Ni group lost some LW (-34 mg) during the period of 24 days.

The metabolizabilities of normal nectar DM resulted 94.6% (Table 4).

The sugars metabolizability of Nt was almost complete and those of CP and of fat were 57% and 44%, respectively. The absorption of minerals equals 41.1%. The digestibility of CF and of cell wall components was negligible. The addition of inulin significantly affected the metabolizability of DM, energy, CP, NFE and ash reflecting the statistically significant change of droppings composition (Table 3). The energy metabolizability of Nt was 94% and 90% that of Ni. Comparing the DM, NFE and energy excretions of the two nectars, we could deduce that, for the most part of inulin, passed through the digestive tract.

Table 1. Dry matter and nutrients content of feeds

Feed (% DM)	**DM**	**CP**	**EE**	**CF**	**Ash**	**NDF**	**NFE**	**GE** (kJ/g)
Nt	93.76	4.28	2.30	0.28	0.95	0.19	92.28	16.63
Ni	93.89	3.99	2.14	0.26	0.88	0.17	92.81	16.66
Inulin	95.64	0.18	-	-	0.02	-	99.80	16.30
Nectar (g/l)								**GE** (kJ/g)
Nt (g)	178.56	7.65	4.11	0.50	1.69	0.33	164.78	2970
Ni (g)	178.81	7.14	3.83	0.46	1.58	0.31	165.96	2979

DM: Dry Matter; CP: Crude Protein; EE: Ether Extract; CF: Crude Fiber; NDF: Neutral Detergent Fiber; NFE: Nitrogen Free Extract; GE: Gross energy; Nt: artificial nectar; Ni: artificial nectar with inulin

Table 2. Average daily intake of nectar (ml) and nutrients (mg/d)

	Nt	**Ni**	**F**	**df**
ml	11.5^{B}	13.13^{A}	40.89	32
DM	2053.98^{B}	2348^{A}	41.76	32
CP	87.99^{B}	93.73^{A}	9.36	32
EE	47.3^{B}	50.23^{A}	8.45	32
Ash	19.48^{B}	20.72^{A}	8.92	32
NDF	3.81^{B}	4.05^{A}	8.45	32
NFE	1895.38^{B}	2179.25^{A}	45.42	32
GE (kJ)	34.16^{B}	39.12^{A}	45.37	32
CF	5.72^{B}	6.08^{A}	8.45	32

DM: Dry Matter; CP: Crude Protein; EE: Ether Extract; CF: Crude Fiber; NDF: Neutral Detergent Fiber; NFE: Nitrogen Free Extract; GE: Gross energy; Nt: artificial nectar; Ni: artificial nectar with inulin; F: Anova test; df: degree freedom. In the same row, mean values with different capital letters are significantly different (P< 0.05)

Table 3. Average daily excretion and chemical composition of droppings (mg/d)

	Nt	**Ni**	**F**	**df**
Fresh	1110.53^{B}	2890.08^{A}	84.87	32
DM	110.53^{B}	221.68^{A}	83.86	32
CP	37.8^{B}	46^{A}	8.47	32
EE	26.52^{B}	33.25^{A}	11.11	32
Ash	11.49^{A}	9.62^{B}	7.49	32
NDF	3.70	3.84	0.31	32
NFE	30.39^{B}	128.35^{A}	219.04	32
GE (kJ)	2.07^{B}	3.74^{A}	67.78	32
CF	6	6	0.33	32

DM: Dry Matter; CP: Crude Protein; EE: Ether Extract; CF: Crude Fiber; NDF: Neutral Detergent Fiber; NFE: Nitrogen Free Extract; GE: Gross energy; Nt: artificial nectar; Ni: artificial nectar with inulin; F: Anova test; df: degree freedom

Table 4. Energy, DM and nutrient metabolizability (%)

	Nt	Ni	F	df
DM	94.62[A]	90.52[B]	59.4460	32
CP	57.08[a]	50.77[b]	4.3355	32
EE	43.96[a]	33.59[b]	6.5733	32
Ash	41.05[B]	53.43[A]	14.0682	32
NDF	2.87	4.75	0.0925	32
NFE	98.39[A]	94.09[B]	191.0953	32
GE (kJ)	93.93[A]	89.99[B]	51.2376	32
CF	1.95	3.74	0.0822	32

DM: Dry Matter; CP: Crude Protein; EE: Ether Extract; CF: Crude Fiber; NDF: Neutral Detergent Fiber; NFE: Nitrogen Free Extract; GE: Gross energy; Nt: artificial nectar; Ni: artificial nectar with inulin; F: Anova test; df: degree freedom

4. DISCUSSION

4.1 Nectar Formulation

Many aspects have to be considered for formulating artificial nectar for long-term maintenance of different species of hummingbirds, but primarily the ingredients to use to achieve suitable palatability, to satisfy requirements and to compose a stable solution/suspension.

Wild hummingbirds meet their energy requirements essentially from nectar sugars, while protein needs are independently satisfied by eating a changeable amount of small arthropods [32-36], other stuffs and possibly pollen [37]. Hummingbirds prefer flowers providing nectars with sucrose, fructose and glucose [6] and with a higher proportion of sucrose to hexoses. Rarely contains a single sugar [38-43]. According to [44] hummingbirds select nectars containing 64.4±18.5%) of sucrose. Less univocal is the sugars preference in captive hummingbirds, which prefer the more concentrated sugar solution [6,45]. Many factors influence sugars preference and the first adaptation to a definite sugar solution or to a frequent variations sugars solution requires several hours or feeding sessions [6,46-48]. Different results indicate that sucrose is generally preferred alone or when paired with glucose [49,50].

The preference for solutions of sucrose and hexoses should be compared at the same concentration on equicaloric basis. At high concentration of sugars (≥ 25%), hummingbirds do not discriminate among sugars; instead, at lower concentration, equicaloric, equimolar and equiweight solutions of diverse combination of sugars rank in a different order of preference [48,51,52].

According to these suggestions, we consider that the simultaneous presence of sucrose and its hydrolytic products at moderate concentration could represent a more valid starting option for hummingbirds living in a confined artificial environment. Then sucrose, glucose and fructose are added to compose 86% of the nectar DM, where sucrose contributes with the 56% to the mix of three sugars and the rest is equally balanced between the two hexoses. The rest of sugars come from pollen and supplements.

There are some grounds for inclining in the direction of artificial nectar having moderate sugars' concentration, which appears to favor hexoses preferences and absorption [45]; firstly, because the sugars' concentration of nectars does not influence the rate of daily energy intake [53]. For these reasons, being in situation where nectar availability is not limiting and foraging does not require a high-energy cost, we considered preferable to try the option of offering hummingbirds relatively diluted nectar. The drinking solution was prepared to keep DM concentration around 18%, i.e. 0.5 M sucrose equivalent.

Energy and main nutrients requirements for maintenance, flight and reproduction have been differently studied and indicated [9-11,54,55]. More often energy requirement is associated to time budget since different activities have different energetic cost [34]. Sugars in the nectar provide birds for the most part of energy. Commonly, the daily sugar intake is assumed to cover the energy requirement. The energy requirements of free-living hummingbirds (FMR) has been measured with different methods or estimated in different environmental situations. FMR results high and largely variable, from 29.1

to 81.7 kJ/d [10,56-59]. Typically, the metabolic rate measured in captivity ranges from 7.3 kJ/d to 10 kJ/d per g LW [5,60].

The overall concentration of amino acids in nectar of flowers lies within the range of 30-400 µg/ml and their profile is variable; these amounts are considered insufficient to meet the requirements of nectivorous [61,62].

For [34], the single DM weight of flies is 1.5 mg. The protein content of arthropods is high, from 54% to 77% on DM, depending upon species, the biological value of arthropods body protein, as results from amino acids profile, is variable and usually rather high [63,64]. [1] observed that arthropods are extensively digested in the stomach within 3-4 hours, following only brief storage in the crop. Evidently not all parts are equally digested; the components of soft body and some constituents of cuticle (e.g. lipid, waxes and soluble protein) are probably quite digestible and fragments of their body are detectable in faeces. Unfortunately, not all this information is supported by values or measures.

In captive birds, a convenient and easy way to offer arthropods could be that of raising *Drosophila* in the same aviary or in a bottle; *Diptera* are also within the preferred preys in free-living hummingbirds. [12,36] observed that when nectar and fruit flies are separately offered, their intakes are independently regulated.

The number of fruit flies a hummingbird has to ingest to cover N maintenance requirements is differently quoted, as 312 by [9], 150 by [12], 283 by [65]. According to [5], a hummingbird of 3 g LW ingests 677 flies/d. In contrast, [35] assume that hummingbirds could meet the daily protein requirement by ingesting only 38 flies each containing 2.1 mg of DM, with a protein content of about 60% and a metabolizability of 80%. [12] observed that *Sephanoides sephaniodes* weighting about 6 g require approximately 150 fruit flies to maintain body mass.

If the animal preys are not available, a protein substitute must enter the nectar composition in a prearranged fixed amount, as well as alternative sources of fat, ashes, and dietary fiber.

Hummingbirds do not specifically necessitate protein but are able to live eating a diet containing free amino acids [66]. As the prudential criteria was to guarantee an adequate intake of protein to cover synthesis of semi essential and non-essential amino acids and possible metabolizability inefficiencies, we opted for a nectar containing 4.28% on DM or 7.65 g/l of CP, from a balanced mix of amino acids with the addition of an extra amount of organic N (hydrolyzed protein).

In some experiments, diets were formulated adding from 2% of crude oil [9], to 3% of corn oil [8] and 3.7% of corn oil [67,68]. On fat utilization by hummingbirds, any specific information has been experimentally obtained. Hummingbirds oxidize carbohydrates preferentially, but when sugar availability is shortened, as after overnight fasting, quickly switch to long chain fatty acids, also sparing the exiguous glycogen reserves [59,69,70]. Body fat content is variable and changes during the day and with season. Highest variations are found in migratory species, which can store more than 40% of body weight as fat [71].

Arthropods should provide hummingbirds also with essential unsaturated fatty acids, being one of the richest animal feed sources [72]. *Drosophila sp.*, are a significant source of fat (8-15%) and of mono (13-51%) and poly unsaturated fatty acids (8-31%) [73,74]. The effective exigencies of unsaturated fatty acids are unknown, but a certain amount of these fatty acids should be available to hummingbirds so that oil has to be added to nectar. This opportunity must be considered especially when captive birds have not the chance of preying live fruit flies. On this basis, it was decided to introduce in the formula 2.5% of a 1:1 mix of olive and soybean oils.

The hummingbirds do not collect pollen and only groom it off their bills [75]. The hummingbirds' attitude for eating pollen and their capability of extracting nutrients from ungerminated pollen is controversial [37,76]. [7] feeding Anna's and Costa's hummingbirds 20% pollen on diet DM found that the digested amount was always nil or very low (<7%), however reported a higher digestibility of pollen in Anna's nestling hummingbirds, but nestling are offered solid and slow passing food, mainly insects [77]. [78] comments that pollen digestibility is limited by the fast transit trough gastrointestinal tract, but it also depends upon bird's age and species.

4.2 Nutrients Intake and Metabolizability

The LW of hummingbird species largely varies and, to compare values, daily intakes are more often referred to metabolic weight. Hummingbirds suckled 886 *vs.* 1004 ml/kg$^{0.75}$ of Nt and Ni nectar respectively, corresponding to 121.5 and 134.4 g/kgLW$^{0.75}$ of DM. When inulin was added to the nectar, the hummingbirds increase the average daily intake by 14.3%.

When concentration of sugars is low, as in this experiment, the intakes should be compared at the same molarity or energy or sugar contents. [53], with 0.75, 0.5, 0.25 M sucrose nectars, measured increasing intakes (8.6, 12.6 and 24.4 ml/d). [49], with reference to captive hummingbirds of different species, determined that intake steadily increase when the sucrose molarity of nectar diminishes and, at 584 mmol/l equals 12 ml/d. Practically, higher the sugar molarity lower is the intake but, *ceteris paribus*, the overall daily consumption and the metabolizability are the same. In our experiment (0.5 M, sucrose equivalent), the intake of Nt was 11.5 ml/d and the sugars (NFE) metabolizability resulted 98.4%, within the range of the ordinary values. In fact, sugar assimilation efficiencies measured in hummingbirds using traditional balance methods are higher than 95% and by and large independent from sugar concentration and from *digesta* retention time [45,49]. [79] measured sucrose digestibility of 98-99%; [80] found an efficiency of sucrose extraction of 97.1±0.3% on *Selaphorus rufus* hummingbirds. [81] reported that sucrose and hexoses are digested to the same high extent (>97%).

In our trial the DM intake of Nt nectar averaged 2.05 g/d (128 g/d/kgLW$^{0.75}$) and metabolizability resulted 94.6 %, as a combination of sugars metabolizability (98.4 %) and that of other nutrients which all together results 49.5 %.

In different experiments, where diverse nectars were given to several hummingbird species, the daily intakes of metabolizable energy varied from 29.1 to 138.5 kJ or from 5.63 to 9.32 kJ/d/g LW [10,12,54,58,82]. The apparent metabolizable energy intake of Nt and Ni were 32.1 and 35.2 kJ/d, which correspond to 7.97 and 8.87 kJ/d/g LW, values which are consistent with other experimental results. The Nt group along the 24 days of trial basically maintained their LW and those receiving Ni nectar, lost a little of LW.

Beside sugars, the Nt nectar contains other ingredients of different palatability and metabolizability, as protein, fat and minerals, which affect DM intake and metabolizability.

In trial, with *Sephanoides sephaniodes* weighing about 5.5 g, [12] compared nectars of different sugars and CP content – from zero to 11.2%. They observed an increasing DM intake from 93 to 139 g/d/kgLW$^{0.75}$ and a declining energy metabolizability (from 0.99 to 0.93); when CP was 4.7%, the DM intake was 119.5 g/d/kgLW$^{0.75}$ and the metabolizability 0.97. From an experiment of [68], who offered a liquid diet containing a rising percentage of CP - from 0 to 3% in 25% solids – we can easily calculate to some extent an increase of DM consumption from 123 to 130 g/kgLW$^{0.75}$. In another experiment, [54], comparing two diets differing for sucrose molarity supplemented with protein, vitamins and minerals (8 g/l), measured a DM intake of 2.22 g/d or 110 g/d/kgLW$^{0.75}$ (0.5 M) and 1.87 or 93 g/d (0.75 M); DM metabolizability in both groups was 0.95±0.02. Thus, when nectar contains protein, the intake is adjusted to sugars content and the DM metabolizability decreases. DM intake and metabolizability of diets comparable to Nt were similar to those we measured on *Amazilia amazilia*.

The average individual intake of CP was 88 mg/d (Nt) which correspond to 5.5 g CP/d/kgLW$^{0.75}$, equivalent to 14.1 mg Nitrogen (N)/d or 880 mg N/d/kgLW$^{0.75}$ (Table 2). In the previously reported experiments, [12] substituting part of sugars with an increasing amount of protein (ProMod–Abbott), observed that CP intakes grew up to 313 mg/d (14.1 g CP/d/kgLW$^{0.75}$). In this experiment and in that of [67], the DM intake increases by means of protein content of DM. [54] measured a daily CP intake of 80 mg (0.5M) and 40 mg (0.75M), corresponding to 3.8 and 1.9 g CP/d/kgLW$^{0.75}$ and observed that the different CP intakes depend upon sugar molarity. These and other results [68,83], lead to assume that, in a complete nectar diet, the frequency of bouts and the DM intakes depend upon the sugar concentration, but not from that of N. N excretion, when measured, results positively and, within a "physiological" level, almost positively correlated with the intake.

To conclude, when protein is offered in the nectar, hummingbirds are not capable to adjust the N intake to their exigencies, so that the CP content of the nectar solution must be decided on the basis the rate of ingestion of DM and the N metabolizability, verifying that the intake correspond to the requirements.

The protein content of nectar DM given hummingbirds to maintain their body condition varies from 2.4% up to 6.5%; usually the level protein considered more appropriate seems to be of about 4.5% CP on DM [9,12,54,67,83].

In our experiment, giving Nt containing 4.3% CP on DM, the average intake of CP was 88 mg/d or 8 mg/d of apparently metabolizable N (3.1 $g/kg^{0.75}/d$ of metabolizable CP).

On protein requirements and feeding of hummingbirds, not much information is available, but it is generally recognized that they have low N requirements compared to non-nectivorous birds [12,55,68,83]. [84] proposed simple equations, frequently used to weigh up the N requirement of passerines and non-passerines. The Minimal N Requirement (MNR) is satisfied by 430 mg $N/kg^{0.75}/d$, being the Total Endogenous Nitrogen Losses (TENL) of 270 mg $N/kg^{0.75}/d$. [9], on the basis of this MNR value, suggest a CP content of 6.5% DM in complete nectar for hummingbirds, assuming a N metabolizability no less than 80%.

[68], in an experiment on *Calypte costae* hummingbirds of 3.8 g LW, measured a similar MNR and TENL of 77.1 and 77.8 mg $N/kg^{0.75}/d$. Also [12], in the previously cited research on N requirement of *Sephanoides sephaniodes*, comparing MNR and TENL, measured giving also a "Zero N" nectar, obtained an identical value of 67.9·$mg/kg^{0.75}/d$. [83] offering three hummingbirds species of different LW a nectar containing casein acid hydrolyzate (Sigma®), gave evidence of a variable MNR. The values went from 85.5 to 158 $mg/kg^{0.75}/d$, and TENL, from 46 to 77.7 $mg/kg^{0.75}/d$; in a previous experiment, on a fourth species, measured a MNR of 62.5 $mg/kg^{0.75}/d$ and a TENL of 45.5 $mg/kg^{0.75}/d$ [85]. [12], shifting the N content in equicaloric nectars from 1.2 up to 11.2%, registered daily N intakes varying from 5.2 to 50.1 mg (from 32 to 316 mg CP) and only the hummingbirds receiving at least 4.7% CP did not lose weight. The digestibility coefficients were inversely related to the protein content and significantly differed among diets, decreasing from 68±7% to 48±3%. [8] also comment that these birds are unable to efficiently assimilate N in high protein diets, because of a protein overload.

On the basis of these results, [55] in a notable discussion paper on N requirements of birds, reassessing Robbins' allometric equation [84], propose for nectivorous-frugivorous birds an average MNR of 152 mg $N/kg^{0.76}/d$ and TENL of 54.1 mg $N/kg^{0.69}/d$. To compare the reported MNR and TENL, we recalculated the values according these metabolic weights, always obtaining lower values than the two averages calculated by [55]. Both MNR and TENL varied giving evidence that this requirement is largely influenced by nectar composition, by hummingbirds' species and by rearing and experimental conditions.

The relationship between MNR and TENL values measured in all these experiments depends also on CP metabolizabilty or retention efficiency, that is the slope of the linear regression between ingested and retained N. Amino acids, if are not quickly deaminated, are used mainly for synthesis of skeletal muscles and, to a lesser extent, of other protein compounds. The metabolizability depends on the biological value of protein. In the wild, hummingbirds complemented the nectar feeding catching arthropods the protein of which is highly metabolizable, 78.8±6.4% according to [12]. The few information of N metabolizability experimentally measured using different protein supplements are always lowers than that from fruit flies. [83], using casein acid hydrolyzate (SIGMA–St. Louis), measured N metabolizability within 49 and 54%. [12], giving different amounts of hydrolyzed collagen (Promod–Abbott, Columbus, USA), calculated a declining metabolizabilty from 68% to 48%, increasing with protein intakes. In our trial, the protein metabolizability was 57%.

All these N sources solubilized or dispersed in the nectar, other than natural supplies, which are separately ingested, should differently modify osmolarity and transit time and *affect* the intakes.

The average lipids intake of hummingbirds during the trial was of 47 mg/d and the excretion of 26.52 mg/d; the fat concentration in droppings DM was higher than 25%, but we had already observed that hummingbirds' *excreta* appear "greasy". In our trial, the lipids metabolizability in hummingbirds receiving normal nectar was 44%, indicating a moderate but effective absorption. Weight for weight the assimilated lipid was less than 1% of that of digested NFE. Under an energetic point of view, this value does not appear relevant to captive hummingbirds, but oil brings also indispensable essential fatty acids.

The intake of fibrous fractions (NDF) is reported in Tables 2. The basal intake was of about 3-4 mg/d. The amount of fibrous fraction of droppings averagely excreted by hummingbirds (Table 3) practically matches the intakes (Tables 2 and 3). The digestibility of CF and of NDF, did not significantly differ between nectars and was always very low and variable.

The addition of inulin to the nectar was followed by an increase of DM intake of about 14% (Table 2). Apart from significantly modifying the intake, the dietary fibrous source added to normal nectar significantly affected the nutrients' excretion and thus their metabolizability: those calculated for DM, NFE and Energy were reduced by about 4%; higher differences were measured for CP and EE, instead the absorption of ash was increased. [86] comments that various hypotheses have been tested to tentatively explain the modifications of the digestion and absorption of macronutrients, especially carbohydrates, triglycerides and minerals, which may occur in animals and humans, either by delaying gastric emptying and/or shortening small intestinal transit time [87]. In fact, the increase of daily DM intake and excretion stands for a shortened transit time. The absorption of ashes was significantly higher with Ni (53.4%) than Nt (41.1%). Similar effect of inulin has been observed in poultry, other animal species and in man [88,89]. The DM excretion was about doubled by inulin addition, but the difference between the DM and NFE contents of droppings collected from birds fed Nt and Ni nectar was lower than the inulin intake. We have tried to interpret our results, on which there is any specific reference, on the basis of existing knowledge and the rationale which premised this experiment. Crossing data of NFE intakes and excretions in Nt and Ni birds, the inulin disappearance through the digestive tract should concern less than one third of the intake (*i.e.* ~ 55 mg/d). In animals, inulin passes through the stomach and duodenum undigested and is easily available to the gut microorganisms. [90] report that hummingbirds possess high density of intestinal bacteria and a degradation activity on urinary uric acid, urea and urates, always there, to ammonia has been observed. So that also a fermentation activity on sugars could occur with production of volatile fatty acids (VFA). The viable presence of undigested inulin in the lower gut, where presumably stays longer, could favor the presence of microorganisms producing also different alloenzymes, *e.g.* inulinase, which modify to a certain extent indigestible carbohydrates. The major limit to this potential activity is the swiftness of transit as fluid passes through a hummingbird's digestive tract in about 0.5 h. [57]. In birds, diverse components of the same meal may move at greatly different rates through the different tracts of digestive system [78]. Soluble and insoluble fraction of dietary fiber have a high water holding capacity and if not digested they provide bulk to the droppings, differently altering the transit time of nectar in the different sections of the whole digestive tract.

5. CONCLUSIONS

In this research, we tested a complete diet, which seems to satisfy the major requirement of hummingbirds. In fact, only sugar quality and concentration in the nectar were studied in depth. Few, but consistent information is available on protein, but not with reference to its quality. The knowledge on exigencies of fat and fatty acids, structural, homeostatic and trace elements, vitamins is scarce. Next to that, little information exists on the digestive interaction among nutrients and simple feeds, which are necessary to obtain a relatively complete and balanced nutrients' input.

Beside nutritional issues, we also have to deal with dietetic and health aspects. In fact, some deaths, due to infections as aspergillosis and candidiasis, happened and occasionally the necropsy revealed an anamnesis resembling the occurrence of steatosis and/or hepatosis of uncertain origin. This induced to add MOS, silimarin and inulin to nectar and to consider even the use of a prebiotic as inulin, evaluating their effect on the nectar digestion.

The attempt of simultaneously measuring the metabolizability of main nutrients of a feed open to a different view on the digestive efficiency of hummingbirds and helps in testing other ingredients than the few generally used. There is much to understand when we consider that typically about 40 single nutrients among amino acids, fatty acids, hydro and liposoluble vitamins and minerals are required in the diets of birds.

The only option given captive hummingbirds was that of varying the intake and preying fruit flies. For practical reasons, only one or two nectar recipes could be prepared and offered to hummingbirds kept in an artificial environment. The results of these trials offered some indications that should be considered for a necessary further improvement or adaptation to different rearing conditions and hummingbirds' species.

The option of using sucrose and hexoses in a ratio resembling that of floral nectars preferred by hummingbirds seems to guarantee a satisfactory volumetric intake and an overall sugars. The average daily intake of DM (2.05 g/d) and the average volumetric consumption of nectar find a good agreement with other experimental evidences obtained with nectar of similar sucrose molarity, digestibility. The intake of sugars and energy, (about 130 g/kg$^{0.75}$LW/d and 2150 kJ kg$^{0.75}$LW/d), seem adequate to cover the maintenance requirements.

When intake values are given as metabolizable N per unit of metabolic weight, we measured an intake (about 500 mg N/ kg$^{0.75}$LW/d) which was about twice the more suitable indication obtained by other Authors for the maintenance. On the basis of existing knowledge and on the result of this trial, assuming a CP metabolizability of 50-60%, the suggestion is to formulate the nectar containing not less than 3.5% of protein or even more if birds have to cover requirement for reproduction and molt.

The addition of dietary fiber significantly reduced the amount of absorbed fat. We try an interpretation of these results considering a possible effect of the inulin on the transit time, but this remains a nutritional aspect, which requires more careful evaluation. Even if the energy requirements of captive hummingbirds can be easily satisfied by sugars, a small amount of vegetal oil is useful to meet essential fatty acids exigency, which is unspecified, but probably not avoidable. If we assume that fat could be anyhow necessary, the low assimilation suggest to add oil in the amount commonly adopted (2-2.5%) to guarantee a small but significant metabolizable quantity.

The main role of fiber was thought to substitute for the indigestible fractions of fruit flies. For inulin a prebiotic effect could also be devised, with some benefit to the health condition of a captive bird living in an aviary. The addition of soluble fiber modified intakes and decreased the metabolizabilities of nutrients. Positive dietetic effects should compensate this negative consequence.

ETHICAL ISSUES

The experiment was conducted according the Ethical recommendations adopted by the University of Udine with reference to "Directive. 86/609/EEC concerning the protection of animals used for experimental and other scientific purposes".

ACKNOWLEDGMENTS

We like to thank Mr. Raniero Avanzi for his help as Animal keeper during the experimental phase of this work (Centre for the Safeguard of Hummingbirds, Trieste) and Mr. Andrea Fabris (University of Udine) for the valid support given in the chemical evaluation.

COMPETING INTERESTS

The authors declare that they have no competing interests.

REFERENCES

1. Remsen JV, Stiles F, Scott PE. Frequency of arthropods in stomachs of tropical hummingbirds. Auk. 1986;103:436-441.
2. Stiles FG. Behavioral, Ecological and morphological correlates of foraging for arthropods by the hummingbirds of a tropical wet forest. The Condor. 1995;97:853-878.
3. Sherman AR. Experiments in feeding hummingbirds during seven summers. Wils Bull. 1913;25:153-166.
4. Cordier AA. A better way to feed hummingbirds. Avic Mag. 1953;58:143-145.
5. Schelthauer W. Thomas Crowell Y. Hummingbirds. New York; 1967.
6. Hainsworth FR, Wolf LL. Nectar characteristics and food selection by hummingbirds. Oecologia. 1976;25:101-114.
7. Brice AT, Dahl KH, Grau CR. Pollen digestibility by hummingbird and psittacines. Condor. 1989;91:681-688.
8. Zanotto FP, Bicudo JEPW. Dietary protein and carbohydrate affect feeding behaviour and metabolic regulation in hummingbirds (*Melanotrochilus fuscus*). Rev Chil Hist Nat. 2005;78:281-294.
9. Peaker M. Nutritional requirements and diets for hummingbirds and sunbirds. Int. Zoo Yb. 1990;29:109-118.
10. Weathers WW, Stiles FG. Energetics and water balance in free-living tropical hummingbirds. Condor. 1989;91:324–331.
11. Mc Whorter TJ, López Calleja MV. The integration of diet, physiology, and ecology of nectar feeding birds. Rev Chil Hist Nat. 2000;73:451–460.

12. López-Calleja MV, Fernández MJ, Bozinovic F. The integration of energy and nitrogen balance in the hummingbird *Sephanoides sephaniodes*. J Exper Biol. 2003;206:3349-3359.
13. Paoletti MG, Norberto L, Cozzarini E, Musumeci S. Role of chitinases in human stomach for chitin digestion: AM case in the gastric digestion of chitin and chit in gastric pathologies. binomium chitin-chitinase. - Eds: Salvatore Musumeci, Maurizio G. Paoletti © 2009 Nova Science Publishers, Inc. 2009;339(9):339-358. Recent Issues ISBN 978-1-60692.
14. NRC. Dietary Reference Intakes: Proposed Definition of Dietary Fiber. National Academy Press; 2001.
15. Thomas S, Durand D, Chassenieux C, Jyotishkumar P. Handbook of Biopolymer-Based Materials: From Blends and Composites to Gels and Complex Networks (eds.) - John Wiley & Sons; 2013.
16. Barton LL, Northup DE. Microbial ecology. Wiley-Blackwell; 2011.
17. Jeuniaux C. Method Enzymol. 1966;8:644-650.
18. Eurich K, Segawa M, Toei-Shimizu S, Mizoguchi E. Potential role of chitinase 3-like-1 in inflammation-associated carcinogenic changes of epithelial cells. World J Gastroenterol. 2009;15:5249-5259.
19. Patton RS, Chandler PT, Gonzales OG. 1975 Nutritive Value of Crab Meal for Young Ruminating Calves J. Dairy Sci. 1975;58(3):404-409.
20. Razdan Pettersson AD. Effect of chitin and chitosan on nutrient digestibility and plasma lipid concentrations in broiler chickens. British Journal of Nutrition. DOI:10.1079/BJN19940029. 1994;72:277-288.
21. Hossain S, Blair R. Chitin utilisation by broilers and its effect on body composition and blood metabolites. Brit Poultry Sci. 2007;48:33-38.
22. Deuchi K, Kanauchi O, Imasato Y, Kobayashi E. Decreasing effect of chitosan on the apparent fat digestibility by rats fed on a high-fat – Biosci Biotechnol Biochem. 2014;58:1613-1616.
23. Flickinger EA, Van Loo J, Fahey GC. Nutritional Responses to the Presence of Inulin and Oligofructose in the Diets of Domesticated Animals: A Review. Crit Rev Food Sci Nutr. 2003;43:19–60.
24. Verdonk JM, Shim SB, van Leeuwen P, Verstegen MV. Application of inulin-type fructans in animal feed and pet food. Br J Nutr. 2005;93:125-138.
25. Koepche M. Las aves del departamento de Lima. Morson, Lima Perù. 1964.
26. González O, Bananaquit (*Coereba flaveola*), a new bird in the ecosystem of the parks of Lima (Peru). Ecology. 1998;1:79-83.
27. Delzenne N, Roberfroid MB. Physiological effects of nondigestible oligosaccharides. Lebensm-Wiss Technol. 1994;27:1–6.
28. Roberfroid MB, Slavin J. Nondigestible oligosaccharides. Crit Rev Food Sci Nutr. 2000;40:461.
29. A.O.A.C. Official Methods of Analysis of the Association of Official Analytical Chemist. 16th ed., Publ. Washington, DC, USA. 1990;931-933.
30. Van Soest PJ. Nutritional Ecology of the Ruminant: Ruminant metabolism, Nutritional Strategies, the Cellulotic Fermentation and the Chemistry of Forages and Plant Fibers. O and B Books, Corvallis; 1982.
31. FAO - Food and Agricultural Organization. Food energy: methods of analysis and conversion factors. Report of a technical workshop. FAO Food and Nutrition paper No. 77. Rome; 2003.
32. Gass CL, Montgomerie RD. Hummingbird foraging behavior: decision-making and energy regulation. In Foraging Behavior: Ecological, Ethological and Psychological Approaches (ed. Kamil AC, TD Sargent), New York: Garland STMP Press. 1981;159-194.
33. Pyke GH. Why hummingbirds hover and honeyeaters perch. Anim. Behav. 1980;29:861-867.
34. Hainsworth FR. Foraging efficiency and parental care in *Colibri coruscans*. Condor. 1977;79:69-75.
35. Montgomerie RD, Redsell CA. A nesting hummingbird feeding solely on arthropods. Condor. 1980;82:463-464.
36. Brice AT. The essentiality of nectar and arthropods in the diet of the Anna's Hummingbird (*Calypte anna*). Comp Biochem Phys. 1992;101:151–155.

37. Wooller RD, Richardson KC, Pagendham CM. The digestion of pollen by some Australian birds. Aus J Zool. 1988;36:357–362.
38. Pyke GH, Waser NM. The production of dilute nectars by hummingbirds flowers. Biotropica. 1981;13:260–270.
39. Freeman CE, Reid WH, Becvar JE, Scogin R. Similarity and apparent convergence in the nectar-sugar composition of some Hummingbird-Pollinated Flowers. Bot Gaz. 1984;145:132-135.
40. Martínez del Rio C. Sugar preferences in hummingbirds: the influence of subtle chemical differences on food choice. Condor. 1990;92:1022-1030.
41. Martínez del Rio C, Baker HG, Baker I. Ecological and evolutionary implications of digestive processes: bird preferences and the sugar constituents of floral nectar and fruit pulp. Experientia. 1992;48:544-551.
42. Stiles FG, Freeman CE. Patterns in floral nectar characteristics of some bird-visited plant species from Costa Rica. Biotropica. 1993;25:191–205.
43. Chalcoff VR. Composición de azúcares de la flora ornitófila del Bosque Templado de Sudamérica Austral: efectos históricos y preferencias presentes de *Sephanoides sephaniodes*. Undergraduate Thesis, Universidad Nacional del Comahue, Argentina; 2001.
44. Nicolson SW, Fleming PA. Nectar as food for birds: The physiological consequences of drinking dilute sugar solutions. Plant Syst Evol. 2003;238:139–153.
45. Schondube JE, Martínez del Rio C. Concentration-dependent sugar preferences in nectar-feeding birds: mechanisms and consequences. Funct Ecol. 2003;17:445-453.
46. Stiles, FG. Taste preferences, color preferences, and flower choice in hummingbirds. Condor. 1976;78:10–26.
47. Roberts WM. Hummingbirds' nectar concentration preferences at low volume: the importance of time scale. Anim Behav. 1996;52:361-370.
48. Schmid R. Sugar preferences of hummingbirds; 2008. Available:http://www.easternct.edu/~elliottp/documents/Sugar_000.pdf
49. Mc Whorter TJ, Martínez del Rio C. Does gut function limit hummingbird food intake? Physiol Biochem Zool. 2000;73:313–324.
50. Schondube J, Martínez del Rio C. Sugar and protein digestion in flower piercers and hummingbirds: a comparative test of adaptive convergence. J Comp Phys. 2004;174:263-273.
51. Brown M, Downs CT, Johnson SD. Sugar preferences of nectar feeding birds. A comparison of experimental techniques. J Avian Biol. 2008;39:479-483.
52. Fleming PA, Bakken BH, Lotz CN, Nicolson SW. Concentration and temperature effects on sugar intake and preferences in a sunbird and a hummingbird. Funct Ecol. 2004;18:223-232.
53. López–Calleja MV, Bozinovic F, Martinez del Rio C. Effects of sugar concentration on hummingbird feeding and energy use. Comp Biochem Phys.1997;118:1291-1299.
54. López-Calleja MV, Bozinovic F. Dynamic energy and time budgets in hummingbirds: a study in *Sephanoides sephaniodes*. Comp Biochem Phys A. 2003;134:283-295.
55. Tsahar E, Arad Z, Izhaki I, Martínez del Rio C. Do nectar- and fruit-eating birds have lower nitrogen requirements than omnivores? An allometric test. Auk. 2006;123:1004–1012.
56. Stiles FG. On the field identification of California hummingbirds. Calif. Birds. 1971;2:41-54.
57. Hainsworth FR, Wolf LL. Crop volume, nectar concentration and hummingbird energetics. Comp Biochem Phys. 1972;42:359-366.
58. Powers DR, Conley TM. Field metabolic rate and food consumption of two sympatric hummingbird species in south-eastern Arizona. Condor. 1994:96:141–150.
59. Suarez RK, Gass CL. Hummingbird foraging and the relation between bioenergetics and behaviour. Comp Biochem Phys. 2002;133:335-343.
60. Mc Whorter TJ, Bakken Hartman B, Karasov WH, Martinez del Rio C. Hummingbirds rely on both paracellular and carrier-mediated intestinal glucose absorption to fuel high metabolism. Biol Lett. 2005;2:131–134.
61. Baker HG. Non-sugar chemical constituents of nectar. Apidologie. 1977;8:349-356.

62. Martínez del Rio C. Nutritional ecology of fruit eating and flower visiting birds and bats. In The Digestive System in Mammals: Food Form and Function (ed. Chivers DJ, Langer P.), Cambridge: Cambridge University Press. 1994;102-127

63. Robel RJ, Press BM, Henning BL, Johnson KW, Blocker HD, Kemp KE. Nutrient and energetic characteristics of sweep net-collected invertebrates. J Field Ornithol. 1995;66:44–53.

64. Barker D, Fitzpatrick MP, Dierenfeld ES. Nutrient composition of selected whole invertebrates. Zoo Biol. 1998;17:23-134.

65. Lotz CN, Martínez del Rio C. The ability of rufous hummingbirds *Selasphorus rufus* to diluite and concentrate urine. J Avian Biol. 2004;35:54-62.

66. Baker HG, Baker I. Sugar ratios in nectar. – Phytochem Bull. 1979;2:43-45.

67. Brice AT, Grau CR. Hummingbird nutrition: development of a purified diet for long-term maintenance. Zoo Biol. 1989;8:233-237.

68. Brice AT, Grau CR. Protein requirements of Costa's Hummingbirds *Calypte costae*. Physiol Zool. 1991;64:611–626.

69. Suarez RK, Lighton JRB, Moyes CD, Brown GS, Gass CL, Hochachka PW. Fuel selection in rufous hummingbirds: ecological implications of metabolic biochemistry. Proc. of the National Academy of Science USA. 1990;87:9207–9210.

70. Carleton SA, Bakken Hartman B, Martínez del Rio C. Metabolic substrate use and the turnover of endogenous energy reserves in broad-tailed hummingbirds (*Selasphorus platycercus*). J Ex. Biol. 2006;209:2622–2627.

71. Odum EP, Connell CE, Stoddard HL. Flight energy and estimated flight ranges of some migratory birds. Auk. 1961;78:515-527.

72. De Foliart GR. Insect fatty acids: similar to those of poultry and fish in their degree of unsaturation, but higher in the poly-unsaturates. Food Insects Newslett. 1991;4:1-4.

73. Green PR, Geer BW. Changes in the fatty acid composition of *Droshophila melanogaster* during development and ageing. Arch Int Physiol Biochim. 1979;87:485-491.

74. Robinson SJW, Zwaan B, Partridge L. Starvation resistance and adult body composition in a latitudinal cline of *Drosophila melanogaster*. Evolution. 2000;54:1819-1824.

75. Castellanos MA, Wilson P., Thomson JD. Pollen transfer by hummingbirds and bumble bees, and the divergence of pollination modes in Penstemon. Evolution. 2003;57:2742–2752.

76. Richardson KC, Wooller RD. Adaptations of the alimentary tracts of some Australian lorikeets to a diet of pollen and nectar. Aust J Zoo. 1990;38:581–586.

77. Carpenter FL, Castronova JL. Maternal diet selectivity in *Calypte anna*. Am Midl Nat. 1980;103:175-179.

78. Klasing KC. Comparative Avian Nutrition. CAB International; ISBN: 0851992196 – UK; 1998.

79. Hainsworth FR. Food quality and foraging efficiency. J Comp Phys. 1974;88:425–431.

80. Karasov WH. Energetics, physiology and vertebrate ecology. Trends Ecol Evol. 1986;1:101-104.

81. Duke GF. Gastrointestinal physiology and nutrition in wild birds. Proc. of the Nutrition Society. 1997;56:1049-1056.

82. Powers DR, Nagy KA. Field metabolic rate and food consumption by free-living Anna's hummingbirds. Physiol Zool. 1988;61:500-506.

83. Mc Whorter TJ, Powers D, Martínez del Rio C. Ammonotely, nitrogen excretion and nitrogen requirements in three sympatric hummingbird species. Physiol Biochem Zool. 2003;76:731–743.

84. Robbins CT. Wildlife Feeding and Nutrition (2nd edition). San Diego: Academic Press; 1993.

85. Mc Whorter TJ. Energy assimilation, protein balance and water absorption in broad-tailed hummingbirds, *Selasphorus platycercus*. M.S. Thesis, University of Wyoming, Laramie. vii +. 1997;89.

86. Roberfroid M, Buddington RK. Inulin and oligofructose: proven health benefits and claims. J Nutr. 2007;137:2489–2597.

87. Oku T, Tokunaga T, Hosoya N. Non digestibility of a new sweetener "Neosugar" in the rat. J Nutr. 1984;114:1574–1581.

88. Ortiz LT, Rodriguez ML, Alzueta C, Rebole A, Trevino J. Effect of inulin on growth performance, intestinal tract sizes, mineral retention and tibial bone mineralisation in

broiler chickens. Brit Poultry Sci. 2009;50:325-332.

89. Awad AW, Ghareeb K, Paßlack N, Zentek J. Dietary inulin alters the intestinal absorptive and barrier function of piglet intestine after weaning. Res Vet Sci. 2013;5:249-254.

90. Preest MR, Folk DG, Beuchat CA. Decomposition of nitrogenous compounds by intestinal bacteria in hummingbirds. Auk. 2003;120:1091–1101.

9

Assessment of Genetic Diversity of Elite Indian Rice Varieties Using Agro-Morphological Traits and SSR Markers

Keshavulu Kunusoth[1*], Krishnasamy Vadivel[2], Raman Meenakshi Sundaram[3], Razia Sultana[1], Passoupathy Rajendrakumar[3], Sheshumadhav Maganti[3], Lella V. Subbarao[3] and Sebastian Reyes Chin-Wo[4]

[1]Department of Seed Science and Technology, Acharya NG Ranga Agricultural University, Hyderabad, TS, India.
[2]Centre for Plant Molecular Biology, Tamil Nadu Agricultural University, Coimbatore, Tamil Nadu, India.
[3]Crop Improvement, Directorate of Rice Research, Hyderabad, TS, India.
[4]Department of Plant Sciences, University of California, Davis, California, USA.

Authors' contributions

Authors KK, KV, RMS designed the experiments. Author KK performed most of the experiments. Authors RMS, RS and PR reviewed experimental design and contributed in experiments and all draft of the manuscript. Author KK, SM, LVS and SRC managed analyses of the study and performed the statistical analysis. All authors read and approved the final manuscript.

Editor(s):
(1) Dalong Guo, College of Forestry, Henan University of Science and Technology, Henan, People's Republic of China.
(2) Juan Yan, Sichuan Agricultural University, China.
Reviewers:
(1) Valentin Ivanov Kosev, Breeding and seed production of forage crops, Institute of Forage Crops – Pleven, Bulgaria.
(2) Guillermo Castañón-Nájera, División Académica de Ciencias Biológicas, Universidad Juárez Autónoma de Tabasco. Villahermosa, Centro, Tabasco, México.
(3) Anonymous, Turkey.

ABSTRACT

Assessment of genetic variability of the crop varieties is essential to assure selection of genetically divergent lines useful for the future breeding programmes. Thus, genetic diversity assessment of 24 elite Indian rice varieties was performed based on 24 agro-morphological traits and 86 SSR markers. The morphological and grain traits exhibiting significant variation are useful for discrimination of the rice varieties and were confirmed by Principal Component Analysis. Genetic

**Corresponding author: E-mail: keshava_72@yahoo.com;*

diversity assessment based on SSR markers displayed genetic similarity coefficients and grouped the varieties into five major clusters. The genetic population structure obtained was predominantly associated with UPGMA clustering and the structure bar plot. Cluster analysis based on both phenotype and SSR marker data did not show perfect congruence between the two measures of genetic diversity. However, the correlation between morphological and molecular diversity was positive and significant ($r = 0.36$ $P<10^{-4}$) indicating usefulness in classification of rice varieties.

Keywords: Genetic diversity; agro-morphological; characterization; Indian varieties; simple sequence repeats; Oryza sativa.

1. INTRODUCTION

Rice (*Oryza sativa* L.) is one of the most important crops that provide food for more than half of the world population. It plays a vital role in the national food security in India and contributes a major source of calories for the urban and rural populace [1]. With changing consumer preferences and the demand by the export market, rice breeding program in India has received significant attention with respect to improvement of yield with grain quality. In spite of more than 900 rice varieties have been developed in India, only considerable number of elite rice varieties have been released to suit different agro-climatic conditions which includes in the present study contributing ~ 29% of the total rice production in the country and being cultivated in many rice growing states (http://agricoop.nic.in). In order to produce improved varieties in future, assessment of the variability among these elite rice varieties is also essential which will enable the identification of genetically divergent varieties useful in future breeding programmes.

Characterization and genetic diversity of rice germplasm is well progressing globally, nevertheless, knowledge regarding the amount of phenotypic and genotypic variation specific to the geographical region which will enable to design effective breeding programs in order to achieve high productivity. Further, it has a great importance for seed production, certification and cultivar propagation and implementation of Protection of Plant Varieties and Farmer's Right Act of India (http://www.planthauthority.org). Traditionally, morphological traits have been used to evaluate distinctness and to establish broad description of rice varieties [2,3]. However, despite the fact that the varieties growing in different agro-climatic regions differ substantially from each other in agronomical traits, they are difficult to characterize using only morphological characteristics recommended by the International Union for the Protection of New Varieties of Plants (UPOV) descriptors [4]. Moreover, rice breeding tends to produce elite varieties which are phenotypically less distinct due to narrow genetic base of the populations and in such situations morphological traits seems inadequate because of environmental influence [2]. Besides, morphological variation does not always reflect real genetic variation because of genotype x environment interactions and the largely unknown genetic control of polygenic morphological and agronomic traits [5].

Molecular marker based approaches provide an opportunity to look into the genotypic basis of the observed phenotypic variation in released varieties [6]. Different types of molecular markers such as RAPDs [7,8], RFLPs [9], AFLPs [10] and ISSRs [7,11] have been used for estimation of genetic diversity and development of molecular fingerprints in rice. Among these, simple sequence repeats (SSRs) or microsatellites are the most popular markers in rice research [12]. SSR markers are becoming markers of choice for molecular characterization and genetic diversity studies in rice [13,14]. Breeding of rice varieties in India has been developed based on few important traits, mainly plant height, flowering, pest and disease resistance, maturity and yield. So far, no comprehensive study based on molecular marker in particular has been documented in the existing popular rice varieties. In view of the above, we attempt to characterize and estimate the genetic diversity in elite rice varieties grown in India using agro-morpho traits and SSR markers along with a correlation between the two diversity estimates.

2. MATERIALS AND METHODS

2.1 Plant materials

Pure seeds (breeder seed) of 24 elite rice varieties collected from the Acharya N G Ranga Agricultural University, Hyderabad, Tamil Nadu

Agricultural University, Coimbatore and Directorate of Rice Research, Hyderabad, TS, India were used for the present study (Table 1).

2.2 Agro-morphological Characterization

The field experiments were carried out in three replications (5×6 m plot size with a spacing of 30 ×20 cm) and all the cultivars were grown under screen-house simultaneously to confirm qualitative traits at Directorate of Rice Research, Hyderabad, India during dry season 2011-12 and 2012-13 for recording stable and uniform morphological characteristics. Ten plants were selected at random and tagged in each variety per replication. The highly variable plant morphological and grain traits in rice cultivars were selected on the basis of variability among closely related varieties [15]. A set of 16 key plant morphological traits, which included qualitative and quantitative characteristics and their states (scores) were recorded as per the National/Indian guidelines for the conduct of tests for Distinctness, Uniformity and Stability (DUS) (http://www.plantauthority.gov.in). The data on quantitative characteristics recorded for two seasons were grouped and the average values were used for statistical analysis. The grain quantitative and qualitative traits namely grain weight, grain length and width, decorticated grain length and width, shape of the de-husked grain, amylose content of endosperm (%) and gelatinization temperature was determined and grouped as per DUS testing protocol (http://www.plantauthority.gov.in).

2.3 SSR Marker Analysis

Total genomic DNA was isolated from 10-15 day old seedlings as per the protocol of Zheng et al. [16]. A total of 86 hyper-variable SSR (microsatellite) primer pairs purchased from Integrated DNA Technologies, USA were used for PCR amplification. The SSR markers were selected based on their polymorphism content and distribution across the 12 rice chromosomes. PCR was carried out in 10 µl reaction containing 0.2 µM of each primer, 200 µM of dNTPs, 1X PCR assay buffer (50 mM KCl,10 mM, Tris HCl (P^H 8.3), 1.5 mM of $MgCl_2$, 0.1% gelatin, 15 ng of DNA and 0.5 unit of *Taq* DNA polymerase. The amplification was performed with initial denaturation at 94°C for 5 min followed by 35 cycles of 94°C for 30 sec, 55°C for 1 min, 72°C for 7 minutes before cooling at 4°C. The PCR products were electrophoretically resolved on 3% superfine grade agarose (Amresco, USA) stained with ethidium bromide and visualized under UV.

Table 1. Details of 24 elite Indian rice varieties used in the study

Variety	Year of release	Parentage
ADT 43	1998	IR 50 x White ponni
ASD 19	1995	Lalnakanda x IR 30
BPT 5204 (Samba mahsuri)	1986	GEB 24 x TN 1 x Mahsuri
GEB 24	1929	Mutant of Konmani
IR 64	1992	IR 5657-33-2-1 x IR 2061-465-1-5-5
JGL 384 (Polasa prabha)	2002	WGL 48684 x BPT 5204
JGL 1798 (Jagtial sannalu)	2002	BPT 5204 x WGL 48684
Mahsuri	1972	65 x Mayang Ebos 6080 x 2
MTU 1001 (Vijetha)	1995	MTU 5249 x MTU 7014
MTU 1010 (Cottondora sannalu)	2000	MTU 2067 x IR 64
MTU 2067 (Chaitanya)	1988	Sowbhagya x ARC 5984
MTU 2077 (Krishnaveni)	1989	Sowbhagya x ARC 5984
MTU 4870 (Deepthi)	2000	MTU 4569 x ARC 6650
MTU 5249 (Vajram)	1986	Sowbhagya x ARC 6650
MTU 7029 (Swarna)	1982	Vasishtha x Mahsuri
RDR 7555(Rudrama)	1991	HR 19 x TN 1
RDR 763 (Indur samba)	1997	BPT 5204 x WGL 13400
WGL 14 (Warangal sannalu)	-	BPT 5204 x ARC 5984 / BPT 3291
WGL 13400 (Surekha)	1976	IR-8 x Siam 29
WGL 20471 (Erra mallelu)	1991	BC 5-55 x W 12708
RPW6-17 (Phalguna)	1977	IR-8 / Siam 29
RNR 23064 (Taramathi)	-	BPT 5204 x Tellahamsa
Tellahamsa	1971	HR 12 x TN 1
Kalajeera	-	Land race

2.4 Data Analysis

Univariate descriptive statistical data for quantitative traits were computed for mean, range, coefficient of variation and standard deviation. Before analysis, the data was standardized by subtracting the mean value and dividing it by the standard deviation. Principal Component Analysis (PCA) was done on observed morpho-grain traits with the standardized data. Euclidean distances were calculated for all the traits considered together, morphological traits and for the SSR. Each dataset was plotted in a dendrogram using the average linkage method cluster analysis across varieties. With the molecular marker data, Jaccard's index was calculated for genetic distance among the varieties to generate a dendrogram. These analyses were performed in R 2.9.1, using the software packages Vegan 1.15-4 [17] and the base package Stats. Only clear and unambiguous SSR amplicons were scored for their presence and absence among the cultivars as '1' and '0', respectively. The binary data matrix was subjected to further analysis. Sequential Agglomerative Hierarchial Non overlapping (SAHN) clustering was done on the squared Euclidean distance matrix and the similarity matrix was generated based on Jaccard's index by utilizing the UPGMA algorithm through the NTSYS pc 2.0 software [18]. Polymorphism Information Content (PIC) for each SSR was calculated according to the formula: PIC =1- ($\sum$ piY) whilo '1' is the total number of alleles detected for SSR marker and 'pi' is the frequency of the ith allele [19]. All computations were performed with appropriate procedures of the NTSYS-pc version 2.1 software [18].

The genetic structure in the elite Indian rice varieties was inferred using the software Structure 2.3.2 [20] based on the amplification pattern of the SSR markers. Structure employs a model based clustering method that accounts for Hardy-Weinberg and linkage disequilibrium and attempts to find population groupings that minimizes linkage disequilibrium. Markers used in this study were distributed across the rice genome and can therefore be assumed to be independent. To estimate the number of sub-populations (the K parameter), the dataset was analyzed allowing the value of K ranging from 1 to 5. Five independent runs were carried out for each K value. All the parameters were set to their default values. An admixture model was chosen as it can account for some individuals having mixed ancestry. The optimal value for number of sub-populations was determined based on likelihood variance values obtained for each run using ΔK, an *ad-hoc* parameter described by Ewens et al. [21]. A bar plot of line membership in each population was generated using structure 2.3.2.

For comparisons between the phenotypic and genotypic data, simple (r) and spearman rank (rs) correlation coefficient was calculated between the Euclidean distances for each variety, across all the datasets. The 'P' value for the correlation coefficient was calculated according to each distribution. To determine the similarity between the dataset, each pair was tested using the Mantel Z statistic [22]. The test was carried out using the R package ade4 [23], which calculate the Z value and the corresponding P value. Significance of Z was determined by comparing the observed Z values obtained and by calculating Z for one matrix with 2000 permuted variants of second matrix.

3. RESULTS AND DISCUSSION

3.1 Morphological Characterization

Characterization and genetic diversity assessment of elite rice varieties are important in view of identification and selecting suitable parental lines to develop new varietal combinations as well as, for protecting genotypes from unauthorized exploitation. Modern breeding makes the use of natural processes but takes full advantage of available genetic diversity to create new genotypes [24] and this needs clear-cut information on the extent of genetic diversity available so that the breeder can exploit the genotypes efficiently. Characterization of 24 elite Indian rice varieties in the present study showed considerable variation in all morpho-grain traits. The details of qualitative and quantitative morpho-grain traits for plant morphology and grain quality traits observed and their distribution among the varieties are shown in the Table 2 and 3. In case of quantitative traits, the coefficient of variation ranged from 8.01% (panicle length of main axis) to 25.67 per cent (stem thickness) among the varieties (Table 4). The highest variability was observed for stem thickness (25.67) followed by leaf length (21.05%) and grain weight (19.47%). The leaf length among the varieties ranged from 19.5 cm (RNR 23064) to 47.50 cm (WGL 13400). The lowest (0.85 cm) and the highest (1.45 cm) leaf width were observed in Tellahamsa and MTU

5249, respectively. Among the quantitative traits, time of heading which ranged from 78 days (RDR 7555) to 122 days (MTU 5249), the lowest stem length (60.50 cm) was observed in RDR 7555, while MTU 1010 showed the highest (99.0 cm). Kalajeera was characterized with the lowest stem thickness (0. 54 cm) against the highest in ASD 19 (1.15 cm). The varieties Tellahamsa and MTU 2077 respectively displayed the shortest (102.66 days) and the longest (147.33 days) time of maturity, respectively. Leaf and stem characteristics are important to distinguish the rice cultivars. The variation in the time of heading has been reported as a useful trait by Nethra [25]. According to national DUS testing guidelines of plant protection Act of India and also as per the present study, the stem length was less suitable for classification of the varieties. One of the most distinguishable features of cultivated rice was variation in the size and shape of grains. Among the 11 qualitative traits, only panicle awn trait displayed no variation among the varieties except WGL 20471. The rice varieties were grouped into five categories based on grain length, viz., very short (Kalajeera), short (ADT 43, BPT 5204, GEB 24, JGL 384, JGL 1798, Mahsuri, MTU 7029 and RDR 7555), medium (ASD 19, MTU 1001, MTU 2067, MTU 2077, MTU 5870, MTU 5249, RDR 763, WGL 14 and RNR 23064) and long (IR 64, MTU 1010, WGL 13400, WGL 20471 and RPW6-17) and very long (Tellahamsa). The grain traits showed considerable variation among all the varieties (Table 3). The seed weight was lowest in JGL 384 (13.5 g), while the variety RPW6-17 showed the highest (26.5 g). The variety Kalajeera with the lowest grain length (6.02 mm) whereas, Tellahamsa had shown having the highest (10.23 mm). Similarly, in BPT 5204, the grain width was very narrow (1.80 mm) and in RDR 7555it was broadest (3.05 mm). Based on the decorticated grain length, the varieties ranged from 4.53 (Kalajeera) to 6.90 mm (WGL 13400), while the decorticated grain width was very narrow (1.78 mm) in WGL 20471and very broad (2.70 mm) in RDR 7555. Panicle and grain characters such as the presence of awns, length and width of grain as well as for decorticated grain shape of grains were useful for variety characterization. In the present study, the length to width ratio ranged from 2.72 (land race Kalajeera) to 4.32 (Tellahamsa). According to Adair et al. [26], grain size and shape are important components of rice quality in developing new varieties and also for determining consumer acceptance. All the short and medium slender varieties were late in flowering and also showed considerable variations between these varieties and other varieties of rice. The amylose content in rice genotypes, influenced by amylose: Amylopectin ratio is a useful measure to characterize rice genotype [27]. In the current study, all the varieties possessed medium amylose content except MTU 1010, RDR 7555 and Tellahamsa which showed low and high amylose content respectively.

The broad morpho-grain trait diversity existed among the varieties in the present study indicated fair distribution of variation across all of the morphological traits. The significance of agro-morphological characters in diversity assessment of rice varieties has been well documented [28]. In agreement with Singhal and Prakash [29], a set of morphological traits are useful to distinguish the varieties from each other. A set of 12 morphological traits viz., altitude of flag leaf blade, time of heading and maturity, stem length and thickness, leaf traits and grain traits like grain weight, length and width of seed, shape of grain, shape of decorticated grain and gelatinization temperature were useful in characterization of the elite Indian rice varieties.

3.2 Principal Component Analysis (PCA)

The principal characters with higher Eigen vectors that delineated the varieties into separate groups in the first seven components are presented in Table 5. The 24 varieties were scattered by their component scores of the first seven principal components (PCs) with Eigen value higher than one and explained a total of 75.8% of their phenotypic variation components. The first PC explained 18.32% of the total variation, mainly related to quantitative variation of leaf length and width, time of heading, stem thickness, time of maturity and qualitative traits like intensity of green leaf colour, pubescence of leaf blade, attitude of culm and gelatinization temperature. In the second PC, which explained 14.59% of the total variation, predominant traits were leaf length and width of blade, time of heading, stem length, time of maturity, flag leaf attitude of leaf blade, panicle curvature of main axis, and grain traits (grain weight, grain and decorticated grain width). The third principal component, which accounted for 12.04% of the total variation, was dominated by traits such as leaf and stem traits, time of maturity, panicle number per plant, grain and decorticated grain length and shape, flag leaf attitude. While leaf width of blade, stem thickness, time of maturity, grain length, leaf pubescence of blade surface and grain quality traits (grain weight, grain width and decorticated grain width, endosperm amylose content and gelatinization temperature) were important delineating traits associated with the fourth principal component, which documented 10.99% of the total variation. Principal components 5, 6 and 7 accounted for 8.32%, 6.23% and 5.39% respectively of the total variation and the main characteristics of these components were positively related to the width of leaf blade, stem length, panicle length of main axis, leaf pubescence of blade surface, panicle awn, stem thickness and flag leaf attitude of blade. The first three PCs, which contributed to 45% of the total variation, were plotted graphically to demonstrate the relationships between varieties (Fig. 1). The plot evidenced considerable diversity among the varieties. However, PCA could considerably differentiate the varieties according to their pedigree and grain types and the varieties did not form clear-cut distinct groups. The varieties considered for the present study have been developed to suit different rice growing ecosystems and consumer preferences based on their grain type. There were eight short and medium slender types and a landrace cultivars *viz.*, ADT 43, BPT 5204, JGL 384, JGL 1798, Mahsuri, WGL 14, MTU 7029, RDR 763 and a landrace Kalajeera, which had higher factor loadings (>1) in PC1 compared to other medium and long grain varieties. These were distinct and formed the most divergent single line group compared to medium and long grain varieties for PC1 while, the medium and long slender and bold varieties released from Andhra Pradesh Rice Research station other than RPW 6-17 and WGL 13400 (MTU 1001, MTU 2067, MTU 2077, MTU 4870, MTU 5249, MTU 7029, RPW 6-17 and WGL 13400) are markedly distinct from the others in the PC2. The rest of the varieties *viz.*, ASD 19, GEB 24, RDR 7555 and RNR 2306 showed higher factor loadings for PC3. The PC1 through PC7 vectors explains progressively less of the variation across the plant morphology and grain quality characteristics.

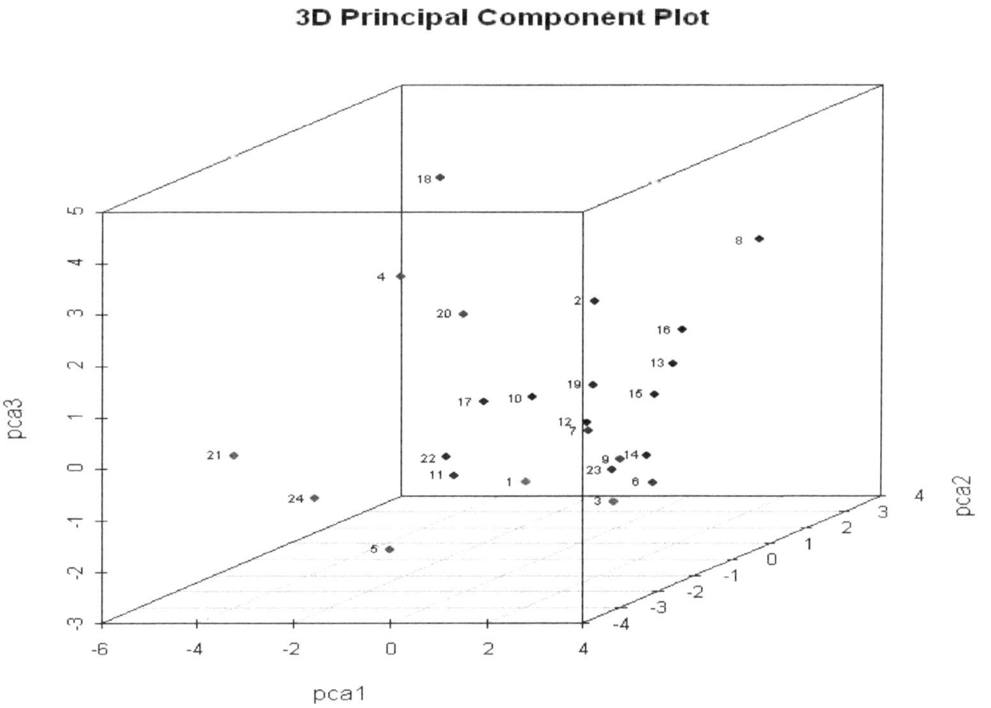

Fig. 1. Scatter plot of the three principal components (PC1, PC2 and PC3) using agro morphological traits (1-ADT 43; 2-ASD 19; 3-BPT 5204; 4-GEB 24; 5-IR 64; 6-JGL 384; 7-JGL 1798; 8-Kalajeera; 9-Mahsuri; 10-MTU 1001; 11-MTU 1010; 12-MTU 2067; 13-MTU 2077; 14-MTU 4870; 15-MTU 5249; 16-MTU 7029; 17-RPW6-17; 18-RDR 7555; 19-RDR 763; 20-RNR 23064; 21-Tellahamsa; 22-WGL 13400; 23-WGL 14; 24-WGL 20471

Table 2. Plant morphological traits among elite Indian rice varieties

Variety	Leaf intensity of green colour	Leaf pubescence of blade surface	Attitude of culm	Flag leaf attitude of blade (early observation)	Flag leaf attitude of blade (late observation)	Panicle curvature of main axis	Panicle awn	Panicle exsertion	Leaf length of blade	Leaf width of blade	Time of heading	Stem length	Stem thickness	Panicle length of main axis	Panicle number per plant	Time of maturity (days)
ADT 43	Light	Weak	Semi-erect	Erect	Erect	Semi-straight	Absent	Exserted	Long	Medium	Early	Very short	Thick	Medium	Medium	Early
ASD 19	Dark	Medium	Semi-erect	Horizontal	Deflexed	Drooping	Absent	Well exserted	Long	Medium	Medium	Very short	Thick	Short	Medium	Medium
BPT 5204	Medium	Medium	Open	Erect	Semi-erect	Semi-straight	Absent	Exserted	Long	Narrow	Late	Very short	Thick	Short	Medium	Late
JGL 384	Medium	Medium	Semi-erect	Semi-erect	Semi-erect	Semi-straight	Absent	Exserted	Long	Medium	Late	Very short	Thick	Long	Medium	Late
JGL 1798	Medium	Medium	Semi-erect	Erect	Horizontal	Semi-straight	Absent	Well exserted	Medium	Medium	Medium	Very short	Thick	Long	Medium	Medium
GEB 24	Light	Medium	Open	Horizontal	Deflexed	Deflexed	Absent	Well exserted	Long	Narrow	Early	Very short	Medium	Long	Medium	Early
Mahsuri	Dark	Medium	Semi-erect	Erect	Semi-erect	Semi-straight	Absent	Exserted	Long	Narrow	Medium	Very short	Thick	Long	Medium	Medium
IR 64	Medium	Strong	Semi-erect	Erect	Semi-erect	Drooping	Absent	Exserted	Long	Medium	Medium	Very short	Thick	Long	Medium	Medium
MTU 1010	Medium	Medium	Erect	Semi-erect	Deflexed	Drooping	Absent	Well exserted	Long	Narrow	Medium	Short	Thick	Long	Medium	Medium
MTU 2077	Medium	Very strong	Erect	Semi-erect	Horizontal	Drooping	Absent	Exserted	Long	Medium	Late	Very short	Thick	Long	Medium	Late
MTU 5249	Dark	Strong	Erect	Semi-erect	Horizontal	Drooping	Absent	Exserted	Long	Medium	Late	Short	Thick	Long	Medium	Late
MTU 4870	Medium	Medium	Semi-erect	Erect	Semi-erect	Semi-straight	Absent	Exserted	Long	Medium	Late	Short	Thick	Medium	Medium	Late
MTU 1001	Medium	Weak	Semi-erect	Erect	Horizontal	Drooping	Absent	Well exserted	Long	Medium	Late	Short	Thick	Long	Medium	Late
MTU 2067	Medium	Medium	Semi-erect	Semi-erect	Horizontal	Semi-straight	Absent	Exserted	Long	Medium	Late	Very short	Thick	Long	Medium	Late
MTU 7029	Dark	Strong	Semi-erect	Semi-erect	Horizontal	Drooping	Absent	Partly exserted	Medium	Medium	Late	Very short	Thick	Long	Medium	Late
WGL 14	Medium	Medium	Semi-erect	Semi-erect	Semi-erect	Drooping	Absent	Exserted	Long	Medium	Late	Very short	Thick	Long	Medium	Late
RPW 6-17	Dark	Medium	Semi-erect	Semi-erect	Horizontal	Drooping	Absent	Exserted	Long	Narrow	Late	Very short	Thick	Long	Medium	Late
WGL 13400	Medium	Medium	Erect	Semi-erect	Semi-erect	Drooping	Absent	Well exserted	Long	Medium	Medium	Very short	Medium	Long	Medium	Late

Variety	Leaf intensity of green colour	Leaf pubescence of blade surface	Attitude of culm	Flag leaf attitude of blade (early observation)	Flag leaf attitude of blade (late observation)	Panicle curvature of main axis	Panicle awn	Panicle exsertion	Leaf length of blade	Leaf width of blade	Time of heading	Stem length	Stem thickness	Panicle length of main axis	Panicle number per plant	Time of maturity (days)
WGL 20471	Medium	Medium	Semi-erect	Semi-erect	Horizontal	Drooping	Present	Exserted	Long	Medium	Early	Very short	Thick	Long	Medium	Early
RDR 7555	Medium	Medium	Semi-erect	Semi-erect	Horizontal	Drooping	Absent	Exserted	Medium	Medium	Early	Very short	Thick	Long	Medium	Very early
RDR 763	Medium	Very strong	Erect	Erect	Horizontal	Drooping	Absent	Well exserted	Long	Medium	Medium	Very short	Medium	Long	Medium	Medium
Tellahamsa	Medium	Absent	Semi-erect	Semi-erect	Horizontal	Semi-straight	Absent	Well exserted	Long	Narrow	Early	Short	Medium	Long	Medium	Very early
RNR 23064	Medium	Medium	Semi-erect	Semi-erect	Horizontal	Drooping	Absent	Well exserted	Short	Medium	Medium	Very short	Medium	Long	Medium	Medium
Kalajeera	Dark	Weak	Open	Semi-erect	Horizontal	Deflexed	Absent	Well exserted	Long	Narrow	Late	Very short	Medium	Long	Few	Late

Table 3. Grain morphological traits among elite Indian rice varieties

Variety	1000 grain weight (g)	Grain length (mm)	Grain width (mm)	Decorticated grain length (mm)	Decorticated grain width (mm)	Decorticated grain shape	Endosperm amylose content	Gelatinization of temperature
ADT 43	Very low	Small	Narrow	Small	Narrow	Medium slender	Medium	High intermediate
ASD 19	Low	Medium	Medium	Small	Narrow	Short slender	Medium	Intermediate
BPT 5204	Low	Small	Narrow	Small	Narrow	Medium slender	Medium	High intermediate
JGL 384	Very low	Small	Narrow	Small	Narrow	Medium slender	Medium	Intermediate
JGL 1798	Low	Small	Narrow	Small	Narrow	Medium slender	Medium	Intermediate
GEB 24	Low	Small	Narrow	Small	Medium	Medium slender	Medium	Intermediate
Mahsuri	Low	Small	Medium	Small	Medium	Medium slender	Medium	High intermediate
IR 64	Medium	Long	Medium	Medium	Medium	Long bold	Medium	High intermediate
MTU 1010	Medium	Long	Narrow	Medium	Medium	Long bold	Low	High intermediate
MTU 2077	Low	Medium	Medium	Small	Medium	Medium slender	Medium	Intermediate
MTU 5249	Low	Medium	Medium	Medium	Medium	Long bold	Medium	Intermediate
MTU 4870	Medium	Medium	Medium	Small	Medium	Medium slender	Medium	Low
MTU 1001	Medium	Medium	Medium	Medium	Medium	Medium slender	Medium	Low
MTU 2067	Low	Medium	Medium	Smal	Medium	Medium slender	Medium	Intermediate
MTU 7029	Low	Small	Narrow	Small	Medium	Shot bold	Medium	Intermediate
WGL 14	Very low	Medium	Medium	Small	Narrow	Medium slender	Medium	Intermediate
RPW 6-17	High	Long	Broad	Medium	Medium	Long bold	Medium	Low
WGL 13400	High	Long	Medium	Medium	Medium	Long slender	Medium	Intermediate
WGL 20471	Medium	Long	Very narrow	Medium	Narrow	Long bold	Medium	Low

Variety	1000 grain weight (g)	Grain length (mm)	Grain width (mm)	Decorticated grain length (mm)	Decorticated grain width (mm)	Decorticated grain shape	Endosperm amylose content	Gelatinization of temperature
RDR 7555	Medium	Small	Very broad	Small	Medium	Short bold	Low	Intermediate
RDR 763	Low	Medium	Narrow	Small	Narrow	Short slender	Medium	Intermediate
Tellahamsa	Medium	Very long	Medium	Medium	Medium	Long slender	High	Low
RNR 23064	Low	Medium	Narrow	Small	Medium	Medium slender	Medium	Intermediate
Kalajeera	Very low	Very small	Narrow	Small	Medium	Short slender	Medium	High intermediate

Table 4. Descriptive statistics for 13 morpho-grain quantitative traits among elite Indian rice varieties

Trait	Mean±SD	Range	C.V
Leaf length of blade (cm)	36.97±7.9	19.5 – 47.50	21.05
Leaf width of blade (cm)	1.16±0.2	0.85 – 1.45	17.31
Time of heading (days)	104.04±15.2	77.66 – 122.33	12.64
Stem thickness (cm)	0.80±0.2	0. 54 – 1.15	25.67
Stem length (cm)	78.91±12.0	60.5 – 99.0	15.22
Panicle length of main axis (cm)	22.12±1.8	19.62 – 25.49	8.01
Panicle number per plant	16.79±2.1	14.00 – 19.00	12.68
Time of maturity (days)	131.56±13.2	102.66 – 147.33	12.20
1000 Grain weight (g)	18.99±3.7	13.5 – 26.5	19.47
Grain length (mm)	8.27±0.9	6.02 – 10.23	11.46
Grain width (mm)	2.29±0.2	1.80– 3.05	10.89
Decorticated grain length (mm)	5.89±0.6	4.53 - 6.90	9.84
Decorticated grain width (mm)	2.09±0.2	1.78 – 2.7	10.22

The combined PCA of the qualitative and quantitative morpho-grain traits grouped most of the varieties into one group, while a few others remained scattered. However, the Eigen values spread up to seven vectors, showed a total of 75.8% of their phenotypic variation components. In concurrence with Meesang et al. [3] this may be due to the use of highly adapted genotypes with less phenotypical distinction for breeding programmes. In PCA of the present study, the first three PCs predominantly explained the total variance that was observed to a considerable degree of distribution based on plant morpho-grain traits. The landrace Kalajeera stood apart from all other varieties in the PCA plot which may be due to its distinct origin, Eastern or North Eastern India (www.mssrf.org/bd/ebooks/ebook8.pdf) and high genetic similarity to *japonica* than *indica*. Further, Oka and Chang [30] also noticed wide variations between landraces and cultivated types in the rice populations of tropical regions. The varieties, RDR 7555, GEB 24, ASD 19, RNR 23064, Tellahamsa, WGL 20471, IR 64 and Kalajeera were distinct from the others and remained scattered across the plot reflecting the important morphological differences among them accounted by leaf, time of heading and maturity, panicle, decorticated grain length and gelatinization temperature. Though the cluster analysis in general was consistent with that of PCA, many of the short and medium slender varieties displayed a tendency to stay together. In conformity with Fuentes et al. [31], the diversity among varieties can be estimated efficiently based on the parentage and phenotypic diversity, permitting a more effective separation of the progenitor set than those obtained solely by phenotypic or genealogical information.

The average intra and inter-cluster distance values for varieties based on Euclidean distances are presented Fig. 2. Intra-cluster Euclidean values ranged from 55.1 (cluster II) to 0.0 (cluster V). Based on the inter-cluster Euclidean values, the highest divergence (100.01) occurred between cluster II and V, while the lowest divergence (47.81) was between cluster III and IV. The inter-cluster distance between I and II and III and IV, were moderately divergent. Cluster analysis based on Euclidean distances of morphological traits classified the varieties in the present study into five major groups and it revealed considerable diversity among the rice cultivars. The varieties JGL 384, MTU 2077, BPT 5204 and MTU 4870 in group 1; varieties RNR 23064 and Kalajeera in group 2; MTU 7029, WGL 13400, JGL 1798, Tellahamsa, ADT 43, MTU 1010, RPW 6-17, MTU 5249, GEB 24 and MTU 1001 in group 3 and the varieties RDR 763, IR 64, Mahsuri, MTU 2067, WGL 20471, ASD 19 and WGL 14 showed the closest morphological similarity and were placed in group 4 while, RDR 7555 variety was in group 5 and was separated from other varieties. The group 2 and 5 were morphologically different with respect to many of the traits characterized. The highest inter-cluster distance observed between II and V clusters (100.01) was due to the variety RNR 23064 and the landrace Kalajeera, while the lowest inter-cluster distance (47.81) between III and IV clusters may be due to the close genetic relationship among the varieties, similar parentages or grain type. The specific groupings of the varieties according to the pedigree and grain type up to some extent indicated the importance of these morphological traits for genetic diversity analysis. For eg., the cultivars BPT 5204 (GEB 24/TN1/Mahsuri) and JGL 384 (Kavya/BPT 5204) and MTU 1001 (MTU 5249 x MTU 7014) and MTU 5249 (MTU 4569 x ARC 6650) showed grouping according to their pedigree and shared at least one common parent in their pedigree.

3.3 Molecular Characterization

A total of 251 alleles were amplified using 86 SSR markers distributed across the genome (primers/chromosome) of 24 rice varieties. The number of alleles amplified by each primer pair ranged from 2 to 6 with an average of 3.0. The polymorphism information content (PIC) for these markers ranged from 0.54 (RM 13) to 0.97 (RM 226). The markers *viz.*, RM 226, RM 11229, RM 16153 and RM 15630 amplified the highest number of alleles (6 each), and was followed by 5 each in RM 204, RM 307, RM 16649, and RM 333. The SSR markers RM 14140, RM 16913, RM 6759, RM 70, RM 324, RM 336, RM 248, RM 72, RM 206, RM 247 and RM 20 amplified 4 alleles each. The number of alleles observed for 24 elite varieties of rice was analyzed in the study. The Jaccard's similarity coefficient ranged from 0.26 (JGL 1798 Vs JGL 384) to 0.76 (Kalajeera Vs RDR 7555) with an average similarity index of 0.52. In pair-wise comparisons, least similarity index was observed between the JGL 384 and JGL 1798 (0.26) whereas, landrace Kalajeera and RDR 7555 showed least similarity index (0.76). The Fig. 3 shows the dendrogram of 24 rice varieties based on SSR amplification pattern, the varieties were observed to be grouped into five major clusters. Major cluster I consisting of JGL 384, JGL 1798, BPT 5204, WGL 14, RDR 763, RNR 23064, ADT 43, ASD 19, GEB 24, Mahsuri, MTU 7029, MTU 7029 and IR 64 had 73.30% similarity, whereas cluster II with varieties Phalguna (RPW6-17), WGL 13400, Tellahamsa and RDR 7555 had average similarity of 66.8% among them. Cluster III consisted of MTU 2077, MTU 2067, MTU 4870, MTU 1001 and MTU 5249 of long grain varieties with a similarity of 42.6%. The variety Erramallelu (WGL 20471) and a landrace, Kalajeera were distinct from each other and also from other cultivars and were separated into cluster IV and cluster V, respectively. In the present study, SSR markers grouped the varieties in each cluster predominantly according to their genetic background. Based on marker data, distinct clusters were formed with related varieties. The varieties JGL 384 and JGL 1798 were grouped together as they were developed using same parents for medium slender grain type. Similarly, medium slender grain type varieties, namely BPT 5204, WGL 14, RNR 23064 and a short slender grain type variety RDR 763 were grouped in the same cluster. But, these cultivars were not grouped together when morphological characteristics were employed. The poly-allelic nature of SSR markers is highly favourable to discriminate the individuals more precisely. SSR markers RM 226, RM 11229, RM 16153 and RM 15630 showed six alleles across the varieties which formed five major groups of different compositions. The varieties' having short, medium or long slender grain types released for commercial cultivation along with other cultivars in diverse geographical locations across India are expected to be genetically diverse, and it become obvious by looking at the range of similarity coefficient between the varieties (0.26 to 0.76). High degree of similarity (76%) observed between JGL 1798 Vs JGL 384 followed by WGL 13400 with RPW6-17 (75%) and BPT 5204 and WGL 14 and Tellahamsa and RDR 7555 (73%), mainly because of their closely related parental lines. Least similarity was observed between Kalajeera and the rest of varieties, as emphasized earlier based on the PCA and may be due to the distinct origin of Kalajeera (Eastern or North-Eastern India). The grouping of MTU 2077, MTU 2067, MTU 4870, MTU 1001, MTU 1010 and MTU 5249 into a cluster with similarity ranging from 0.34 to 0.71 is due to their close pedigree. Further, they were released from Regional Agricultural Research Station, Maruteru, Andhra Pradesh State, India using similarly adapted genotypes as parent material and the possibility of few varieties with one of the genotypes as a common parent. The short slender variety Mahsuri and short bold variety MTU 7029 were clustered in a group as the Mahsuri is the one of the parente for MTU 7029. Similarly, Tellahamsa and RDR 7555 were grouped in one cluster showed similarity as one of the parental lines was common among them and interestingly both of them possessed long and bold grain types. However, the varieties RDR 763 and BPT 5204 did not group together though they have one common parent. Also the varieties, IR 64, MTU 2077 and MTU 1010 with closely related parentage were not a part of the same cluster, which might be due to differential selection pressure and selection criteria adopted for the development of these varieties. Ravi et al. [32] demonstrated the efficacy of SSR markers in an accurate determination of relationship between closely related genotypes and to analyze varieties with close pedigrees. In the present study, out of 86 SSR markers analyzed, 14 most informative primers (RM 70, RM 204, RM 11229, RM 16153, RM 15630, RM 206, RM 217, RM 226, RM 16649, RM 248, RM 307, RM 324, RM 333 and RM 336) were identified for the fingerprinting of closely related elite rice varieties.

3.4 Population Structure

A model based statistical analysis STRUCTURE was performed using the data of 251 polymorphic loci to recognize the genetic structure and to determine if an alternative structure could be elucidated with molecular marker data among the 24 elite varieties as explained in the materials and methods. The study provided significant evidence for existence of a population structure in these rice varieties. In most of the cases, consistent estimates of P(X/K) were obtained across the independent runs. The optimum population structure was inferred based on ΔK, an *ad hoc* parameter and three clusters were identified among the 24 varieties. The lines have been listed by their primary membership in each of these populations, where admixture values, Q, for the population of primary membership were <0.50, line was considered as mixed within their population of primary membership. Each line's estimated membership in each of the three populations is depicted in a STRUCTURE bar plot (Fig. 4) and exact membership proportions are provided in Table 6. Cluster 1 composed of 10 fine-grain (short and medium types) varieties except GEB 24, Mahsuri and Kalajeera; cluster 2 contained six coarse grain (long bold types) varieties except MTU

2077 and MTU 4870 medium slender varieties; while the cluster 3 comprised eight fine grain (short and medium slender) varieties. Population structure based on Q values with k=3 predominantly matched very well with UPGMA tree from the distance based analysis (Fig. 3). Differentiation estimated by F-statistics values between clusters determined by the software STRUCTURE is 0.0813, 0.0805 and 0.2079 in between cluster 1 and 2; 2 and 3 and 1 and 3 respectively, indicating to high differentiation. Genetic structure of populations has been previously documented in rice [33,34]. Based on the cluster analysis, Nelson et al. [34] identified six primary genetic clusters represented by genotypes. The present study, exhibited a structure with three clusters. Overall, the varieties were assigned into corresponding clusters with the membership proportion (Q) more than 0.53 except ADT 43 (0.47) in cluster 3. This variety also shared a secondary membership with cluster 1 (0.38). However, the short slender varieties GEB 24, MTU 7029, Mahsuri and landrace Kalajeera were grouped in cluster 1 and shared secondary membership with cluster 3 group (short and medium grain types). In some of these cases, the shared ancestry could be because of shared pedigree relationships and a perusal of the pedigree of these varieties indicates close relationship among the varieties. The simulations in *structure* population indicated that these varieties In the sub-population 1 could have received gametes from sub-population 3 as these are predominantly short and medium slender types.

3.5 Comparison of Phenotypic and Genotypic Markers Based on Diversity Estimates

The correlation between genotypic and phenotypic similarity value was moderately significant (r = 0.36 $P<10^{-4}$; rs = 0.12 $P<10^{-2}$). The Z-test statistic was also significant between GS and MS matrices (Z=0.39; P=0.0002). The results indicate a positive correspondence between the two similarity measures (Fig. 5) with the clusters of 24 cultivars of rice which were observed to be nearly similar in both the analyses. However, some discrepancies between the morphological traits (Fig. 2) and molecular dendrogram (Fig. 3) were observed, especially with respect to the varieties GEB 24, Tellahamsa, RPW 6-17, ASD 19, IR 64, MTU 7029 and RDR 7555. The three clusters based on genetic structure Q values and five clusters based on UPGMA with molecular markers for the 24 varieties corresponded largely to each other with respect to their pedigree relationship and morphological traits. However, contrary to the UPGMA clustering, the variety WGL 20471 and landrace Kalajeera were divided in subgroup 1 showing unknown linkage with *structure* population in this study.

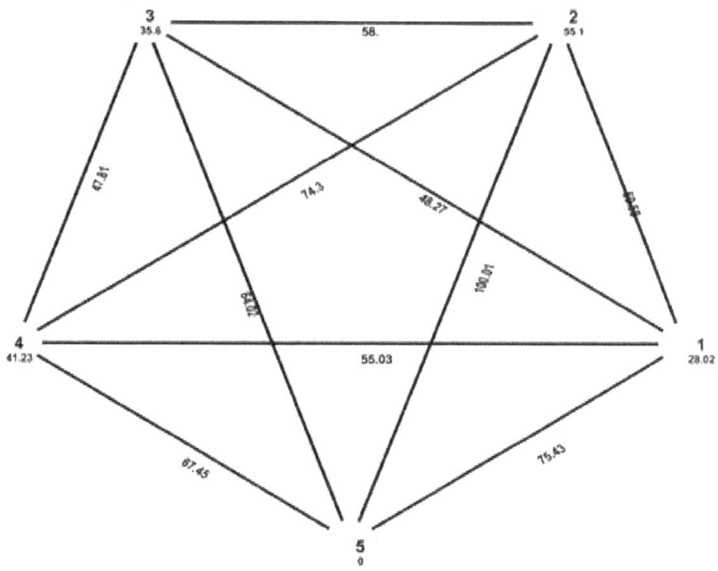

Euclidean² Distance (Not to the Scale)

Fig. 2. Euclidean distances among elite Indian rice varieties

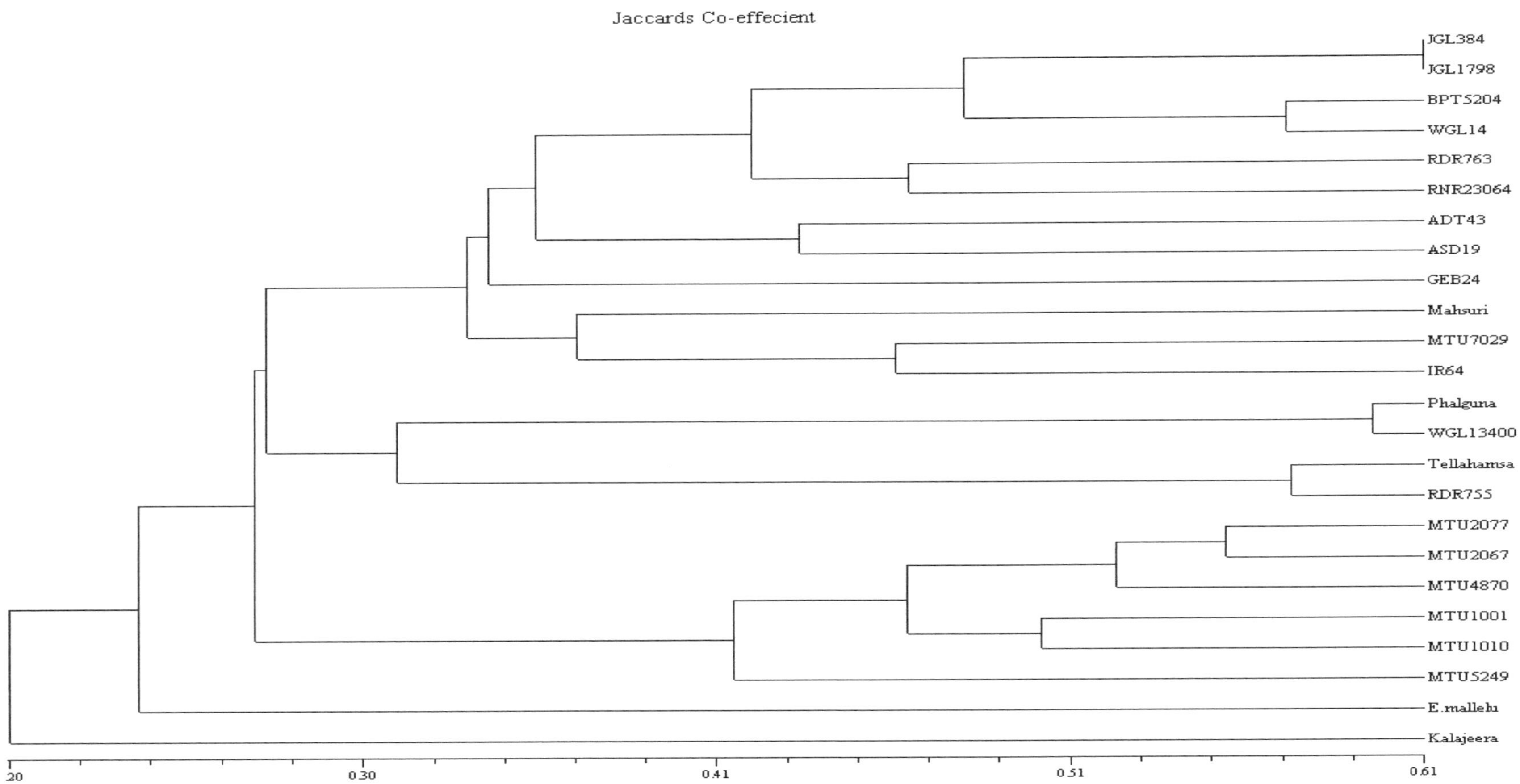

Fig. 3. **Dendrogram based on 256 SSR alleles derived from UPGMA cluster analysis using Jaccard's coefficient**

Table 5. Contribution percentage and agro morphological traits associated with the seven principal components of elite Indian rice varieties and their Eigenvectors

Factor	**1**	**2**	**3**	**4**	**5**	**6**	**7**
Contribution percentage (%)	18.32	14.59	12.04	10.99	8.32	6.23	5.39
Characters	**Eigenvector**						
Length of leaf blade (cm)	1.738	1.342	1.342	-1.815	-2.161	-1.023	0.518
Width of Leaf blade (cm)	1.059	1.527	1.400	1.991	2.345	-1.719	-0.661
Time of heading (days)	3.739	1.785	1.562	-0.927	-0.608	-0.417	0.093
Stem length (cm)	-0.406	0.909	1.289	-1.686	-1.772	1.136	-0.253
Stem thickness (cm)	1.959	-2.893	1.210	1.840	0.263	2.050	-1.238
Panicle length of main axis (cm)	-0.946	-1.347	0.134	-1.240	1.076	-1.495	1.616
Panicle number per plant	-0.233	-3.140	1.845	0.523	-1.226	0.212	0.135
Time of maturity (days)	4.043	1.351	1.332	-0.985	-0.226	-0.937	0.108
1000 grain weight(g)	-2.703	1.324	0.798	1.268	-0.251	0.386	0.772
Grain length (mm)	-3.401	0.197	1.968	-0.538	0.344	0.653	-0.504
Grain width (mm)	-1.135	2.521	-1.137	2.545	-0.781	0.602	0.434
Decorticated grain length (mm)	-3.135	0.342	2.312	-0.541	0.583	0.654	-0.084
Decorticated grain width (mm)	-0.199	2.720	-1.214	2.603	-0.916	0.163	0.962
Intensity of green leaf Colour	2.266	1.303	-0.197	0.760	-0.378	1.242	-1.274
Pubescence of leaf blade surface	2.001	0.406	0.253	0.298	3.636	0.526	0.441
Attitude of culm	0.760	-3.345	-1.647	0.325	-2.050	-2.137	0.363
Flag leaf attitude of blade early observation	-0.442	0.590	-2.633	-1.414	0.345	0.034	0.474
Flag leaf attitude of blade late observation	-0.858	1.320	-2.579	-1.781	0.460	1.557	-0.796
Curvature of main panicle axis	0.082	0.921	-2.866	-1.832	0.818	-0.277	1.069
Panicle awn	-1.170	-1.381	-0.742	-0.549	2.145	-1.972	-2.227
Panicle exsertion	-1.014	-0.441	-1.561	-2.109	-1.113	0.335	-2.540
Decorticated grain shape	-2.690	-0.390	2.487	-0.801	-0.059	-0.683	0.691
Endosperm amylose content	-1.214	-0.283	-1.519	3.426	-1.303	-1.536	-1.016
Gelatinization of temperature	2.652	-3.143	-0.793	0.471	0.321	1.617	0.732

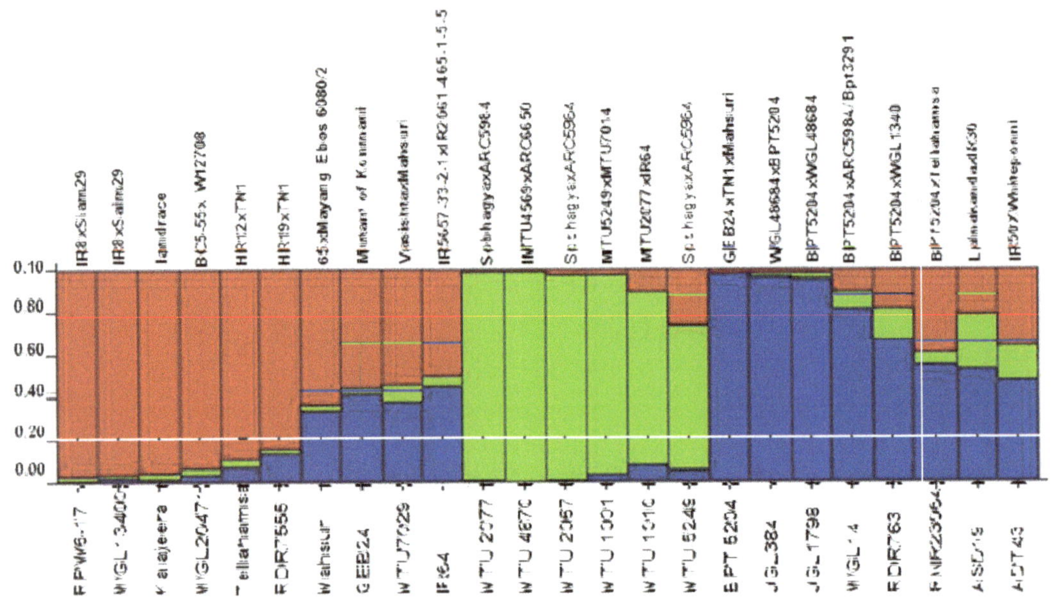

Fig. 4. Estimation of population structure based on 86 SSR markers; each variety in a population is represented by bar plot proportionally to Q values of 24 elite Indian varieties up to three colours into three subpopulation based on their pedigree

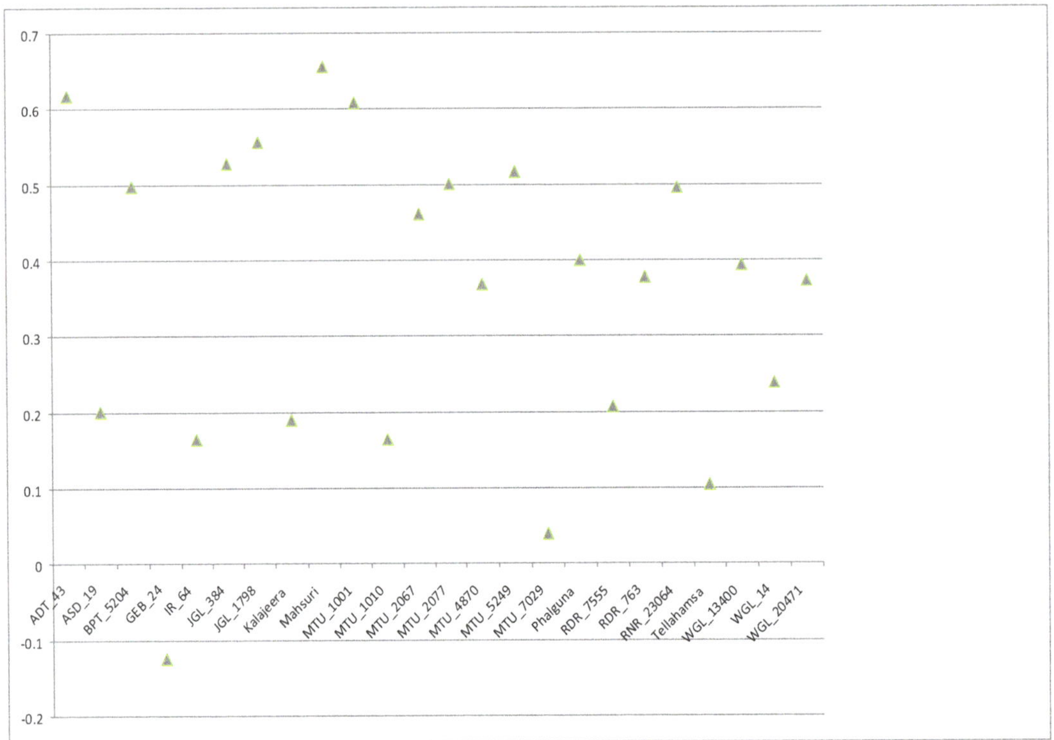

Fig. 5. Correlation between of SSR marker and morphological traits among elite Indian rice varieties

Table 6. Estimated membership inference cluster values among rice varieties

Variety	Cluster 1	Cluster 2	Cluster 3
JGL 384	0.016	0.013	0.971
JGL 1798	0.017	0.024	0.959
BPT 5204	0.014	0.005	0.981
WGL 14	0.100	0.078	0.822
RDR 763	0.190	0.146	0.664
RNR 2306	0.402	0.051	0.548
MTU 2077	0.010	0.981	0.009
MTU 4870	0.013	0.98	0.007
MTU 1001	0.027	0.934	0.039
MTU 2067	0.026	0.963	0.011
MTU 7029	0.558	0.077	0.366
MTU 1010	0.123	0.811	0.066
MTU 5249	0.277	0.667	0.055
RPW6-17	0.971	0.02	0.009
WGL 134	0.965	0.008	0.027
WGL 20471	0.946	0.021	0.033
Tellahamsa	0.894	0.034	0.071
RDR 7555	0.838	0.018	0.144
ADT 43	0.385	0.141	0.474
ASD 19	0.253	0.235	0.512
GEB24	0.576	0.023	0.401
Mahsuri	0.655	0.024	0.321
IR 64	0.526	0.044	0.430
Kalajeera	0.961	0.027	0.012

The correlation between morphological and SSR marker data exhibited similar grouping among the varieties. Three consistent groups were found in both the diversity classifications. The first group consisted, the bold and long slender grain cultivars including a landrace Kalajeera except WGL 14 and ASD 19, the second group consisted of all the medium and long slender MTU varieties developed by the Andhra Pradesh State Rice Research Institute of India. The third group was composed by the short slender varieties *viz.*, JGL 384, JGL 1798, BPT 5204, ADT 43 and Mahsuri. The other varieties were so diverse with respect to their genetic background and therefore, no consistent relationship could be found between the data derived from SSRs and morphological traits. Similarly, high correspondence between phenotypic and genotypic diversities has been documented among rice varieties [31]. The significant correlation observed in the present study between the two diversity measures could be possibly explained by the SSR markers that are being associated with chromosomal regions which could have selected for particular environments and morphological traits relevant to adaptation.

4. CONCLUSION

SSR markers provide a greater allelic discrimination for the characterization of elite Indian rice varieties as compared to morpho-grain traits. Agro-morphological traits can be reliably deployed along with SSR markers for the characterization and genetic diversity assessment of rice varieties and can also be used to design better experiments in DUS testing to assess distinction.

ACKNOWLEDGEMENTS

We are thankful to Dr. B C Viraktamath, Project Director, Directorate of Rice Research, Hyderabad, India for the research facilities provided. Gratefully acknowledge the support of Drs. A. Padmaraju and Late B. Muralimohan Reddy and the assistance of Mr. K. Muralikrishna and Mr. Mohammad Siraj in conducting the field experiments.

COMPETING INTERESTS

Authors have declared that no competing interests exist.

REFERENCES

1. Sasaki T, Burr B. International rice genome sequence project, the effort to complete the sequence of rice genome. Curr Opi Plant Biol. 2000;3.138-141.
2. Cooke RJ. New approaches to potato variety identification. Potato Res. 1999;42:529-539.
3. Meesang N, Ranamukharachi SL, Peterson MJ, Anderson SB. Soybean cultivar identification and genetic purity analysis using microsatellite DNA markers. Seed Sci Technol. 2001;29:637-645.
4. Bonow S, Von Pinho EVR, Vieira MGC, Vosman B. Microsatellite markers in and around rice genes, Applications in variety identification and DUS testing. Crop Sci. 2009;49:880-886.
5. Smith JSC, Smith OS. Fingerprinting crop varieties. Advances Agron. 1992;47:85-140.
6. Govindaraj P, Vinod KK, Arumugachamy S, Maheswaran M. Analyzing genetic control of cooked grain traits and gelatinization temperature in a double haploid population of rice by quantitative

trait loci mapping. Euphytica. 2009;166:165-176.
7. Bhuyan N, Borah BK, Sarma RN. Genetic diversity analysis in traditional lowland rice (*Oryza sativa* L.) of Assam using RAPD and ISSR markers. Curr Sci. 2007;93:967-972.
8. Anshi Z, Jiuwen X, Limin Z, Zhifeng X, Huiyan W. RAPD analysis of classification and genetic relationship among Northern japonica rice. Mol Plant Breed. 2009;5:885-889.
9. Mochida K, Furuta T, Ebana K, Shinozaki K, Kikuchi J. Correlation exploration of metabolic and genomic diversity in rice. BMC Genomics. 2009;10:568.
10. Aggarwal RK, Brar DS, Nandi S, Huang N, Khush GS. Phylogenetic relationship among Oryza species revealed by AFLP markers. Theorl Appl Genet. 1999;98:1320-1328.
11. Bao J, Corke H, Sun M. Analysis of Genetic Diversity and Relationships in Waxy Rice (*Oryza sativa* L.) using AFLP and ISSR Markers. Genet Resour Crop Evol. 2006;53:323-330.
12. McCouch S, Teytelman L, XU Y, Lobos K, Clare K, Walten M, et al. Development of 2,240 new SSR markers for rice (*Oryza sativa* L.). DNA Res. 2002;9:199-207.
13. Wong SC, Yiu PH, Bong STW, Lee HH, Neoh PNP, Rajan A. Analysis of Sarawak Bario Rice Diversity Using Microsatellite Markers. Am J Agron Biol Sci. 2009;4:298-304.
14. Thomson MJ, Thomson PR, Prasetiyono J, Trijatmiko KR, Silitonga TS, McCouch SR. Genetic Diversity of Isolated Populations of Indonesian Landraces of Rice (*Oryza sativa* L.) Collected in East Kalimantan on the Island of Borneo. Rice. 2009;2:80-92.
15. Keshavulu K, Sambasivarao P, Reddy KB, Ganesh M, Manohareddy NM, Reddy BM et al. Morphological, chemical and electrophoretic descriptors of rice (*Oryza sativa* L.) varieties. Technical Descriptors No 2, ICAR- ANGRAU, Hyderabad, India. 2007;15-172.
16. Zheng K, Huang N, Bennett J, Khush GS. PCR-based marker assisted selection in rice breeding. IRRI Discussion Paper Series No 12. International Rice Research Institute, Manila, the Philippines; 1996.
17. Oksanen J, Kindt R, Legendre P, O'Hara B, Simpson GL, Solymos P, Stevens MH, Wagner HV. Community Ecology Package; 2009. Available:http.//cran.project.org/web/packages/vegan/index. htm
18. Rohlf FJ. NTSYS-PC, Numerical Taxonomy and Multivariate Analysis System. Version 3.21, Exeter Publications, New York, USA; 2000.
19. Nei M. Analysis of gene diversity in subdivided populations. Proc Natl Acad Sci, USA. 1973;70:3321-3323.
20. Pritchard JK, Stephens M, Donnelly P. Inference of population structure using multilocus genotype data. Genet. 2000;155:945-959.
21. Ewens WJ, Spielman RS. The transmission/disequilibrium test, history, subdivision, and admixture. Am J Human Genet. 1995;57:455–464.
22. Mantel NA. The detection of disease clustering and a generalized regression approach. Cancer Res. 1967;27:209-220.
23. Chessel D, Dufour A, Dray S. Ade4, Analysis of Ecological Data, Exploratory and Euclidean methods in Environmental sciences; 2009. Available:http.//cran.univ-lyon1.fr/web/packages/ade4/index.html
24. Xu W, Virmani SS, Hernanadez JE, Sebastian LS, Redona ED, Li Z. Genetic diversity in the parental lines and heterosis of the tropical rice hybrids. Euphytica. 2002;127:139-148.
25. Nethra N. Studies on varietal characterization based on morphological biochemical and molecular markers in rice (*Oryza sativa* L.). M Sc. (Ag) Thesis. University of Agricultural Sciences, Bangalore, Karnataka, India; 2003.
26. Adair CR, Beachell HM, Jodan NE, Johnstan TH, Thysell JR, Jr. Green VE, Webb BD, Atkins JG. Rice breeding and testing methods in the US In: Rice in the US, Varieties and production. USDA Agricultural Research Hand book. 1996;289:19-64.
27. Sanjeevarao AR, Murthy V, Subrahmanya RS. The amylose and amylopectin content of rice and their influence on the cooking quality of the cereal. Proc Indian Acad Sci Section B. 1952;36:70-80.
28. Bajracharya J, Steele KA, Jarvis DI, Sthapit BR, Witcombe JR. Rice landrace diversity in Nepal, Variability of agro-morphological traits and SSR markers in landraces from a high-altitude site. Field Crops Res. 2006;95:327-335.
29. Singhal NC, Prakash S. The characterization and identification of wheat cultivars - integrated approach. In,

Proceeding Indo-British Workshop on Plant Breeders Rights, Seed Certification and Storage, Division of Seed Sci and Technol, IARI, New Delhi. 1992;140-146.

30. Oka HI, Chang WT. Hybrid swarms between wild and cultivated rice species. *Oryza perennis* and *O. sativa*. Evaluation. 1961;15:418-430.

31. Fuentes JL, Cornide MT, Alvarez A, Suarez E, Borges E. Genetic diversity analysis of rice varieties (*Oryza sativa* L.) based on morphological, pedigree and DNA polymorphism data. Plant Genet Resour Crop Evol. 2005;3:353-359.

32. Ravi M, Geethanjali S, Sameeya F, Maheswaran M. Molecular marker based genetic diversity analysis in rice (*Oryza sativa* L.) using RAPD and SSR markers. Euphytica. 2003;133:243-252.

33. Agrama HA, Eizenga GC, Yan W. Association mapping of yield and its components in rice cultivars. Mol Breed. 2007;19:341-356.

34. Nelson PT, Coles ND, Holland JB, Bubeck DM, Smith S, Goodman MM. Molecular characterization of maize inbreds with expired US plant variety protection. Crop Sci. 2008;48:1673-1685.

10

Genetic Variation, Heritability and Relationship between Inbred Lines Performances under Drought Stress and Irrigation Conditions in Rice

Charles A. Joseph[1,2*]

[1]Institute of Crop Sciences/National Key, Facility for Crop Gene Resources and Genetic Improvement, Chinese Academy of Agricultural Sciences, 12 South Zhong-Guan-Cun St., Beijing 100081, China.
[2]Department of Botany, University of Dar es Salaam Box 35091, Dar es Salaam, Tanzania.

Author's contribution

The sole author designed, analyzed and interpreted and prepared the manuscript.

Editor(s):
(1) Özge Çelik, Department of Molecular Biology and Genetics, Istanbul Kultur University, Turkey.
Reviewers:
(1) Anonymous, India.
(2) Anonymous, Brazil.

ABSTRACT

Aims: Genetic variability and heritability for agronomic traits are key component in broadening the gene pool and as selection criteria in any rice breeding programs. In present study I estimated genetic variation and heritability of traits, examined relationships among traits and their effects on grain yield under water stressed and non-stressed conditions and distinguished cultivars that are drought tolerance, high yield potential and those that combined both high yield potential and drought tolerance.

Study Design: Field experiment was arranged in a completely randomized block design.

Place and Duration of Study: Study was carried in field research center Hainan of the Institute of Crop Science Chinese Academy of Agricultural Sciences China PR. from 2012-2014.

Methodology: Two parents Minghui 63 and 02428 were crossed to develop 717 lines. Based on spikelet fertility ≥ 50,140 lines were selected for present study. Under drought stress field was drained at panicle initiation stage and maintained dry until maturity. In irrigation water was applied whenever necessary until maturity. At maturity, ANOVA, Dunnetts't tests, correlation and path analysis were performed for 24 traits.

Results: Correlation coefficient of trait performances between water regimes was positive and

**Corresponding author: E-mail: jnyengo@yahoo.co.uk;*

ranged between 0.3 to 0.9. Drought tolerance, yield potential and both drought tolerance and yield potential lines were distinguished. Trait genetic variation, heritability and their contribution to the final grain yield varied over genetic background, water regimes and year. Heritability in *japonica* type was higher under irrigation compared to drought stress. In *indica* background heritability was higher under drought stress in 2013, and varied in 2014. Lower values of residue effects were recorded for *japonica* type compared to *indica* background and under drought stress compared to irrigation. High filled grain number, high 1000-grain weight, moderate tiller number, short flowering time and high plant height could be selection criteria for breeding high grain yield under drought conditions. While high spikelet number coupled with high spikelet fertility, high tiller number, long panicle length, moderate plant height, and moderate flowering time are necessary for breeding high yield potential.

Conclusion: These results provide a platform for dissecting genes for high yield potential, genes for drought tolerance and genes for both yield potential and high drought tolerance.

Keywords: Heritability; genetic variations; Oryza sativa L.; drought tolerance; relative trait performance reduction; yield potential; residual effects.

1. INTRODUCTION

Rice is grown all over the tropics, subtropics, semi-arid tropics and temperate regions of the world. It is anticipated that by 2025, 1.8 billion people will be living in countries or regions with absolute water scarcity, and two-thirds of the world background could be under conditions of water stress [1]. This water scarcity threatens the food security of thousand millions of people who entirely depend on rice as the primary source of nutrition. Yet very little yield gain occurs in rainfed rice systems that occupy much of the 63.5 million hectares of rice planting areas most of which is in tropical Asia, Africa, and Latin America [2,3]. Breeding approaches to develop rice varieties with better yields in drought-prone environments should focus on broadening genetic pool which has narrowed because of the massive international exchange of germplasm [4]. Hence it is imperative that a detailed understanding of genetic information of yield traits such as their heritability, genetic and environmental effects should be addressed properly before performing any genetic improvement programs [5]. Putative traits that enhance drought stress tolerance are many and some do not show enough genetic variation and heritability. In addition these traits interact among themselves. In other words causal factors are not acting independently. Under limited resource both financial and manpower it is important to separate correlation coefficients into components of direct (influence of causal effect directly on dependant variable) and indirect effects (influence of causal effect on dependant via another causal effect) to identify only traits of large effect on grain yield as selection criteria and for genetic manipulation. Moreover many researchers neglect the empirical fact that drought does not occur every season in a target environment. Therefore drought tolerant cultivar should not be one that is drought tolerant but low in yield potential in favorable season, instead cultivar that respond to favorable conditions and produce economically useful yield under drought conditions as well. It is also worth noting that greatest yield loss due to drought stress occurs during reproductive stage. In vegetative stage plant may acclimatize by producing osmolytes and heat shock proteins, however, reproductive organs particularly pollen and stigma have no such ability [6]. In that regard drought resistant (DR) in this study is viewed as the ability of a cultivar to maintain higher grain yield in drought stress as well as normal irrigation conditions compared to recurrent parent, while yield potential is the upper limit of yield not constrained by water, nutrients, or pests [7]. Studies have shown that there is a low but positive correlation between yield under stress and yield potential. A genetic correlation of 0.8 between yield under stress and non-stress have been reported [8]. This indicates that much of the yield under drought is accounted for by yield potential. And varieties with high yield potential had an advantage over varieties with lower yield potential even with moderate drought stress that resulted up to only 50% loss in grain yield [7]. Therefore for successful genetic improvement of rice variety for each category (drought tolerance, yield potential and both drought tolerance and yield potential), it is first essential to identify traits with major effects on grain yield across water regimes and genetic background for rice variety on each category and then undertake a comprehensive study to understand heritability, genetics and environment effects of each trait. In present study we evaluated the performances of BILs under normal irrigated conditions as well as under drought stress to distinguish cultivars with genes for tolerance from those that simply have high yield potentials and cultivars that combine both stress tolerance with responsiveness to favorable conditions. In addition we estimated genetic variation and heritability of traits and examined relationships among traits and their effects on grain yield under water stressed and non-stressed conditions

2. METHODOLOGY

2.1 Study Material

Minghui 63 (MH63) an *indica* was crossed with 02428 a *japonica*. The F_1 hybrids were simultaneously backcrossed twice to both parents to produce two BC_2F_1 populations. At the same time, the F_1 hybrids were self pollinated to generate F_2 populations. Both the BC_2F_1 and F_2 were advanced by single seed descent from year to year until BC_2F_8 and F_6 respectively. Finally, two sets of reciprocal inbred lines, consisting of 226 lines (BC_2F_8) with the MH63 background (MH63_BILs), 229 lines (BC_2F_8) with the 02428 background (02428_BILs) and 262 ($F_{2:8}$) recombinant inbreed lines (RBILs) were selected for the study.

2.2 Methods

2.2.1 Field management and drought stress application

The preliminary screening of the bulk 717 BILs and RIL backgrounds for drought tolerance (DT) was conducted in 2011-12 in Hainan during the dry season. Based on spikelet fertility (SF) ≥ 50, 140 lines mainly from BILs were selected for present study. All selected BILs and their parents were planted under both normal irrigation (IR) and drought stress conditions (DS) in Hainan for two years 2013 and 2014. Field experiment was arranged in a completely randomized block design with 3 and 2 replications for drought stress in 2013 and 2014 respectively, and 2 replications under normal irrigation for both 2013 and 2014 years. In each plot 10 plants were transplanted in a single row with spacing of 25 by 20 cm. At each plot parents were inserted in every 30 lines for MH63_BILs and 40 lines for 02428_IL in each plot as checks. The fields was managed normally with a total of 260 kg N ha−1 applied (about 80 kg N ha−1 as the basal fertilizer and about 60 kg N ha−1 applied each time at 14, 30, and 60 d after transplanting. For drought stress field were drained when plants enter the stage of panicle initiation stage (roughly after 30 days after transplanting). Field was maintained dry until all panicles excert completely. After all the plants had panicle excerted completely irrigation was applied only once to rescue plants that were dying. Any rain that fallen during drought stress application period was recorded. In the irrigated control, irrigation was applied whenever necessary until all lines matured.

2.2.2 Phenotyping

In both drought stress (DS) and well watered conditions (IR) at maturity, 5 representative plants in 2013 and 8 plants in 2014 in each row were measured for Heading Date (HD), Plant Height (PH), Spikelet number per panicle (SNP), filled grain number per panicle (FGN) spikelet fertility (SF), thousand grain weight (TGW) grain yield per plant (GY) grain length (GL), grain width (GW), grain length width ratio (GLW), grain thickness (GT) grain length thickness ratio (GLT) grain width thickness ratio (GWT), Flag leaf length (FLL), leaf width (FLW) and flag leaf rolling (FLR) for well watered plants in 2013. For drought stress in addition to HD, PH, TN, PL, SNP, FGN, SF, GY, TGW, GL, GW, GLWR, GLT, GWT leaf tip drying, leaf rolling and recovery after first day of re-watering was measured. In 2014 HD, PH, TN, PL, SNP, FGN, SF, GY, TGW, GL, GW, GLWR, FLL, FLW and FLR were measured in both drought stress and normal irrigation.

2.2.3 Analysis

Analysis of variance (ANOVA) using SAS PROC GLM (SAS Institute 2002) analyzed the data from two seasons to exam trait variation attributable to differences between the DT treatments, among different BILs within each background, and among BILs from different backgrounds. Dunnetts't tests was used to compare difference between BILs and parents. Relative trait performance reduction (RR) was calculated according to the formula: average trait of each IL under irrigation (Xi) and average trait of each line under drought (Xd); RR = (Xi - Xd)/Xi*100. Total phenotypic variability (R^2) explained by each source of variation for each trait was calculated according R^2 = SS/CT*100.
Where SS=sum of square and CT corrected total. Heritability was calculated according to the

formula: h2m= σ^2_g / σ^2_g + σ^2_e. Where genetic variance σ^2_g = (MS_g-MS_e)/r, variance due to errors σ^2_e = MS_e/r, r = number of replications. MS_g is considered as total phenotypic variation and MS_e is an estimate of non-genetic variation. To investigate the inter-relationships among the variables and their direct and indirect contributions to yield performance under well-watered and water stressed conditions, entry means of variables that showed high genetic variation were subjected to correlation and path analyses according to [9].

3. RESULTS AND DISCUSSION

3.1 Drought Resistant Genotypes

Drought resistant line in this present study means individual line that had high grain yield per plant (GY) compared to the recurrent parent under terminal drought stress for two consecutive seasons. In the year 2013 and 2014, MH63_IL an *indica* background showed that 27 and 24 lines were drought tolerant respectively. 14 lines DQ_15, DQ_34, DQ_68, DQ_82, DQ_146, DQ_212 DQ_214, DQ_215, DQ_219, DQ_225, DQ_227, DQ_230, DQ_231 and DQ_232 were drought tolerant for two consecutive seasons. Out of these 14 lines, only 4 lines (28%) DQ_34, DQ_82, DQ_219 and DQ_227 had high yield in both drought stress and normal irrigation conditions for two consecutive seasons. Hence these are the lines that had genes for both yield potential and yield under drought stress. Lines DQ_146, DQ_212, DQ_230 and DQ_232 did not record higher yield potential for two seasons but had higher yield under drought for two consecutive dry seasons. These lines can be regarded as having genes for high yield under drought but no genes for high yield potential. Under normal irrigation conditions (IR) in 2013, 12 lines had higher yield compared to recurrent parents. Of these, 8 lines (66%) were drought tolerant as well. In 2014 under well watered conditions 17 lines had higher yield compared to recurrent parents. Of these, 12 lines (70%) were drought tolerant as well. 8 Lines DQ_12, DQ_13, DQ_19, DQ_34, DQ_82, DQ_207, DQ_219 and DQ_227 had higher yields for two consecutive seasons under normal irrigation. Out of these, 4 lines (50%) DQ_12, DQ_13 DQ_19 and DQ_207 had high yield potential for two consecutive years but were not drought tolerance hence they had genes for yield potential but not for drought tolerance. The correlation coefficient between the yield potential and the yield under drought was 0.3 and 0.5 in 2013 and 2014 respectively. Other traits such as HD, PH, TN, PL, SNP, FGN, SF, TGW, GL, GW, GLW, the correlation coefficient between water regimes were 0.9vs0.8, 0.8vs0.8, 0.5vs0.3, 0.6vs0.4, 0.6vs0.7, 0.7vs0.6, 0.5vs0.2, 0.6vs0.5, 0.8vs0.8, 0.8vs0.4 and 0.8vs0.4 in 2013 and 2014 under drought and normal irrigations respectively. Correlation coefficients were, 0.3, 0.8 and 0.3 for leaf length, width and rolling in 2003 and 0.6, 0.7 and 0.6 for grain thickness, length thickness ratio and width thickness ratio respectively in 2014. All drought tolerant lines in *indica* background had early flowering (HD) and high plant height, compared to recurrent parent a mechanism called drought escape. Escape strategy relies on successful reproduction before the onset of severe stress.

In 02428_IL a *japonica* background 20 and 17 lines were drought tolerant in the year 2013 and 2014 respectively. 7 lines DQ_237, DQ_288, DQ_386, DQ_389, DQ_390, DQ_393 and DQ_461 were drought tolerant for two consecutive seasons. These 7 lines except DQ_389 had higher yield compared to recurrent parent in both drought stress as well as normal irrigations conditions for two consecutive seasons. Thus they were the lines with both yield potential and drought tolerance. In normal conditions in 2013, 51 lines had higher yield potential compared to recurrent parent. Of these, 16 lines (31%) were drought tolerant as well. In 2014 in normal conditions 41 lines had higher yield compared to recurrent parents whereas 14 lines (34%) were drought tolerant as well. 35 lines had high yield potential compared to recurrent parent for two consecutive seasons. Of these 35 lines 16 lines (45%) DQ_257, DQ_258, DQ_267, DQ_325, DQ_345, DQ_356, DQ_399, DQ_414, DQ_422, DQ_ 430, DQ_433, DQ_440, DQ_441, DQ_443, DQ_460 and DQ_488, had no higher yield under drought stress thus contains genes for yield potential but not for drought tolerance. The correlation coefficient of specific traits performance between well watered and drought stress conditions for GY, TGW, HD, TN, PH, PL, SNP, FGN, SF, GL, GW and GLW were 0.5vs0.5, 0.3vs0.4, 0.5vs0.7, 0.3vs0.5, 0.7vs0.5, 0.5vs0.5, 0.7vs0.6, 0.9vs0.6, 0.4vs0.6, 0.8vs0.9, 0.8vs0.9 and 0.8vs0.9 in 2013 and 2014 respectively. Other traits such as grain thickness, grain length thickness ratio, grain width thickness ratio were 0.7, 0.8, 0.6, in 2013 respectively. Correlation coefficient between drought and irrigation conditions were 0.7, 0.7 and 0.6 for leaf length, leaf width and leaf rolling respectively in 2014. In this background all of the drought tolerant lines had late flowering time (HD) and moderate high plant height compared to recurrent parent. These results reveal that there is positive relationship between traits performance under water stress and under well water conditions for all traits. Correlations coefficients differed among the traits and between the years ranging from 30% to over 90%. This suggests that plants with high yield potential can perform better than low yield potential under drought stress. Same result were reported by [7,8]. These homologous lines selected in each category can act as genetic source for breeding or understanding genetic analysis underlying drought tolerance, yield potential and both drought tolerance and yield potential varieties

3.2 Genetic Variation under Drought Stress and Normal Irrigation Conditions

Traits with high genetic variation, heritability and strong correlations with grain yield under drought and low seasonal variability are desirable for making progress in plant breeding for drought tolerance [5]. From our study total variation for all measured traits were significant under both water regimes and seasons (Table 1). ANOVA indicated the presence of high genetic variability. Suggesting that selection on the basis of phenotype alone could be effective for the improvement of these traits. Similar results have been reported by [10-15]. When data were analyzed separately as treatment per season, using one-way analysis of variance, all traits showed significant variations in both water regimes and in both years. Such results have been reported by [16-18]. Variation was higher under irrigation conditions compared to drought stress for all traits in *japonica* type but lower for TN, SNP, FGN, GY, and GW in *indica* type.

3.3 Heritability and Performance of BILs under Different Water Regimes

Adequate water supply is critical to plant survival and growth whereas its deficit suppresses leaf expansion and tillering, reduces photosynthetic rate and leaf area due to early senescence and ultimately reduce grain yield [19]. Plant genotypes vary in response to water stress. Variations are due to the combined actions of genetic and environments. The proportion of variance observed among BILs that is due to genetic differences is called heritability. The estimates of heritability act as predictive instrument in expressing the reliability of phenotypic value. Therefore, high heritability helps in effective selection for a particular character. Heritability was classified as low (below 0.3), medium (0.3-0.6 and high (above 0.6) according to [20]. Trait-specific heritability obtained from the study under irrigation and water stresses over the two seasons are presented in Table 2 and Table 3. In general traits studied in the present investigation expressed medium to high heritability estimates ranging from 0.3 to over 0.9. In MH63_IL background, in 2013 higher heritability for all traits for both treatments were recorded. Equal heritability under both drought stress and normal irrigation were shown by HD, PH, PL, FGN and TN. Traits such as TN, SNP, SF, GL, GW, GLW, GT, GLT and GWT, had higher heritability under drought stress compared to normal irrigation. In 2014, the lowest heritability was showed by grain width (h^2_m= 0.3) and tiller number TN (h2 m = 0.4) under drought, while spikelet fertility h^2_m= 0.4) and tiller number (h^2_m = 0.5) had the lowest heritability under irrigation. Heritability was higher under normal irrigation condition compared to drought stress for HD, PH, PL, SF, GL, GLW and FLL, while lower under normal irrigation than under drought stress for TN, SNP, FGN, GY and GW. In *indica* background, heritability was generally higher in drought stress compared to normal irrigation and slightly higher in 2013 compared to 2014. These would suggest heritability were affected by treatment and seasons.

In 02428_IL background presented in Table 3, heritability was higher in 2013 compared to 2014 for all traits. In 2013 the lowest heritability was showed by grain width (GY) (h^2_m= 0.5) and tiller number TN (h^2 m = 0.6) under both water regimes, while for 2014 lowest heritability was showed by GY h^2_m= 0.3, 0.4) under drought and irrigation. In contrast to *indica* type, heritability in *japonica* type was higher under normal irrigation for all traits in all seasons compared to drought tress. Higher heritability recorded would mean these traits could be assessed precisely in the genotypes evaluated. Of all traits in both backgrounds GY and FGN and SNP were most affected by drought. Relative reduction (RR%) for the grain yield in *indica* type was 67 vs. 61 in 2013 and 76 vs. 71% for 2014 for the parent and BILs respectively. Likewise RR for the grain yield in *japonica* background was 69 vs 75% in 2013 and 42 vs. 56 for 2014 for the parent and BILs respectively. Variation on yield reduction between 2013 and 2014, in both backgrounds probably came from difference in stress intensity, duration, timing, or developments stage at which stress was applied. It might also have a risen from the year variations per se such as diurnal, temperature, humidity etc or even from experimental errors. In general relative reduction (RR) for the grain yield was high in 2013 and low in 2014 for *japonica* but vice versa in *indica* type. Moreover *indica* expressed early water deficit through leaf tip drying while *japonica* expressed early water deficit through leaf rolling (Fig1). These differences might imply that these two subspecies might have different drought tolerance mechanisms as it was for heading date and plant height mentioned in section 3.1 above of present study. One way ANOVA indicates that genetic variance (σ^2_g) (Table 2&3) for both treatments and in both seasons was higher than variance due to (error σ^2_e) for all traits indicating a lower experimental error across the backgrounds, treatment and seasons. To determine putative traits that enhance yield potential and yield under drought we performed both correlation analysis and path analysis.

Table 1. Total variations (R^2) explained by season (S), Treatment (T), genotype (G), genotype X season interaction (G*S) and genotype X treatment interaction (G*T) for the traits evaluated derived from sum of square (SS) and corrected factor (CT). R^2 less than 5% is insignificant, S.V; source of variation

S. V	02428_IL (*Japonica* type)					MH63_IL (*Indica* type)				
	S	T	G	G*S	G*T	S	T	G	G*S	G*T
HD	8.0	1.9	33.5	8.8	11.0	25	16.4	41	5.8	3.6
PH	0.3	68.1	15.5	2.7	1.9	3.1	64.0	19.7	1.9	1.2
PL	8.2	45.4	20.2	2.1	5.2	2.8	36.3	30.9	4.2	11.1
TN	11.1	43.6	15.4	3.8	5.8	33.4	8.2	23.5	3.8	6.3
SNP	12.6	49.8	19.2	2.3	3.9	5.8	26.1	40.4	6.7	8.1
FGN	5.9	48.1	22.2	2.4	5.3	4.3	33.8	40.0	5.1	7.3
SF	3.2	10.3	37.2	4.5	10.5	1.2	24.3	31.9	7.7	11.8
TGW	0.0	30.6	24.1	6.7	9.4	5.5	33.2	26.0	6.6	5.0
GY	0.4	62.3	12.2	3.6	4.4	0.7	65.6	14.4	4.1	5.0
GL	2.6	0.5	76.8	3.2	3.7	15.4	10.3	53.3	5.4	2.8
GW	1.0	5.9	68.2	2.4	4.1	4.4	10.2	47.7	5.2	8.3
GLW	1.5	0.6	81.4	2.0	3.1	10.3	0.0	59.3	7.8	5.1

Fig. 1. Shows variation in expressing early water deficit between the two subspecies. MH63_IL (*indica* background) relays water deficit through leaf tip drying while 02428_IL (*japonica* background) expresses water deficit through leaf rolling

Table 2. MH63_IL *Indica* type mean performances of recurrent parent (P) and backcross introgression inbred lines (BILs) genetic variance (σ^2_g), non-genetic variances (σ^2_e) heritability (h^2_m) and relative trait performance reduction (RR%) of measured traits in two contrasting treatments(T) drought stress (DS) and normal irrigation(IR) in 2013 and 2014. HD; heading date, PH; plant height (cm), PL; panicle length (cm), TN; tiller number, SNP; spikelet number per panicle, FGN; filled grain number per panicle, SF spikelet fertility(%), TGW 1000-grain weight(g), GY; grain weight per plant (g/plant) delayed leaf senescence, GL; grain length(μ), GW; grain width (μ), GLW; grain length width ratio GT grain thickness (μ), GLT; grain length thickness ratio, GWT; grain width thickness ratio, FLL; flag leaf length (cm) FLW flag leaf width (cm) and FLR; flag leaf rolling

MH63_IL performance under DS & IR IN 2013										
FACTORS		P	BILs			RR%		VARIANCES		
Trait	T		mean	Min	Max	P	BILs	σ^2_g	σ^2_e	h^2_m
HD	DS	121	116	97	127	-11	-9	60.3	0.5	0.9
	IR	109	106	92	118			39.6	0.6	0.9
PH	DS	65.8	65	50	90	22	22	48.6	1.7	0.9
	IR	84.8	84	67	107			31.8	3.7	0.9
PL	DS	19.4	19	14.1	26	16	12	3.68	0.1	0.9
	IR	23.1	22	17.7	26.1			2.3	0.1	0.9
TN	DS	7.0	6.7	4.0	10	23	14	1.1	0.1	0.8
	IR	9.1	7.8	4.0	12			1.2	0.7	0.6
SNP	DS	72.3	78	42	216	44	36	917	16	0.9
	IR	129	121	70.4	328			623	189	0.7
FGN	DS	41.7	54	19	142	58	43	760	13	0.9
	IR	98.9	94	50.2	166			539	19	0.9
SF	DS	58	67	31	92	25	15	124	6.7	0.9
	IR	76.9	78	32.7	94.7			94.7	12	0.8
TGW	DS	24.4	24	10.5	33.8	16	18	10.7	3.7	0.7
	IR	29	30	23.6	38.8			6.1	0.2	0.9
GY	DS	7.4	7.3	3.2	17.5	67	61	6.5	0.2	0.9
	IR	22.3	19	8.1	32.3			17.6	0.9	0.9
GL	DS	9.7	9.4	7.7	10.3	0.0	2.0	0.1	0.0	0.9
	IR	9.7	9.6	8.1	10.3			0.1	0.0	0.8
GW	DS	2.6	2.7	2.4	3.1	4.0	4.0	0.02	0.0	0.9
	IR	2.7	2.8	2.4	3.7			0.04	0.0	0.8
GLW	DS	3.6	3.5	2.8	4	3.0	2.0	0.07	0.0	0.9
	IR	3.7	3.4	2.7	4.1			0.07	0.0	0.8
GT	DS	2.0	2.0	1.7	2.2	0.0	0.0	0.01	0.0	0.9
	IR	2.0	2	1.8	2.5			0.01	0.0	0.7
GLT	DS	5.0	4.8	3.7	5.9	0.0	0.0	0.12	0.0	0.9
	IR	5.0	4.8	3.4	5.4			0.07	0.0	0.8
GWT	DS	1.3	1.4	1.1	1.7	7.0	0.0	0.01	0.0	0.9
	IR	1.4	1.4	1.1	1.8			0.01	0.0	0.8

MH63_IL PERFORMANCE UNDER DS & IR IN 2014										
FACTORS		P	BILs			RR%		VARIANCES		
Trait	T		Mean	Min	Max	P	BILs	σ^2_g	σ^2_e	h^2_m
HD	DS	129	122	97	137	-7	-3	57.7	7.4	0.8
	IR	121	118	112	125			11.9	0.8	0.9
PH	DS	67	67	97	103	30	27	32.5	10	0.7
	IR	95.1	92	74	114			42.9	2.5	0.9
PL	DS	18.5	19	15	27	24	16	2.84	0.7	0.7
	IR	24.5	23	18	27			3.42	0.4	0.8
TN	DS	10	9.2	3	16	7.0	15	1.49	1.4	0.5
	IR	10.8	11	6.5	17			0.89	1.1	0.4
SNP	DS	59	67	33	135	28	26	470	34.9	0.9
	IR	81.9	91	45	139			215	56.2	0.7
FGN	DS	46	46	16	110	37	38	339	30.7	0.9
	IR	72.9	75	32	128			249	46.6	0.8
SF	DS	77.4	68	42	87	13	16	30.7	39.5	0.4
	IR	89	82	49	97			53.7	9.1	0.8
TGW	DS	22.2	22	6.4	29	22	18	10	1.2	0.8
	IR	28.6	27	22	34			4.23	0.5	0.8
GY	DS	6	6.5	0.7	20	76	71	17.1	2.9	0.8
	IR	25.5	23	6.6	43			21.5	9.72	0.6
GL	DS	9	8.8	7.3	9.7	8.0	6.0	0.16	0.02	0.8
	IR	9.8	9.4	7.9	10			0.18	0.01	0.9
GW	DS	2.7	2.8	2.4	3.3	4.0	3.0	0.01	0.0	0.7
	IR	2.8	2.9	2.6	3.6			0.01	0.02	0.3
GLW	DS	3.5	3.2	2.7	3.8	0.0	3.0	0.03	0.01	0.8
	IR	3.5	3.3	2.7	3.7			0.05	0.0	0.9
FLL	DS	17.1	20	13	36	38	28	5.19	3.7	0.5
	IR	27.4	27	20	37			9.0	1.7	0.8
FLW	DS	1.9	1.8	1.5	2.3	5.0	5.0	0.02	0.0	0.8
	IR	2	1.9	1.6	2.4			0.02	0.0	0.8
FLR	DS	1.7	1.7	1.1	2.2	11	11	0.02	0.0	0.8
	IR	1.9	1.9	1.5	2.3			0.02	0.0	0.8

Table 3. 02428_IL *Japonica* type mean performances of recurrent parent (P) and backcross introgression inbred lines (BILs) genetic variance (σ^2_g), non-genetic variances (σ^2_e) heritability (h^2_m) and relative trait performance reduction (RR%) of measured traits in two contrasting treatments(T) drought stress (DS) and normal irrigation(IR) in 2013 and 2014. HD; heading date, PH; plant height (cm), PL; panicle length (cm), TN; tiller number, SNP; spikelet number per panicle, FGN; filled grain number per panicle, SF spikelet fertility(%), TGW 1000-grain weight(g), GY; grain weight per plant (g/plant) delayed leaf senescence, GL; grain length(μ), GW; grain width (μ), GLW; grain length width ratio GT grain thickness (μ), GLT; grain length thickness ratio, GWT; grain width thickness ratio, FLL; flag leaf length (cm) FLW flag leaf width (cm) and FLR; flag leaf rolling

02428_IL population performances under DS & IR 2013

FACTORS		P	BILs			RR%		VARIANCES		
Trait	T		Mean	Min	Max	P	BILs	σ^2_g	σ^2_e	h^2_m
GY	DS	5.7	5.0	1.8	12	69	75	1.0	0.9	0.5
	IR	19	20	11	30			10	3.1	0.8
TGW	DS	19	20	0.8	33	16	19	18	1.9	0.9
	IR	22	24	19.2	34			5.5	0.4	0.9
HD	DS	100	106	87	127	2.9	-3.7	52	16	0.8
	IR	103	102	88	119			17	1.6	0.9
PH	DS	60	61	22	88	31	30	44	7.6	0.9
	IR	87	87	67	110			45	3.7	0.9
TN	DS	5.0	4.6	0.8	8.4	47	49	0.5	0.3	0.6
	IR	9.5	9.0	5.0	14			2.1	0.7	0.7
PL	DS	18	17	13	22	17	20	1.5	0.6	0.7
	IR	21	21	17	25			1.8	0.4	0.8
SNP	DS	107	92	44	205	39	49	261	118	0.7
	IR	174	182	107	309			1015	261	0.8
FGN	DS	80	64	2	163	49	58	281	121	0.7
	IR	158	153	53	276			1334	200	0.9
SF	DS	72	68	3.4	91	21	19	156	51	0.8
	IR	91	83	36	98			95	11	0.9
GL	DS	6.8	7.2	6.0	9.2	-2	1.2	0.3	0.0	0.9
	IR	6.9	7.3	6.1	9.2			0.4	0.0	0.9
GW	DS	3.4	3.3	2.1	3.7	2.2	2.9	0.1	0.0	0.9
	IR	3.5	3.4	2.2	4.0			0.1	0.0	0.7
GLW	DS	2.0	2.2	1.8	4.0	-4	-1.5	0.1	0.0	0.9
	IR	2.0	2.2	1.7	4.1			0.1	0.0	0.9
GT	DS	2.2	2.1	1.4	2.5	-3	-1.0	0.0	0.0	0.8
	IR	2.1	2.1	1.8	2.6			0.0	0.0	0.8
GLT	DS	3.2	3.4	2.6	5.3	1.3	2.1	0.2	0.0	0.9
	IR	3.2	3.5	2.7	5.2			0.2	0.0	0.9
GWT	DS	1.6	1.5	1.2	2.3	5.0	3.8	0.0	0.0	0.6
	IR	1.6	1.6	1.1	1.8			0.0	0.0	0.8

02428_IL population performances under DS & IR 2014

FACTORS		P	BILs			RR%		VARIANCES		
Trait	T		Mean	Min	Max	P	BILs	σ^2_g	σ^2_e	h^2_m
GY	DS	11	8.2	0.5	23	42	56	9.2	4.7	0.3
	IR	18	18	4.2	41			20	12	0.4
TGW	DS	18	19	12	26	20	24	4.9	0.2	0.7
	IR	22	25	17	39			6.3	1.3	0.7
HD	DS	108	108	100	117	0.1	0.1	13	1.4	0.9
	IR	108	108	89	120			10	2.0	0.8
PH	DS	66	65	43	82	20	25	24	22	0.5
	IR	83	86	68	109			33	9.3	0.8
TN	DS	10	7.2	3.1	14.2	7.8	27	1.4	0.8	0.6
	IR	10	9.8	5.8	15.1			2.0	0.7	0.8
PL	DS	20	19	15	23	11	13	1.4	0.7	0.7
	IR	22	22	18	26			2.3	0.3	0.9
SNP	DS	135	130	58	228	28	33	606	217	0.7
	IR	187	195	95	319			1178	346	0.8
FGN	DS	122	104	17	205	26	36	748	323	0.7
	IR	165	162	60	287			1085	384	0.7
SF	DS	88	78	20	98	2.2	4.9	145	52	0.7
	IR	90	83	47	97			67	18	0.8
GL	DS	6.4	7.0	6.1	8.8	0.9	1.3	0.2	0.1	0.8
	IR	6.5	7.1	6.1	9.2			0.3	0.0	0.9
GW	DS	3.4	3.3	2.4	3.7	4.4	4.9	0.0	0.0	0.8
	IR	3.5	3.5	2.4	3.8			0.1	0.0	0.9
GLW	DS	1.9	2.1	1.8	3.6	3.5	3.7	0.1	0.0	0.9
	IR	1.9	2.2	1.7	3.7			0.1	0.0	1.0
FLL	DS	22	23	14	44	26	24	6.5	3.2	0.7
	IR	30	30	21	36			8.8	1.3	0.9
FLW	DS	1.7	1.7	1.1	2.1	13	12	0.0	0.0	0.7
	IR	1.9	1.9	1.4	2.9			0.0	0.0	0.7
FLR	DS	1.6	1.5	0.8	2.8	2.2	5.6	0.0	0.0	0.6
	IR	1.6	1.6	0.8	2.1			0.0	0.0	0.8

3.4 Pearson's Linear Correlation

Correlation is a measure of the degree of relationship between two variables. Pearson's linear correlation test for the grain yield (GY) under drought stress and normal irrigation is presented in Table 4. The test revealed that flowering time and plant height were more important determinants of grain yield under drought stress than under normal irrigation. For example correlation coefficients for HD under drought stress were -0.36,-0.22 -0.32, and 0.37 against -0.21, -0.07, -0.13 and -0.14 under normal irrigations. The negative correlation for the heading date observed means that early flowering plants had higher yield compared to late flowering plants. This suggest plants completed their reproductive life cycle before drought stress become severe. A drought resistance mechanism called escape [21]. Higher plant height had influence on GY, the influence was higher under drought stress compared to well watered conditions in 2013 and but vise versa in 2014. Under drought stress plant height could be associated with deep roots. Deep and effective root systems enable plants to penetrate hard soils thereby gain access to deeper water. This phenomenon coupled with closure of stomata and leaf rolling is referred to as dehydration avoidance [21,22]. Conversely PN had moderate positive effect on GY in *indica* type (0.26 vs 0.24) under normal irrigation but not under drought stress (0.06, 0.03). For *japonica* type PN had positive influence on GY under drought stress in 2013. It is known that drought stress at reproductive stage has no effect on tiller number, consequently leading to less detrimental effect on yield [23]. Yield component traits, SNP, FGN, SF TGW were the primary determinants of grain yield under both water regimes, seasons and backgrounds. Their correlation coefficients tended to be higher under drought stress compared to normal irrigation. For instance the correlation coefficients for FGN under drought were 0.75, 0.70, 0.78 and 0.76 against 0.40, 0.50, 0.61, and 0.56 under normal irrigation. Grain size traits (thickness, length and width), generally had moderate effect on grain yield under drought stress but not under normal irrigation. Flag leaf is where large percentage of carbohydrates in the grains are produced. This is because the yield is a product of source-sink relationship. Green tissues (Source) such as leaf, stem, or even own makes photosynthetic assimilates carbon compounds (carbohadrates) through interactions with light. Light makes green tissues to convert mineral carbon to organic carbon. Then the assimilates move towards sink (grain). Flag leaf is the most exposed green tissue to the sun light thus contribute to over 70% of accumulated carboharates in grain. In this research flag leaf length (FLL), flag leaf width had influence on GY under drought as well as under normal irrigation. Pearson correlation indicated complex inter-relationships among other traits. These inter-relationships among traits, each may contribute positively or negatively to the observed coefficients between GY, thus we split coefficient of correlations into direct and indirect components using path analysis.

Table 4. Correlation coefficients between GY and traits performance under drought stress and normal irrigation conditions for 2013 and 2014 in MH63_IL and 02428_IL populations

	2013					2014			
Trait	MH63-*Indica*		02428-*Japonica*		Trait	MH63-*indica*		02428-*Japonica*	
	Ds	IR	DS	IR		DS	IR	DS	IR
HD	-0.36****	-0.21*	-0.22*	-0.07	HD	-0.32***	-0.13	-0.37****	0.14
PH	0.52****	0.32***	0.45****	0.22**	PH	0.51****	0.58****	0.12	0.4****
PL	0.41****	0.51****	0.49****	0.14	PL	0.69****	0.3***	0.52****	0.41****
TN	0.06	0.26***	0.54****	0.05	TN	0.03	0.24**	0.01	0.12
SNP	0.59****	0.19*	0.46****	0.37****	SNP	0.76****	0.53****	0.58****	0.42****
FGN	0.75****	0.4****	0.70****	0.5****	FGN	0.78****	0.61****	0.76****	0.56****
SF	0.75****	0.36****	0.62****	0.46****	SF	0.36****	0.46****	0.68****	0.49****
TGW	0.54****	0.28***	0.49****	0.11	TGW	0.29****	0.42****	0.49****	0.21**
GL	0.16*	0.22***	-0.02	0.08	GL	0.17	0.15	-0.05	-0.14
GW	-0.26***	0.16	0.23***	-0.09	GW	0.001	0.04	0.26***	0.02
GLW	0.28****	-0.03	-0.09	-0.05	GLW	0.12	0.06	-0.12	-0.03
GT	0.47****	0.02	0.36****	0.1	FLL	0.41****	0.33***	0.24**	0.25**
GLT	-0.2**	0.11	-0.21**	-0.09	FLW	0.09	0.31***	0.23**	0.2**
GWT	-0.52****	0.16	-0.14*	0.001	FLR	-0.1	0.22**	0.09	0.02

3.5 Path Analyses for Direct and Indirect Effects

Path analysis for grain yield GY under both drought stress and normal irrigation is presented in Table 5&6. According to linear correlation analysis presented in Table 4; GL, GW, GLWR, GLTR, GWTR, FLW, and FLR did not show consistent correlations with the GY across water regimes, seasons and backgrounds thus were omitted for path analysis. Heritability estimates using standardized data did not differ greatly from the actual heritability estimates for any trait, or season. Same result was reported by [5]. Path analysis for MH63_IL an *indica* type presented in Table.5 reveals that under both water regimes, all the traits except HD had positive correlations with the GY. All traits had strong influence on GY except panicle number. Partitioning correlation coefficient into direct and indirect effects showed that FGN followed by TN PL and TGW had direct positive influence on GY, while SNP and SF had negative direct influence on GY under drought dress. Under irrigation conditions in 2013, SNP, SF, PN and PL had positive direct effect on GY while in 2014, GY was strongly directly influenced by FGN and TN followed by PH and TGW but directly negatively affected by SNP and SF. Under drought stress all traits had significant indirect influence on GY via FGN except HD and PN that had negative indirect effect on GY via FNG. In 2013, under irrigation, no significant indirect effects traits had on GY via FGN, but in 2014 all traits except HD and TN had indirect positive influence on GY via FGN. In this background, these results generally illustrates that TN and PL are important in both water regimes. Under drought stress high TGW and FGN are the primary determinants of grain yield while many SNP and low SF negatively affect grain yield. Under irrigation high SNP coupled with higher SF are the primary determinants of grain yield. Thus high grain yield under drought stress can be achieved by few tiller number, moderate panicle length, heavier grains and highly filled grains. All drought tolerant plants expressed drought escape, thus short flowering time, high plant height and ability to resist leaf tip drying Fig. 1 are also important traits for selection. For high grain yield under well watered conditions many tiller number, long panicle length, high spikelet number coupled with high spikelet fertility and moderate plant height, moderate heading date are important.

Path analysis for 02428_IL a *japonica* type presented in Table 6 reveals that under drought stress all traits had positive influence on GY except heading date that affected GY negatively. In 2013, GT, TN, HD, and TGW did not have large impact on grain yield under irrigation. In 2014, all traits except PH, HD and TGW had strong effect on grain yield. Partitioning correlation coefficients into direct and indirect effects, showed that FGN had the strong positive direct effect on grain yield followed by panicle number and plant height. SNP and SF had negative effect and remaining traits affected GY indirectly via FGN. Under irrigation, in 2013, SF, SNP, TN and TGW had strong positive direct influence on GY. In 2014, positive direct influence on GY came from FGN followed by PH, and PL. SF and SNP had negative direct influence on GY. All traits except plant height and 1000-grain weight had negative indirect influence on GY via FGN. All traits had negative indirect effect on GY via SNP except plant height and 1000-grain weight. Generally path analysis partitioning revealed that all measured traits had significant positive correlations coefficients with the GY under drought as well as irrigation conditions, in both seasons except flowering date (HD) which had negative correlations with GY (Table 5&6). Filled grain number FGN followed by effective tiller number are the primary positive determinant of grain yield under both drought and irrigation conditions. Under irrigation, direct negative influence of SNP and SF on GY is counterbalanced by strong positive association with FGN (Table6). Many SNP required high SF in order to bring positive influence on GY. Low SF requires high FGN to influence GY. Grain thickness and flag leaf length are more important under drought stress compared to irrigation conditions in influencing GY, where both influence GY indirectly via strong association with FGN. The proportion of grain yield or residual effects (R) not accounted for by the variables studied under drought and irrigation conditions for two seasons and in both backgrounds are presented as 'R' in Table 5&6. In these studies residual effect were 0.378 and 0.473 under drought stress 0.517 and 0.415 under irrigation for the year 2013 and 2014 in MH63_ILbackgrounds indicating that 63, 53, 49 and 59% percentages of the factors influencing GY were accounted for by these 10 factors studied. Residue effect for the 02428_ILbackground were 0.216, 0.185 under drought and 0.362 and 0.316 under irrigations implying that 79, 82, 64 and 69% respectively of the factors influencing yield were contributed by these 10 traits covered in this research [24,25].

Table.5 Path analysis partitioning genotypic correlation with grain yield into direct (bold) and indirect effects of yield enhancing traits under drought stress and normal irrigation for 2013 and 2014 in MH63_IL *indica* type population, R= residual effect

Drought stress 2013 (R.0.378)											Drought Stress 2014 (R.0.473)										
Trait	**GT**	**TGW**	**SF**	**FGN**	**SNP**	**TN**	**PL**	**PH**	**HD**	**GY**	**Trait**	**FLL**	**TGW**	**SF**	**FGN**	**SNP**	**TN**	**PL**	**PH**	**HD**	**GY**
GT	**0.087**	0.097	-0.037	0.74	-0.283	0.006	0.006	-0.024	-0.086	0.506	FLL	**0.037**	-0.029	-0.018	0.46	-0.086	0.002	0.235	0.04	0.009	0.650
TGW	0.027	**0.319**	-0.062	0.849	-0.334	0.018	0.056	-0.155	-0.029	0.689	TGW	0.011	**-0.096**	-0.021	0.19	-0.029	-0.002	0.173	0.023	0.005	0.253
SF	0.029	0.177	**-0.112**	1.196	-0.343	-0.005	0.027	-0.132	-0.045	0.792	SF	0.006	-0.018	**-0.114**	0.48	-0.052	0.003	0.084	0.013	0.011	0.413
FGN	0.038	0.161	-0.079	**1.685**	-0.722	-0.077	0.044	-0.21	-0.055	0.785	FGN	0.022	-0.024	-0.072	**0.768**	-0.122	-0.038	0.223	0.033	0.015	0.806
SNP	0.032	0.138	-0.05	1.574	**-0.773**	-0.091	0.047	-0.214	-0.043	0.62	SNP	0.025	-0.022	-0.047	0.737	**-0.127**	-0.046	0.244	0.037	0.014	0.815
TN	0.002	0.024	0.002	-0.559	0.305	**0.232**	-0.01	0.04	0.008	0.045	TN	0.001	0.001	-0.001	-0.144	0.029	**0.201**	-0.039	-0.008	0.006	0.046
PL	0.006	0.203	-0.034	0.841	-0.417	-0.026	**0.088**	-0.239	0.018	0.438	PL	0.026	-0.049	-0.029	0.508	-0.092	-0.024	**0.337**	0.05	0.007	0.734
PH	0.007	0.17	-0.051	1.218	-0.571	-0.032	0.072	**-0.291**	0.02	0.544	PH	0.019	-0.03	-0.02	0.338	-0.062	-0.02	0.223	**0.075**	-0.003	0.520
HD	-0.052	-0.065	0.035	-0.647	0.23	0.014	0.011	-0.041	**0.143**	**-0.373**	HD	-0.009	0.012	0.032	-0.301	0.046	-0.033	-0.059	0.006	**-0.038**	-0.344
Normal irrigation 2013 (R.0.517)											Normal irrigation 2014 (R.0.415)										
Trait	**GT**	**TGW**	**SF**	**FGN**	**SNP**	**TN**	**PL**	**PH**	**HD**	**GY**	**Trait**	**FLL**	**TGW**	**SF**	**FGN**	**SNP**	**TN**	**PL**	**PH**	**HD**	**GY**
GT	**0.072**	0.008	-0.021	-0.011	0.039	-0.106	-0.061	-0.013	0.144	0.05	FLL	**0.127**	0.039	-0.025	0.187	-0.117	0.106	0.019	0.007	0.004	0.347
TGW	0.032	**0.018**	0.054	-0.006	-0.015	-0.055	0.105	0.02	0.138	0.293	TGW	0.019	**0.269**	-0.15	0.324	-0.068	-0.043	0.012	0.106	0.005	0.474
SF	-0.003	0.002	**0.500**	-0.044	-0.021	-0.065	0.004	0.031	-0.006	0.398	SF	0.008	0.100	**-0.404**	0.956	-0.231	-0.092	0.009	0.165	-0.015	0.495
FGN	0.011	0.001	0.298	**-0.073**	0.233	-0.216	0.059	0.074	0.024	0.411	FGN	0.016	0.060	-0.267	**1.446**	-0.686	-0.155	0.005	0.267	0.014	0.701
SNP	0.009	-0.001	-0.033	-0.053	**0.320**	-0.199	0.074	0.069	0.026	0.211	SNP	0.02	0.025	-0.124	1.322	**-0.751**	-0.141	0.003	0.249	0.025	0.627
TN	-0.016	-0.002	-0.067	0.032	-0.131	**0.488**	0.059	-0.037	-0.006	0.321	TN	0.033	-0.028	0.091	-0.545	0.357	**0.511**	0.004	-0.124	-0.006	0.292
PL	-0.011	0.005	0.005	-0.011	0.059	0.072	**0.402**	0.067	-0.049	0.54	PL	0.08	0.104	-0.118	0.257	-0.077	0.053	**0.03**	0.07	-0.024	0.375
PH	-0.007	0.003	0.104	-0.037	0.151	-0.122	0.185	**0.146**	-0.077	0.347	PH	0.002	0.076	-0.177	1.028	-0.497	-0.136	0.006	**0.376**	-0.010	0.669
HD	-0.037	-0.009	0.01	0.006	-0.029	0.011	0.069	0.04	**-0.283**	-0.222	HD	-0.005	-0.012	-0.059	-0.198	0.185	0.025	0.007	0.036	**-0.100**	-0.121

Table.6 Path analysis partitioning genotypic correlation with grain yield into direct (bold) and indirect effects of yield enhancing traits under drought stress and normal irrigation for 2013 and 2014 in MH63_IL *indica* type population, R= residual effect

Drought 2013 (R. 0.216)											Drought stress 2014(R 0.185)										
Trait	**GT**	**SF**	**FGN**	**SNP**	**PL**	**PH**	**TN**	**HD**	**TGW**	**GY**	**Trait**	**FLL**	**SF**	**FGN**	**SNP**	**PL**	**PH**	**TN**	**HD**	**TGW**	**GY**
GT	**0.152**	-0.131	0.335	-0.059	-0.002	-0.002	0.131	0.012	0.008	0.445	FLL	**-0.028**	0.025	0.250	-0.036	-0.037	0.102	-0.032	0.005	0.025	0.274
SF	0.050	**-0.396**	0.887	-0.090	0.006	0.047	0.183	-0.015	0.016	0.688	SF	-0.008	**0.094**	0.573	-0.048	-0.020	0.101	0.018	-0.027	0.048	0.730
FGN	0.044	-0.304	**1.156**	-0.361	0.019	0.070	0.097	-0.006	0.010	0.726	FGN	-0.009	0.071	**0.760**	-0.125	-0.042	0.179	-0.011	-0.037	0.022	0.808
SNP	0.019	-0.074	0.871	**-0.479**	0.023	0.062	-0.021	0.007	0.002	0.409	SNP	-0.007	0.031	0.653	**-0.145**	-0.045	0.184	-0.027	-0.042	0.003	0.605
PL	-0.008	-0.059	0.497	-0.258	**0.043**	0.105	0.043	-0.002	0.006	0.368	PL	-0.012	0.022	0.370	-0.075	**-0.086**	0.236	-0.006	-0.019	0.018	0.448
PH	-0.002	-0.131	0.572	-0.211	0.032	**0.142**	0.092	-0.024	0.008	0.479	PH	-0.009	0.029	0.420	-0.082	-0.063	**0.324**	0.042	-0.055	0.013	0.620
TN	0.049	-0.179	0.278	0.025	0.005	0.032	**0.404**	0.001	0.010	0.625	PN	0.005	0.010	-0.051	0.023	0.003	0.081	**0.170**	-0.050	0.022	0.214
HD	-0.033	-0.108	0.131	0.058	0.002	0.061	-0.008	**-0.056**	0.005	0.052	HD	-0.001	-0.022	-0.239	0.052	0.014	-0.153	-0.072	**0.117**	-0.059	-0.364
TGW	0.059	-0.289	0.559	-0.037	0.013	0.054	0.194	-0.014	**0.022**	0.559	TGW	-0.003	0.020	0.074	-0.002	-0.007	0.019	0.017	-0.030	**0.227**	0.315
Irrigation 2013 (R. 0.362)											Irrigation 2014 (R. 0.316)										
Trait	**GT**	**SF**	**FGN**	**SNP**	**PL**	**PH**	**TN**	**HD**	**TGW**	**GY**	**Trait**	**FLL**	**SF**	**FGN**	**SNP**	**PL**	**PH**	**TN**	**HD**	**TGW**	**GY**
GT	**0.127**	0.017	-0.002	-0.037	-0.003	-0.002	-0.063	-0.048	0.112	0.101	FLL	**0.015**	-0.086	0.710	-0.389	0.113	-0.081	0.037	0.026	-0.019	0.325
SF	0.005	**0.472**	0.098	0.229	0.000	0.004	-0.218	0.016	-0.060	0.545	SF	0.004	**-0.367**	1.152	-0.218	0.021	-0.050	0.023	0.017	-0.036	0.547
FGN	-0.002	0.327	**0.141**	0.629	0.007	0.027	-0.408	-0.005	-0.126	0.591	FGN	0.005	-0.209	**2.017**	-1.090	0.052	-0.159	0.055	0.009	-0.072	0.609
SNP	-0.007	0.155	0.128	**0.695**	0.010	0.034	-0.423	-0.014	-0.127	0.450	SNP	0.005	-0.066	1.830	**-1.202**	0.055	-0.168	0.055	0.004	-0.062	0.450
PL	-0.010	-0.005	0.026	0.173	**0.039**	0.039	-0.110	0.021	0.010	0.183	PL	0.008	-0.037	0.492	-0.309	**0.215**	-0.030	0.065	0.034	0.027	0.465
PH	-0.003	0.019	0.038	0.234	0.015	**0.100**	-0.206	0.039	-0.019	0.217	PH	-0.003	0.046	-0.813	0.513	-0.016	**0.394**	-0.019	-0.029	-0.001	0.072
TN	-0.012	-0.152	-0.086	-0.436	-0.006	-0.031	**0.674**	-0.010	0.002	-0.057	TN	0.004	-0.058	0.754	-0.452	0.095	-0.052	**0.146**	0.010	0.004	0.452
HD	-0.038	0.047	-0.005	-0.062	0.005	0.025	-0.043	**0.159**	-0.010	0.077	HD	0.004	-0.058	0.175	-0.042	0.067	-0.106	0.014	**0.108**	-0.011	0.150
TGW	0.047	-0.094	-0.059	-0.295	0.001	-0.006	0.005	-0.005	**0.301**	-0.105	TGW	-0.002	0.072	-0.791	0.407	0.032	-0.002	0.003	-0.006	**0.183**	-0.103

Lower values for the residue effect for *japonica* type compared to *indica* and in 2013 compared to 2014 would suggest that the extent of contribution of a particular trait on grain yield was affected by environment and perhaps genetic background.

4. CONCLUSION

This study shed light on the genetic information of grain yield and related traits such as their heritability, genetic and environmental effects in the two sets of BILs from a cross between MH63(an *indica*) and 02428 (a *japonica*). Drought tolerant lines with better yields in *indica* had early heading date and higher plant height whereas delayed heading date and moderate plant height in *japonica*. Moreover *japonica* background had low residual effect compared to *indica* suggesting these two population had different drought resistance mechanism or responded differently to environment and seasonal effects. Because of GXE interactions some traits of drought tolerant lines in one season performed poorly the following season. Therefore, selection should focus on different traits in multiple years under water deficit stress and non-stress conditions. High filled grain number, high 1000-grain weight, moderate tiller number, short flowering time and high plant height could be selection criteria for breeding high grain yield under drought conditions. High spikelet number coupled with high spikelet fertility, high tiller number, long panicle length, moderate plant height, and moderate flowering time are necessary for breeding high yield potential plants. Further investigations will focus on identifying QTLs with effect on grain yield under drought, irrigation conditions and both drought and irrigations separately in these two reciprocal backcross introgression inbred lines.

ACKNOWLEDGEMENTS

I acknowledge Dr. Zhi-kang Li and Dr. Tian-qing Zheng of the Institute of Crop Science Chinese Academy of Agricultural Sciences for proposing the study providing study location and funds for data collection. Lastly I acknowledge Chinese Scholarship council for the scholarship.

COMPETING INTERESTS

Author has declared that no competing interests exist.

REFERENCES

1. Jacqueline A, John C, Erica G, Kevin P, and Christoph Z. State-and-Trends of the environment: 1987–2007. 2007;119-155.
2. Narciso J, Hossain M. World Rice Statistics. IRRI. 2002;29.
3. Lafitte HR, Ismail, A. Bennett J. Abiotic stress tolerance in rice for Asia: Progress and the future in "New directions for a diverse planet". Proceedings of the 4th International Crop Science Congress. Brisbane, Australia; 2004.
4. Khush GS, Aquino RC, Breeding tropical *japonica*s for hybrid rice production. In: S. S. Virmani (ed) Hybrid rice technology: New Developments and future prospects, International Rice Research Institute, Los Banos, Philippines. 1994;33–36.
5. Sabu KK, Abdullah MZ, Lim LS and Wickneswari R. Analysis of heritability and genetic variability of agronomically important traits in *Oryza sativa* x O. rufipogon cross. Agron Res. 2009;7(1):97-102.
6. Valliyodan, Nguyen HT. Understanding regulatory networks and engineering for enhanced drought tolerance in plants. C. Opin. P Biol. 2006;9:189-195.
7. Fischer KS, Lafitte R, Fukai S, Atlin G, Hardy B, (editors). Breeding rice for drought-prone environments. Los Baños (Philippines): International Rice Res Inst. 2003;98.
8. Kumar R, Venuprasad R, Atlin G. Genetic analysis of rainfed lowland rice drought tolerance under naturally occurring stress in eastern India: Heritability and QTL effects. F Crop Res. 2007;103:42-52.
9. Dewey DR, Lu KH. A correlation and path-coefficient analysis of components of crested wheatgrass seed production. Agron J. 1959;51:515–518.
10. Kumar R, Krishnapal R, Mandal SK, Ramashankar PS, Rai R. Genetic study of major characters in upland rice. Environment and Ecology. 1994;12(2):363-365.
11. Chaudhary PK, Singh RP. Genetic variability, correlation and path analysis of yield components of rice. Madras Agricultural Journal. 1994;81(9):468-470.
12. Satya AG, Kandasamy, Ramalingam J. Heterosis in hybrid rice, Crop Res. 1999;18(2):243-246.

13. Shivani D, Sree N, Reddy R. Variability, heritability and genetic advance for morphological and physiological traits in certain rice hybrids, Oryza. 2000;37(3):231-233.
14. Iftekharuddaula KM, Hassan MS, Islam MJ. Badshah MA, Islam MR, Akhter K. Genetic evaluation and selection criteria of hybrids rice in irrigated ecosystem of Bangladesh. Pakistan J of Biol Sci, 2001;4(7):790-792.
15. Sao A. Studies on combining ability and heterosis in F1 rice hybrids using cytoplasmic male sterile lines. M. Sc. (Ag.) Thesis, IGAU, Raipur; 2002.
16. Jayasudha S, Sharma D. Genetic parameters of variability, correlation and path- coefficient for grain yield and physiological traits in Rice (*Oryza sativa L.)* under shallow lowland situation. Electronic J. Plant Breed. 2010;1(5):33-38.
17. Idris AE, Justin FJ, Dagash YM, and Abual A IGenetic Variability and Inter Relationship between Yield and Yield Components in Some Rice Genotypes A. J. Exp. Agric. 2012;2(2):233-239.
18. Kuixian J, Yangyang W, Weining S, Qiaojun L, Hanwei M, Shihua S and Hui C. Drought-responsive mechanisms in rice genotypes with contrasting drought tolerance during reproductive stage J. 2012;169:336–344.
19. Araus JL, Slafer GA, Reynolds MP, Royo C. Plant breeding and drought in C3 cereals: What should we breed for? Ann. Bot. 2002;89:925–940.
20. Babu RV, Shreya K, Kuldeep SD, Usharani G and Nagesh P Genetic Variability Studies for Qualitative and Quantitative traits in Popular Rice (*Oryza sativa* L.) Hybrids of India. Int. J. Sci. & Res Pub. 2012;2:(6)1-12.
21. Turner NC, Wright GC, Siddique KH. Adaptation of grain legumes (pulses) to water-limited environments, Adv. Agron. 2001;71:123–231.
22. Kavar T, Maras M, Kidric M, Sustar-Vozlic J, Meglic V. Identification of genes involved in the response of leaves of Phaseolus vulgaris to drought stress, Mol. Breed. 2007;21:159–172.
23. Boojung H, Fukai S. Effects of soil water deficit at different growth stage on rice and yield under optimum conditions phenology biomass production and yields. F. Crop. Res. 1996;48:47-55.
24. Biswash PL, Ghosal S Kamal H, Ujjal KN Genetic variability, correlation and path analysis, in maintainer lines of hybrid rice (*Oryza sativa* L) Bangladesh. J Prog. Sci. & Tech. 2010;8(1)115-118.
25. Ghosal S, Syed MA, Mkhatun, Biswash PL, Aditya TL. Evaluation of selection criteria in premium quality rice genotypes through genetic variability, charcater association and path analysis. Bangladesh Rice J. 2012;16:59-62.

11

Assessing the Effects of Water Management Regimes and Rice Residue on Growth and Yield of Rice in Uganda

T. Awio[1], B. Bua[2*] and J. Karungi[1]

[1]Department of Agricultural Production, College of Agriculture and Environmental Sciences, Makerere University, P.O.Box 7062, Kampala, Uganda.
[2]Department of Agriculture, Faculty of Vocational Studies, Kyambogo University, P.O.Box 1, Kyambogo-Kampala, Uganda.

Authors' contributions

This work was carried out in collaboration between all authors. Author TA conducted the field and screen house trials, did data analysis and drafted the manuscript. Author BB and JK guided in design of the trials, data collection and analysis. All authors read and approved the final manuscript.

Editor(s):
(1) T. Muthukumar, Root & Soil Biology Laboratory Department of Botany, Bharathiar University, India.
(2) Daniele De Wrachien, State University of Milan, Italy.
Reviewers:
(1) Fermín Orona Castro, INIFAP (National Research Institute of Forestry, Agricultural and Livestock), Mexico.
(2) Muhammad Rafi Qamar Hashmi, Agronomy, University of Agriculture Faisalabad, Pakistan.

ABSTRACT

Aim: This study was conducted to assess the influence of different water and rice straw management practices and rice genotypes on growth and yield of rice in Uganda.
Study Design: Field experimental design was a Randomized Complete Block Design while the screen house study design was a Completely Randomized Design.
Place and Duration of Study: The study was conducted in the field at National Crops Resources Research Institute (NaCRRI) Namulonge and in the screen house at Kyambogo University during the period of February-July 2013.
Materials and Methods: Ten rice genotypes obtained from the cereals program at NaCRRI Namulonge were grown under different water management regimes, with and without rice straw incorporation both in the field and screen house. Water management regimes used were alternate wetting and drying (AWD), continuous flooding (CF) and continuous drying (CD).

**Corresponding author: E-mail: boscobua@yahoo.com;*

Results: A significant variation in grain yield was observed among rice genotypes and under different water management regimes ($P<0.001$). Use of rice straw influenced rice yield in the screen house ($P<0.001$) but not in the field ($P=0.23$); interactions of water management x genotype and water management x rice straw x genotype influenced rice yield in the field ($P=0.003$) but not in the screen house ($P=0.5$). Higher yield gain was observed under the water-saving technology alternate wetting and drying compared to continuous flooding or drying.
Conclusion: This study has indicated significant variations in field performance of rice under different water management regimes and rice straw usage. These findings are therefore important because they suggest that efficient management of water resources and rice residues from rice fields coupled with the use of drought tolerant rice varieties could be an effective integrated approach to improve rice yield and an adaptation strategy to the observed climate variability.

Keywords: Alternate wetting and drying; continuous drying; continuous flooding; rice straw.

1. INTRODUCTION

Rice is a major cereal crop in the sub-Saharan Africa, providing up to 21% of the calorie intake. As such, rice is an important strategic component of food security and crucial element in the staple food economies of the region where it has gained prominence in the farming systems and diets in Africa [1,2]. Traditionally, rice has the potential to improve nutrition, boost food security, foster rural development and support sustainable land use in Africa. In Uganda, rice has also become an important food staple and source of income for a big portion of the populace [3,4].

Although much of the global rice acreage is in developing countries, rice yield per unit area is still very low, averaging 1.5t/ha for both lowland and upland rice compared to yield potential of 8 t/ha and 5 t/ha for lowland and upland rice, respectively [5]. The major reasons for these wide yield gaps is due to environmental stress, low inherent yielding varieties, poor management of rice fields, technology incompatibility and socio-economic factors. According to Mostajeran et al. [6], one of the main problems of rice cultivation and production in Africa is lack of water resources, especially during periods of low rainfall which affect the vegetative growth rate and yield. As such, drought is the most serious problem for rain fed dry land rice cultivation. Therefore, water management in rice production; whether in upland or lowland systems is an important aspect in ensuring increased rice production and environmental integrity. Additionally, crop residues are good sources of plant nutrients and are important components for the stability of agricultural ecosystems. In rice production, a large quantity of rice residues is left in the field after harvesting, which can be recycled for nutrient supply [7]. For some rice farmers, the rice residues are incorporated in to the soil during land preparation by ploughing while some farmers burn the rice residues during land preparation [7,8]. However, in Uganda there is no information on whether rice residue incorporation in to the soil results in any improvement in the growth and yield of rice. Moreover, there is also need to understand how different rice genotypes interact with water and rice residue management in influencing rice performance. Therefore, this study aimed at evaluating the effects of water management, rice straw use and rice genotypes on the growth and yield of rice in Uganda.

2. MATERIALS AND METHODS

This study was conducted both in the field at National Crops Resources Research Institute (NaCRRI) Namulonge, (1200 m above sea level, 0°32°N 32°34°E, 1300 mm average annual rainfall) and in the screen house at the Department of Agriculture, Kyambogo University, (1189 m above sea level, 0°20°N 32°35°E). Ten rice genotypes (namely; K5, NERICA-4, IR09A136, P27H14, P27H2, P33H3, P49H3, P55H17, P55H20 and P59H9) were used to assess the effects of water management régimes, rice straw use and rice genotypes on growth and yield of rice of rice both in the field and screen house.

2.1 Field Experiment

The study was conducted during the period of February to July 2013. The experimental design was a randomized complete block design arranged as a split-split plot with water regimes being the main plots, rice straw the sub-plots and rice genotypes made the sub-sub plots with three replications. During land preparation, all the surface residues (organic materials) were collected and disposed off without incorporating it into the soil. Rice straw was chopped into smaller pieces of about 1cm and then later incorporated into the soil two weeks prior to planting at a depth of 10-15 cm and at a rate of 8t/ha as described by Tadashiro et al. [9]. Treatments were as follows:

Main plots (water regimes)	Subplots (rice straw)	Sub-subplots (rice genotypes)
Alternate wetting & drying	8t/ha	All 10
Continuously dry (no flooding)	8t/ha	All 10
Continuous flooding	8t/ha	All 10
Alternate wetting & drying	None	All 10
Continuously dry (no wetting)	None	All 10
Continuous flooding	None	All 10

In alternate wetting and drying (AWD), irrigation water was applied by pumping water in the field using a water pump to flood the field for 7 days after which the field was drained and left dry for another 7 days. The field was alternately flooded and non-flooded for 7 days throughout the growth period of the rice crop.

Under continuous drying (CD), the field was solely dependent on rainfall and this was a rainy season which had atleast one rain per week. Where the rain was high and caused flooding of the field, the field was drained to keep it dry throughout the crop growth period. Under continuous flooding (CF), irrigation water was continuously pumped into the field to flood the field throughout the growth period of the crop.

During flooding, the water level was kept at 5cm above the soil surface. Plastic sheets were fixed around each treatment by digging 20 cm deep to restrict water movement from one treatment to another and a distance of 1.5 m was left between the treatments for easy monitoring. Rice straw rate of 8t/ha was used following the trials and recommendation of Tadashiro et al. [9].

Planting was by direct seeding, with a spacing of 20 cm x 20 cm. In the sub-sub plots, each genotype was planted in four rows, each row with 11 stands and each stand with two plants making a total of 88 plants per genotype. Agronomic data was collected on total number of tillers, number of productive tillers (panicles), crop height at harvest and grain yield. Data on total number of tillers, number of productive tillers and crop height were collected from 14 plants in the two middle rows, 7 plants from each row by counting all the tillers, counting all tillers that had panicles and measuring the height of the plants using a measuring tape, respectively and obtaining the average. Grain yield data was obtained by harvesting the grains from all the plants within a plot (treatment) and the total grain weight for each treatment was obtained (in grams) using an electronic weighing scale and later expressed in t/ha. All the data collected was subjected to Analysis of Variance (ANOVA) of the Genstat computer program (14th edition) and differences in means were determined using Least Significant Difference (LSD) at 5% probability level.

2.2 Screen House Experiment

This experiment was conducted concurrently with the field trial using 10 rice genotypes. The trial design was a completely randomized design with three replications. Rice paddy soil collected from the plough layer of a paddy field at NaCRRI as per Tadashiro et al. [9] was used. Plastic pots containing 8kg of the moistened soil samples were prepared for the treatments. Rice straw was chopped and mixed with the 8 kg of soil before filling in the pots at the same rate as in the field (32 g / 8 kg soil) [9]. Treatments were as follows:

Water regimes	Rice straw	Rice genotypes
Alternate wetting & drying	32 g	All 10
Continuously dry (no flooding)	32 g	All 10
Continuous flooding	32 g	All 10
Alternate wetting & drying	None	All 10
Continuously dry (no wetting)	None	All 10
Continuous flooding	None	All 10

For alternate wetting and drying, the pots were alternately flooded and non-flooded for 7 days throughout the growth period of the rice crop. In continuous drying, watering was done once every week throughout the trial to wet the soil while under continuous flooding, the pots were kept flooded throughout the crop growth period and water level was maintained at 5cm above the soil surface. For all treatments, each pot was then direct seeded with five rice seeds and thinning was later done to two rice plants per pot at the tillering stage. The pots were maintained in the screen house at temperature of 26°C.

Weeding of pots was done twice during the trial, no artificial fertilizer was applied and all plants in each pot were harvested by cutting the panicles to obtain the grains. Data collection and analysis was the same as in the field experiment. The trial lasted from February to July, 2013.

3. RESULTS

3.1 Field Experiment

Water management significantly influenced plant height, number of productive tillers and yield (*P<0.001*) but not total number of tillers. All the growth and yield parameters measured varied significantly for rice genotypes (*P<0.001*); whereas incorporation of rice straw significantly (*P<0.001*) influenced number of productive tillers and total number of tillers, but not plant height and yield. The interactions of water management x genotype, water management x rice straw, rice straw x genotype and water management x rice straw x genotype was not significant for any of the growth parameters (P=0.5). Grain yield on the other hand was significantly affected by the water management x rice genotype (P<0.001), and the water management x rice straw x rice genotype interactions (P=0.003).

Genotypes K5 and IR09A136 had the highest average number of productive tillers per plant and the highest average total number of tillers per plant (Table 1). More productive tillers were recorded in AWD and CD than in the CF (Table 1). With regard to rice straw usage, there were more productive tillers recorded in plots without rice straw incorporated compared to plots which had rice straw incorporated (Table 1).

In regard to grain yield, K5 and P33H3 had the highest grain yield compared to other rice genotypes (Table 2). The highest grain yield was recorded under the CD water regime followed by AWD, and the least was from the CF regime, with no significant difference between AWD and CD (*P>0.05*; Table 2). Genotypes P33H3, NERICA-4 and K5 were among those that performed better under AWD and, K5 still performed well under CF. All the genotypes performed well under CD except for NERICA-4, IR09A136 and P49H3 (Table 2). With regard to rice straw usage, higher grain yields were recorded in plots without rice straw incorporation compared to those plots which had rice straw incorporated (Table 3).

3.2 Screen House Experiment

Results indicated that rice genotype significantly affected number of productive tillers and plant height (*P<0.001*) but not total number of tillers and grain yield. Water management regimes significantly (*P<0.001*) influenced all the studied variables except number of productive tillers. Rice straw incorporation significantly influenced yield and plant height (*P<0.001*) but not tillering. The interactions of water management x rice straw, water management x rice genotypes, rice straw x rice genotypes, and water management x rice straw x rice genotypes was not significant for any of the studied growth and yield variables in the screen house.

Table 1. Average number of productive tillers and total number of tillers under different water management regimes and rice straw usage in the field at NaCRRI

Genotypes	No. of productive tillers								Total no. of tillers
	AWD	CD	CF	LSD (5%)	With straw	Without straw	LSD (5%)	Mean	
K5	20	20	12		14	20		17	21
NERICA-4	10	9	5		7	9		8	11
IR09A136	16	20	9		13	17		15	19
P27H14	10	10	7		8	11		9	13
P27H2	12	13	7		8	14		11	14
P33H3	12	10	7		8	11		10	12
P49H3	13	9	9		9	12		11	14
P55H17	10	9	6		8	9		8	12
P55H20	9	8	7		7	9		8	11
P59H9	11	9	6		7	11		9	11
Mean	12	12	8	1.33	9	12	0.98	11	14
LSD (5%)								2.48	2.62

*(**AWD**=Alternate wetting and drying **CD**=Continuous drying **CF**=Continuous flooding)*

Table 2. Yield of 10 rice genotypes grown under different water management regimes in the field at NaCRRI

Genotypes	Yield (t/ha)				
	AWD	CD	CF	Mean	LSD (5%)
K5	2.47	2.03	2.07	2.19	
NERICA-4	2.18	0.78	0.47	1.14	
IR 09A136	0.92	1.57	1.52	1.34	
P27H14	1.66	2.57	0.38	1.54	
P27H2	1.71	2.66	0.66	1.68	
P33H3	2.14	2.61	0.92	1.89	
P49H3	1.35	1.86	0.64	1.28	
P55H17	0.51	2.25	0.69	1.15	
P55H20	1.47	2.34	0.53	1.45	
P59H9	1.59	2.09	0.40	1.36	
Mean	1.60	2.08	0.83	1.50	0.84
LSD (5%)				0.47	

***(AWD**=Alternate wetting and drying **CD**=Continuous drying **CF**=Continuous flooding)*

Table 3. Yield of 10 rice genotypes grown under different water management regimes and rice straw amendment in the field at NaCRRI

Genotype	Yield (t/ha)									
	AWD with straw	AWD without straw	Mean	CD with straw	CD without straw	Mean	CF with straw	CF without straw	Mean	LSD (5%)
K5	2.08	2.86	2.47	1.71	2.35	2.03	1.31	2.83	2.07	
NERICA-4	1.54	2.82	2.18	0.75	0.81	0.78	0.54	0.41	0.48	
IR 09A136	1.20	0.65	0.93	1.34	1.81	1.58	1.04	2.00	1.52	
P27H14	1.38	1.94	1.66	1.17	3.98	2.58	0.44	0.32	0.38	
P27H2	2.40	1.02	1.71	1.75	3.56	2.66	0.70	0.62	0.66	
P33H3	1.48	2.80	2.14	1.98	3.24	2.61	0.72	1.12	0..92	
P49H3	1.18	1.51	1.35	2.10	1.63	1.87	0.88	0.40	0.64	
P55H17	0.65	0.36	0.51	1.67	2.84	2.26	0.80	0.58	0.69	
P55H20	1.13	1.82	1.47	1.41	3.27	2.34	0.63	0.44	0.54	
P59H9	1.04	2.14	1.59	2.13	2.05	2.09	0.28	0.52	0.40	
Mean	1.41	1.79	1.60	1.60	2.55	2.08	0.73	0.93	0.83	1.50
LSD (5%)			0.82			0.82			0.82	

***(AWD**=Alternate wetting and drying **CD**=Continuous drying **CF**=Continuous flooding)*

As in the field trial, K5 and IR09A136 had the highest number of productive tillers in the screen house (Table 4). Plant height varied significantly among genotypes with the tallest genotype, K5, measuring 0.9 m while the shortest genotypes, NERICA-4, P27H14, P33H3, P55H17 and P59H9 measured 0.7 m (Table 4).

With respect to grain yield and water management, the highest yield was recorded under CF followed by AWD and the least under CD, with no significant difference in grain yield between AWD and CD but CF differed significantly from AWD and CD (Table 5). Contrary to the field trial, higher grain yields were recorded in treatments amended with rice straw compared to those which were not (Table 5). The pooled yield data for rice genotypes over the different water management regimes and rice straw usage showed that genotypes K5, P33H3, P27H14 and P27H2 generally recorded the highest grain yields (Table 5).

4. DISCUSSION

The difference in yield observed among rice genotypes could have been due to the difference among genotypes in number of productive tillers (panicles). According to Hayashi [10], many high-yielding rice varieties are short and high tillering; characteristics exhibited by some of the high yielding genotypes such as P33H3, P27H2 and P27H14 used in this study. These results suggest that the most important factor for obtaining high yield is to get high number of productive tillers (panicles) per unit area.

Consequently, the difference in number of productive tillers / panicles among genotypes could be the main reason that contributed to the yield difference observed among genotypes. Additionally, the difference in grain yield could be attributed to the difference in water use efficiency among the rice genotypes and this could explain the significant interaction of genotypes and water regime on grain yield. Efficiency in water use by rice crops are factors that positively contribute to high yield of rice [11].

The high yield recorded under AWD during field trial compared to CD and CF may be attributed to the high number of productive tillers, moisture retention and efficient moisture use under AWD during the growing periods. The observed effect of AWD on grain yield is in agreement with several prior studies. According to Roderick et al. [12] and Tejendra et al. [13], AWD uses less water without significant reduction in grain yield, harvest index, number of productive tillers and plant height compared with conventional irrigation. Uphoff and Randriamiharisoa [14] and Uphoff [15] noted that keeping rice fields moist but not continuously saturated gives better results, both agronomically and economically, than flooding rice throughout its crop cycle. Bhuiyan and Tuong [16] concluded that a standing depth of water throughout the season is not needed for high rice yields. They added that about 40–45% of the water normally used in irrigating the rice crop in the dry season was saved by applying water in small quantities only to keep the soil saturated throughout the growing season without sacrificing rice yields.

The fact that most of the genotypes used in this study like P27H14, P27H2, P33H3, P55H20, P55H17 and P59H9 yielded relatively well under CD water regime implies that these genotypes can be grown under upland conditions. Since most of the genotypes used in this trial are being developed for upland rice environments, this shows suitability of these genotypes to be used as breeding materials for further breeding work for upland conditions or drought tolerant rice varieties. Developing drought tolerant rice varieties would help farmers adapt to the unpredictable weather patterns currently being observed.

The results of this study also indicated that rice straw significantly influenced rice yield in the screen house but not in the field during the trial. The results of this field study showed no effect of rice straw on rice yield. This contrasts previous studies [7,17] where application of organic materials such as rice straw, rice husks and plant residues significantly improved rice yields and soil productivity. Supapoj et al. [17] further noted that the residual effect of these materials also had a large impact on the growth and yield of rice, with rice straw having a higher residual effect on rice yields than rice husks and wheat straw. Additionally, Pandey [18] noted that using organic materials such as crop residues, green manure and organic wastes improve soil properties, give better crop yield and reduce costs. The result of this study on rice straw utilization shows that for high yield realization, there is need to appropriately manage water in rice fields and use the right rice genotypes.

Table 4. Number of productive tillers, total number of tillers and plant height of 10 rice genotypes grown in the screen house at Kyambogo University

Genotypes	No. of productive tillers	Total no. of tillers				Plant height (m)
		AWD	CD	CF	Mean	
K5	7	8	8	9	9	0.9
NERICA-4	6	9	7	11	9	0.7
IR 09A136	7	9	8	9	9	0.8
P27H14	6	11	9	13	11	0.7
P27H2	6	9	9	9	9	0.8
P33H3	6	9	8	9	9	0.7
P49H3	6	9	9	11	10	0.8
P55H17	6	10	7	10	9	0.7
P55H20	6	11	8	11	10	0.8
P59H9	6	9	8	10	9	0.7
Mean	7	10	8	10	9	0.8
LSD (5%)	0.6				1.4	0.06

*(**AWD**=Alternate wetting and drying **CD**=Continuous drying **CF**=Continuous flooding)*

Table 5. Yield of 10 rice genotypes under different water management regimes and rice straw amendment in the screen house at Kyambogo University

Genotypes	Yield (t/ha)					
	AWD	CD	CF	With straw	Without straw	Mean
K5	1.86	1.33	2.31	2.02	1.64	1.83
NERICA-4	1.14	1.25	1.64	1.49	1.20	1.34
IR 09A136	1.72	1.47	1.52	1.79	1.36	1.57
P27H14	1.89	1.87	1.72	1.97	1.68	1.83
P27H2	1.55	1.62	2.26	1.93	1.69	1.81
P33H3	1.82	1.71	2.40	2.05	1.89	1.97
P49H3	1.63	1.55	1.57	1.71	1.45	1.58
P55H17	1.77	1.35	1.89	1.76	1.58	1.67
P55H20	1.61	1.56	1.77	1.98	1.31	1.64
P59H9	1.57	1.40	1.92	1.87	1.39	1.63
Mean	1.66	1.51	1.90	1.86	1.52	**1.69**
LSD (5%)	0.21			0.17		**0.38**

*(**AWD**=Alternate wetting and drying **CD**=Continuous drying **CF**=Continuous flooding)*

Growing rice varieties that efficiently utilize the available nutrients released from rice straw decomposition and produce high grain yield could result to achieving high yield from rice fields with rice straw being used as the only nutrient source for the rice crops, a common practice with most of the rice farmers in rural areas who lack access to artificial fertilizers.

The significant interaction of water management by rice straw by genotype on rice yield suggests that yield under rice straw usage depends on the water management regime as well as the rice genotype. Varieties efficient in utilizing the available nutrients from straw decomposition under specific water regimes could have helped maintain rice productivity and yields. This could explain the consistent better performance of K5 across water regimes, P33H3 and NERICA-4 under AWD and, P27H14 and P27H2 under CD.

5. CONCLUSION

This study has shown a significant variation in growth and yield parameters among rice genotypes, water management regimes and rice straw usage. Rice yield varied significantly under different water management regimes and the two-way water management by genotype interaction; and the three-way water management by straw by genotype interaction significantly influenced rice yield. This implies that grain yields obtained from the rice genotypes depended on the water management regime and rice straw usage.

The study has demonstrated the possibility of AWD being a promising method of water management in rice cultivation since it offers the benefits of maintaining rice productivity compared with the conventional system of rice production (continuous flooding). This benefit is likely to be a critical factor that could make farmers and other stakeholders adopt the use of AWD in rice production. However, it is very important that other comparative experimental field studies be conducted in different environments or regions with different soil types, topography and climatic conditions in the country to validate these results. Additionally, genotypes used in this study need to be further evaluated for their suitability as genetic materials in breeding for drought tolerant or drought resistant rice varieties.

ACKNOWLEDGEMENTS

We acknowledge the Rockefeller Foundation through its Climate Change Unit Makerere University College of Agriculture and Environmental Sciences for funding this study. We also acknowledge Kyambogo University, National Crops Resources Research Institute

(NaCRRI), Namulonge-Uganda, all the scientists and field technicians who contributed to this work.

COMPETING INTERESTS

Authors have declared that no competing interests exist.

REFERENCES

1. Japan International Cooperation Agency (JICA) in collaboration with Sasakawa Africa Association Uganda. Final survey report on the status of rice production, processing and marketing in Uganda. 2006;90.
2. Oonyu J. Upland rice growing: A potential solution to declining crop yields and the degradation of the Doho wetlands, Butaleja district-Uganda. African Journal of Agricultural Research. 2011;6(12):2774-2783.
3. Bigirwa G, Kikafunda J, Lamo J, Ochen J, Opio F, Tsuboi T. Upland rice growing in Uganda: Reasons for the rapid spread. African Crop Science Conference Proceedings. 2005;7:151-154.
4. National Agricultural Research Organisation (NARO). Rice Diseases and Insects. Promotion of Rice Development (PRiDe) project, NaCRRI. 2012;23.
5. FAOSTAT. Rough rice yield (t/ha), by country and geographical region, 1961-2004. FAO, Rome; 2005.

   Available:http://www.irri.org/science/ricestat/index.asp
6. Mostajeran A., Rahimi-Eichi V. Effects of drought stress on growth and yield of rice (*Oryza sativa* L.) cultivars and accumulation of proline and soluble sugars in sheath and blades of their different ages leaves. American-Eurasian J. Agric. & Environ. Sci. 2009;5(2):264-272.
7. Ogbodo EN. Effect of Crop Residue on Soil Chemical Properties and Rice Yield on an Ultisol at Abakaliki, Southeastern Nigeria. World Journal of Agricultural Sciences. 2011;7(1):13-18.
8. Krishna GM, Arun KM, Kuntal MH, Kali KB, Prabir KG, Manoranjan M. Rice residue-management options and effects on soil properties and crop productivity. Food, Agriculture & Environment. 2004;2(1):224-231.
9. Tadashiro K, Makoto K. Methane emission fluxes from rice paddies and the methane productive potential in the soils. Nutrient Cycling in Agroecosystems. 1997;49:129-138.
10. Hayashi K. Efficiencies of solar energy conversion and relating characteristics in rice varieties. Proceedings of the Crop Science Society of Japan. 1969;38:495-500.
11. Yoshida S. Fundamentals of rice crop science. The international rice research institute Los Baños, Philippines. 1981;284.
12. Roderick MR, Florencia GP, Divina G, Rodriguez P, Ruben ML, Bouman BAM. Impact of the alternate wetting and drying (AWD) water-saving irrigation technique: Evidence from rice producers in the Philippines. Food Policy. 2011;36:280-288.
13. Tejendra C, Riseman A, Yamaji E. Achieving more with less water: Alternate wet and dry irrigation (AWDI) as an alternative to the conventional water management practices in rice farming. Journal of Agricultural Science. 2010;3(3):1-11.
14. Uphoff N, Randriamiharisoa R. Reducing water use in irrigated rice production with the Madagascar System of Rice Intensification (SRI). In: Bouman BAM, Hengsdijk H, Hardy B, Bindraban PS, Tuong TP, Ladha JK, (Eds.). Water-wise rice production. International Rice Research Institute, Philippines. 2002;356. ISBN 971-22-0182-1.
15. Uphoff N. The system of rice intensification (SRI) as a methodology for reducing water requirements in irrigated rice production. Paper for International Dialogue on Rice and Water: Exploring Options for Food, Security and Sustainable Environments, held at IRRI, Los Baños, Philippines; 2006.
16. Bhuiyan SI, Tuong TP. Water use in rice production: Issues, research opportunities and policy implications. Paper presented at the Inter-Centre Water Management Workshop, Colombo, Sri Lanka: International Irrigation Management Institute, Geneva. 1995;5.
17. Supapoj N, Naklang K, Konboon Y. Using organic material to improve soil productivity in rain-fed lowland rice in Northeast Thailand. In: Rain-fed lowland

rice: Advances in nutrient management research, International Rice Research Institute, Los Baños, Philippines. 1998:23-30.

18. Pandey. A primer on organic-based rice farming. International rice research institute, Manila, Philippines. 1991;31-38. ISBN: 971-104-189-8.

12

Temperature Use Efficiency and Yield of Groundnut Varieties in Response to Sowing Dates and Fertility Levels in Western Dry Zone of India

R. S. Meena[1*], R. S. Yadav[2], M. L. Reager[2], Nirmal De[1], V. S. Meena[1], J. P. Verma[3], S. K. Verma[1] and B. C. Kansotia[2]

[1]Institute of Agricultural Sciences, Banaras Hindu University, Varanasi (U.P.) - 221 005, India.
[2]Swami Keshwan and Rajasthan Agricultural University, Bikaner (Rajasthan) - 334 006, India.
[3]Institute of Environment & Sustainable Development, Banaras Hindu University, Varanasi (U.P.) - 221 005, India.

Authors' contributions

All authors contribute for designed the study, performed the statistical analysis and wrote the first draft of the manuscript. Author RSM carried out edited corrections and facilitated all e-mail correspondences .All authors read and approved the final manuscript.

Editor(s):
(1) Edward Wilczewski, University of Technology and Life Sciences in Bydgoszcz, Poland.
(2) Juan Yan, Sichuan Agricultural University, China.
(3) Lanzhuang Chen, Laboratory of Plant Biotechnology, Faculty of Environment and Horticulture, Minami Kyushu University, Miyazaki, Japan.
(4) Peter A. Roussos, Lab. Pomology, Agricultural University of Athens, Greece.
Reviewers:
(1) Anonymous, Iran.
(2) Karistsapol Nooprom, Department of Agricultural Technology, Songkhla Rajabhat University, Thailand.
(3) Mukhtar Musa, Department of Crop Science, Usmanu Danfodiyo University Sokoto, Nigeria.
(4) Anonymous, Iran.
(5) Anonymous, Turkey.

ABSTRACT

A field experiment was conducted in loamy sand soil during *kharif* season of 2009 and 2010 to evaluate the temperature use efficiency and yield of groundnut varieties in response sowing dates and fertility levels. The treatments consisted of four sowing dates (20 April, 15 May, 9 June and 4 July) and two varieties of groundnut (HNG 10 and TG 37A) as main-plot treatments and four fertility levels of nitrogen and phosphorus application (0, 20 N-40 P_2O_5, 30-60 P_2O_5 and 40 N-80 P_2O_5 kg ha^{-1}) as sub-plot making a total of 32 treatment combinations. The experiment was laid out in

Corresponding author: rsmeenaagro@gmail.com;

split-plot design with three replications and crop growth rate, yield parameters, harvest index and temperature use efficiency were recorded on crop. The results showed that significantly higher crop growth rate (CGR) at 30-60 and 60-90 days after sowing (DAS) was observed in HNG 10 variety. Further delays in sowing from 9 June significantly reduce crop growth rate. However, the growth rate of TG 37A variety was observed to be statistically at par among the sowing dates. In HNG 10 variety higher pod yield (3773, 3743 and 3738 kg ha^{-1}), kernel, haulm and biological yield were observed at par to each other from 20 April to 9 June sowing, respectively. Maximum heat unit efficiency (3.23 kg ha^{-1} degree- day-1) was observed on 9 June sowing date while the highest yield (2628 kg ha-1) was recorded in TG 37A variety with 4 July sowing. Harvest index in both the varieties was observed to be significantly higher in 4 July sowing date. All the fertility treatments produced significantly higher crop growth rate and yields than the treatments receiving no fertilizer application. Application of 30 kg N-60 kg P2O5 ha^{-1} significantly enhanced the crop growth and yield over 20 kg N-40 kg P2O5 ha^{-1} and control but at par with 40 kg N-80 kg P2O5 ha^{-1}. However, the crop uses maximum temperature in the month of June for better growth and yield. Thus, it could be concluded that the suitable sowing date of HNG 10 and TG 37A varieties in western dry zone of India are around 9 June and 4 July respectively. Application of 30 kg N-60 kg P2O5 ha^{-1} was optimum for growth and yield of the crop.

Keywords: CGR; groundnut; sowing date; fertility levels; heat unit efficiency; yields.

1. INTRODUCTION

Groundnut is an important oil crop in India. During 2011-2012, India could produce groundnut of 6.94 mt [1]. In dry agricultural area, farmers have limited time for crop sowing, but in regular irrigation is one of the most important non-monetary inputs affecting yield of crops. Sowing time of groundnut is well documented in other regions [2]. In Bikaner region, groundnut sows is under irrigated conditions. It started two decades ago in the command area of Indra Gandhi Nahar Pariyojana (IGNP) and later on it spread to tube well irrigated area of the region. At that time, dust storms were common in the region with minimum vegetation during optimum sowing time of May and June months leading to poor crop establishment due to which the farmers started sowing of groundnut in early summer of April and May for better crop establishment with its harvesting in between October to November. In western dry zone of India some workers [3] reported application of 60 kg N/ha along with equal level of phosphorus as the appropriate fertilizer level while others recommended 20 kg N and 32 kg P_2O_5 ha^{-1} for groundnut. Due to the variation in the optimum fertilizer requirement for groundnut reported by research workers it has also become necessary to determine fertilizer requirement for the crop varieties of different maturity under different sowing dates. This research was conducted using two varieties of groundnut to find suitable sowing dates and fertilizer levels on growth and yield of groundnut in dry western zone of India.

2. MATERIALS AND METHODS

2.1 Experimental Site

The experiment was conducted during *kharif* season of 2009 and 2010 at Agronomy Research Farm, College of Agriculture, Bikaner (Rajasthan) under hyper arid condition, which is situated at a 28° 01'N latitude and 73° 22' E longitudes at an altitude of 234.70 meters above mean sea level (*Arabian* Sea).

2.2 Soil

The soil of the experimental site was loamy sand and having 156.33 kg ha^{-1} alkaline permanganate oxidizable N [4], 16.05 kg ha^{-1} available P [5], 221.0 kg ha^{-1} 1 *N* ammonium acetate exchangeable K [6] and 0.80% organic carbon [7]. The pH of soil was 8.4 (1:2.5 soil and water ratio). Field capacity, permanent wilting point and bulk density recorded were 8.4.0% (w/w), 1.1.83% (w/w) and 1.66 Mg m^{-3}, respectively in 0-30 cm soil depth.

2.3 Temperature

On daily basis maximum and minimum daily air temperature data were recorded from Agro meteorological observatory, Agricultural Research Station, Bikaner. Temperature at which there is no growth which is also called base temperature, it is taken as 10°C [8] for groundnut growth and calculated temperature use efficiency. Interaction effect was observed in different sowing dates and varieties on use of temperatures. The temperature use efficiency went on increasing from vegetative growth to pod filling and physiological maturity of the crop Nuttonson [9].

2.4 Cultural Practices

The seed treated with chloropyriphos at the rate of 2.5 ml/kg before sowing, HNG 10 and TG 37A were sown at the seed rate of 80 kg, 70 kg/ha, respectively. Row spacing of HNG 10 and TG 37A was 25 cm and 30 cm with 5 cm depth. Nitrogen and phosphorus were applied as per treatment using urea (46% N) and SSP 16% P_2O_5 as nutrient sources. The fertilizers were applied as basal application in the furrows just before sowing at a depth of 8-10 cm by "*pora*" method (indigenous). Gap filling and thinning was also done at appropriate stage. In addition to rainfall received during the both cropping seasons average 22.5, 19.5, 16.5 and 12.5 irrigations in HNG 10 and 16.5, 14.5, 13.4 and 11.5 life saving irrigations in TG 37A were given in 20 April, 15 May, 9 June and 4 July sowing respectively, irrigation through sprinkler system from time to time to ensure optimum growth, development and yield of groundnut. Hand weeding was done manually with the help of hand hoe to keep the field weed free. At the time of second weeding earthing up along the rows was also done for improvement in pegging. Groundnut was grown as per recommended practices and was harvested as per days to maturity both the years of experimentation. The crop was harvested manually by uprooting the whole plant when leaf veins started yellowing and about more than 90 per cent pods became fully mature. The plants from each net plot were harvested separately and stacked plot wise for sun drying and subsequent threshing.

2.5 Statistics

The experiment was laid out in split-plot design with three replications, assigning 32 treatments consisting of four sowing dates (20 April, 15 May, 9 June and 4 July) and two varieties (HNG 10 and TG 37A) as main plot treatments and four fertility levels of nitrogen and phosphorus application (0, 20 N: 40 P_2O_5, 30 N: 60 P_2O_5 and 40 N: 80 P_2O_5 kg ha^{-1}) as sub-plots and crop growth rate, yield parameters, harvest index and temperature use efficiency were recorded on crop. All the data obtained from groundnut for two consecutive years of trails were statistically analyzed using the *F*-test [10] Critical difference (CD) values at *P*=0.05 were used for determine the significance of differences during mean values of treatments.

3. RESULTS AND DISCUSSION

3.1 Crop and Weather

Maximum rainfall of 51.50 mm and 21.00 mm were recorded in July of 2009 and August 2010, respectively (Table 1). Highest temperature was observed in of May of both the years. Growth of groundnut was affected by different uncontrollable environmental conditions. Interaction effect was observed in different sowing dates and varieties on thermal temperatures. The heat unit efficiency went on increasing from vegetative growth to pod filling and physiological maturity of the crop. Among the different sowing dates, heat unit efficiency was found to be higher in 9 June than all other three sowing dates (Fig.1). Significantly higher heat unite efficiency was observed in HNG 10 variety compared with TG 37A variety HNG 10 variety used most heat units to reach maturity in vegetative growth [11]. These results are also cnclosed agreement with the findings [12].

3.2 Temperature Use Efficiency

Efficiency of utilization of temperature in terms of dry matter accumulation depends on crop type, genetic factors and sowing. Duration of crop phenophases is an essential component of weather based dynamic crop growth rate and yield simulation models. Crop phenology can be used to specify the most appropriate sowing date of specific development process. The duration of each phenophase determines the accumulation and partitioning of dry matter in different organs reported that the duration of a particular crop growth stages was directly related to temperature and this duration of particular species could be predicted using the sum of daily air temperature [11].

3.3 Crop Growth Rate

Results showed that of tested varieties in the study, in HNG 10 variety higher crop growth rate was observed than TG 37A variety (Fig. 2), primarily due to the differences in the genetic constitution and growth habit of these varieties as evidenced from the similar variation in crop growth rate of these varieties. The variable behaviour of these varieties could be explained in genetic constitution and variation in growth habit of these varieties the similar interaction effect recorded on heat unit efficiency of these varieties. HNG 10 variety is semi-spreading variety which seems to have indeterminate growth habit the considerable. This variety sown early was exposed to longer duration for vegetative growth due to its indeterminate growth habit. On the other hand, the variety sown on later dates was exposed to shorter duration for vegetative growth and there for recorded less crop growth at these dates. In contrast, the TG 37A variety was a Spanish bunch type and probably seems to be a variety of short duration maturity and determinate growth habit. Thus, the variation in growth parameters of the two varieties with variation in sowing dates resulted due to interaction between genetic constitution of the varieties and growth factors like manures, fertilizers, moisture availability *vis-a-vis* environmental factor like temperature, day length, relative humidity, rainfall, wind velocity and biotic and abiotic factors which have a considerable bearing on important plant functions such as photosynthesis, respiration, transpiration, nutrient and water absorption etc. [11].

3.4 Yields

HNG 10 variety was recorded higher and almost similar pod yield (3773, 3743 and 3738 kg ha-^{1}), and kernel yields under the first three sowing dates beginning from 20 April to 9 June, respectively (Table 3). Delayed sowing of this variety beyond 9 June on 4 July resulted in significant reduction in these yield (3444 kg ha^{-1}). On the contrary, yield of TG 37A variety progressively increased under later dates of sowing with the highest yields recorded under the last sowing date of 4 July (2628 kg ha^{-1}). Since this is a short duration variety of determinate growth habit, it recorded the highest yields in the last date of sowing as this sowing date provided optimum maturity period requirement to this variety. The early sowing dates provided very longer than optimum periods and harsh environments to this variety of determinate growth habit, these dates recorded significantly lower yields. Since the pod and kernel yields are cumulative function of the yield attributes, the variation in pod and kernel yields of these varieties as affected by sowing dates. These results are also in close agreement with the findings of an experiment conducted at Aydın (Turkey) on groundnut to determine the appropriate planting date for peanut under the ecological conditions of Aydın and to investigate the effect of different sowing dates on some agronomic traits, yield, and yield components of 3 commercial peanut cultivars (Gazipaşa, Florispan, and NC-7) and the local cultivar. Peanut growth is affected by planting date in semi-arid regions, such as Aydın, where high temperatures and drought stress occur during late crop development and maturation periods [13]. Similar field study conducted on groundnut with four varieties (TPG-41, TKG-19A, TG-37A and GG2) and recorded significantly higher pods per plant, kernel per pod and 100 kernel-weights in GG2 than other three varieties due to the varietal and environmental characteristics seed or kernel size and recovery of sound mature kernels are two important traits exhibiting considerable variability among the yield of groundnut verities [14].

Sowing dates also had significant interaction effects with crop varieties on haulm and biological yields of groundnut (Table 3). The HNG 10 variety was recorded the highest haulm and biological yields when planted on the earliest sowing date of 20 April. Sowing of this variety on latter dates resulted in significant reduction with each delay in sowing with the lowest haulm yield of 4674 kg ha^{-1} and biological yield of 8118 kg ha^{-1}. This could be explained due to the similar variation in the maturity periods made available to this variety for overall growth [2]. Thus the variation in the haulm yield recorded under different dates of sowing with the minimum haulm yield recorded in last date of sowing. The variety HNG 10 also recorded almost similar variation in biological yield as that observed in haulm yield due to the effect of sowing dates. Since biological yield was a cumulative function of haulm and pod yields, the variation in biological yield with sowing dates could be mainly attributed to the haulm yield with minor contribution from the pod yield, as the pod yield of this variety was almost similar in the three initial dates with reduction only under the last date of sowing. Similarly TG 37A variety was recorded significant reduction in haulm yield with every delay in sowing but the reduction was found to be considerably lower than that recorded in HNG 10 variety due to probably to its determinate growth habit. This type of growth habit did not allow the variety to widen the variation in haulm yield with variation in maturity periods as affected by sowing dates. In contrast to haulm yield, the TG 37A variety was recorded a reverse trend in pod yield as affected by sowing dates with the lowest yield recorded under the first date of 20 April sowing and the highest yield recorded under the last date of 4 July sowing. Since the biological yield is a cumulative function of haulm and pod yields, TG 37A variety was recorded almost similar biological yields with minor variations under various sowing dates due to certainly to opposing trends in haulm and pod yields recorded as affected by sowing dates. Thus, the cumulative effect of the variation in growth and yield attributes of the two varieties recorded with variation in sowing dates resulted variation in various yields of groundnut namely pod, kernel, oil, haulm and biological yields as a result of interaction effect between genetic constitution of the varieties and environmental factors, These results was also in close agreement with the findings [15].

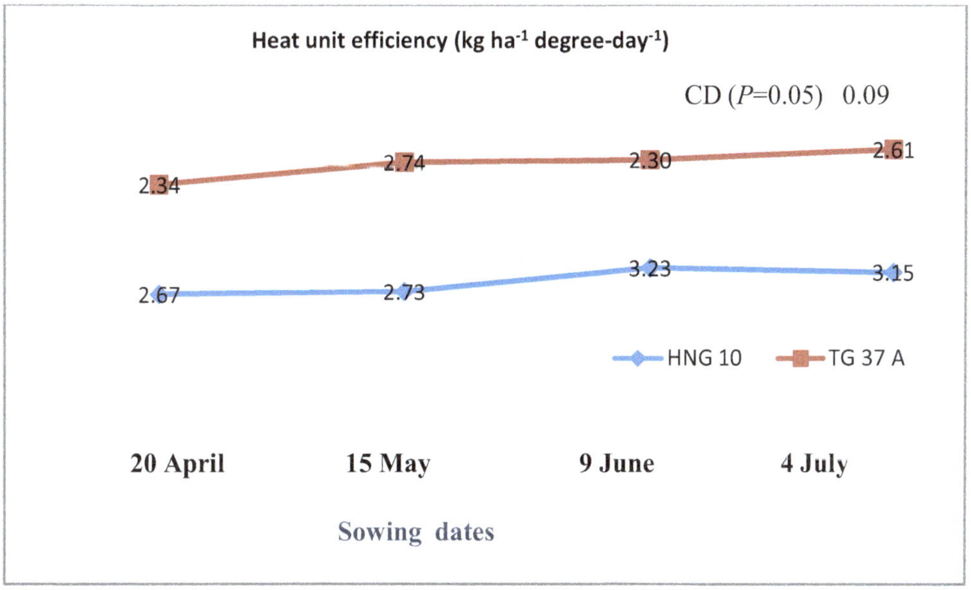

Fig. 1. Effect of sowing dates on heat unit efficency of groundnut (Pooled data of 2 years)

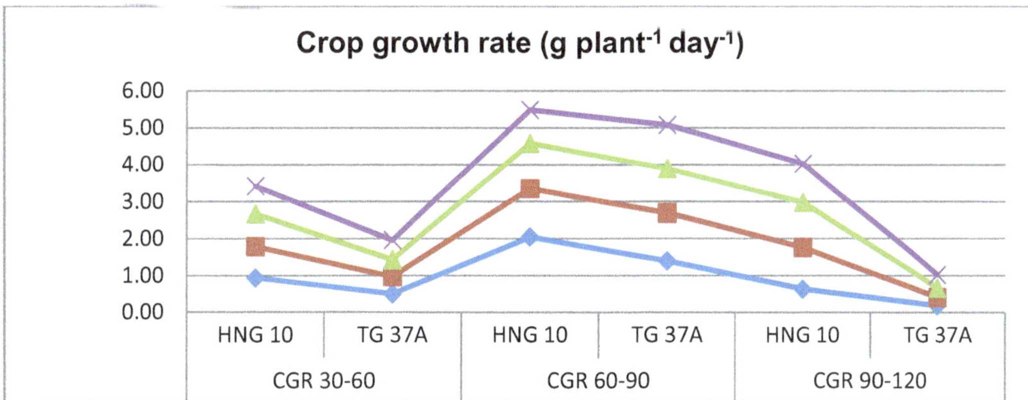

	CGR 30-60		CGR 60-90		CGR 90-120	
	HNG 10	TG 37A	HNG 10	TG 37A	HNG 10	TG 37A
4-Jul	0.75	0.53	0.91	1.18	1.05	0.36
9-Jun	0.87	0.45	1.23	1.21	1.22	0.25
15-May	0.85	0.47	1.31	1.29	1.12	0.22
20-Apr	0.93	0.50	2.03	1.39	0.63	0.18

Fig. 2. Effect of sowing dates on crop growth rate of groundnut (Pooled data of 2 years)

3.5 Effect of Fertility Levels

Application of nitrogen and phosphorus up to 30 kg N-60 kg P_2O_5 ha^{-1} was recorded higher pod yield, kernel yield, haulm yield and biological yield (Table 2) over control and 20 kg N-40 kg P_2O_5 ha^{-1}. Further increase in the fertility level up 40 kg N-80 kg P_2O_5 ha^{-1} was statistically at par with 30 kg N-60 kg P_2O_5 ha^{-1}. Early and plentiful availability of nitrogen and phosphorus to plants favourably influenced the kernel development and kernel size, which ultimately resulted in, increased pod and seed index and shelling percentage. Improved overall growth and profuse branching due to nitrogen and phosphorus fertilization led to increased net photosynthesis on one hand and greater mobilization of photosynthates with suitable temperature requirement towards reproductive structures on the other might have increased the yield attributes significantly result was agreement with those of [16].

Table 1. Temperature and rainfall data during growing seasons of groundnut (2009-2010)

Months	Temperature (°C)				Rainfall			
	Maximum	Minimum	Maximum	Minimum	Total Rainfall (mm)	Rainy days	Total Rainfall (mm)	Rainy days
	2009	2009	2010	2010	2009	2009	2010	2010
April	39.70	24.65	42.15	28.25	0.00	0.00	0.00	0.00
May	43.75	28.95	43.00	28.63	0.50	0.00	0.88	0.25
June	41.58	27.26	41.12	29.44	11.42	0.80	15.10	0.60
July	38.38	27.23	38.18	29.40	51.50	2.00	19.33	0.75
August	37.78	28.32	36.30	27.44	9.00	0.60	21.00	1.40
September	38.90	25.65	34.95	24.50	4.45	0.75	27.00	0.50
October	36.50	22.40	36.35	22.48	0.00	0.00	0.00	0.00
November	33.70	17.85	31.80	16.30	0.00	0.00	0.00	0.00
Monthly average	38.79	25.29	37.98	25.81	9.61	0.52	10.41	0.44

Table 2. Effect of different sowing dates and fertility levels on yields and harvest index of groundnut (pooled data of 2 years)

Treatment	Pod yield (kg ha^{-1})	Kernel yield (kg ha^{-1})	Haulm yield (kg ha^{-1})	Biological yield (kg ha^{-1})	Harvest index(%)
Sowing date					
20 April	2971	1891	6227	9198	32.53
15 May	3084	1962	5332	8416	38.00
09 June	3231	2100	4942	8173	40.86
04 July	3344	2125	4090	7434	45.53
CD (*P*=0.05)	1.62	1.07	1.81	1.93	1.57
Varieties					
HNG 10	3687	2327	6326	10013	37.64
TG 37A	2628	1712	3969	6597	40.82
CD (*P*=0.05)	1.15	0.76	1.28	1.36	1.11
Fertility levels					
Control	2804	1742	4711	7515	38.35
20kgN-40 kg P_2O_5 /ha	3200	2035	5000	8200	40.22
30kgN-60 kg P_2O_5 /ha	3353	2170	5365	8718	39.78
40kgN-80 kg P_2O_5 /ha	3274	2131	5515	8789	38.57
CD (*P*=0.05)	1.05	0.70	1.58	1.91	1.14

Table 3. Interaction effect of different sowing dates and varieties on yields of groundnut (pooled data of 2 years)

Treatment	Pod yield (kg ha^{-1})		Kernel yield (kg ha^{-1})		Haulm yield (kg ha^{-1})		Biological yield (kg ha^{-1})		Harvest index (%)	
	HNG 10	TG 37A	HNG 10	TG 37A	HNG 10	TG 37A	HNG 10	TG 37A	HNG 10	TG 37A
sowing date										
20 April	3773	2170	2365	1416	7830	4623	11603	6793	32.60	32.45
15 May	3743	2426	2346	1578	6563	4101	10305	6527	37.02	38.99
09 June	3789	2673	2444	1756	6238	3646	10027	6319	38.61	43.12
04 July	3444	3244	2151	2099	4674	3506	8118	6751	42.33	48.73
Mean	3687	2628	2327	17.12	6326	3969	10013	6597	37.64	40.82
CD (P=0.05)		2.30	1.52		2.56		2.73		2.22	

4. CONCLUSION

The present study indicates that suitable sowing date for HNG 10 variety to be around 9 June and TG 37A variety for 4 July and recommended fertilizing dose 30 kg N-60 kg P_2O_5 ha^{-1} for better growth and higher productivity of groundnut in western dry zone of India. However, groundnut crop use maximum temperature in June moth for better growth of HNG 10. These provide very clear picture of the amount, pattern and efficiency of heat energy consumption of the crops. These can also be used very effectively for forecasting the occurrence of different phenophases of the crops.

COMPETING INTERESTS

Authors have declared that no competing interests exist.

REFERENCES

1. ASG. Agricultural statistics at a glance. Department of Agriculture and Co-operation, Ministry of Agriculture Government of India. 2012;128-185.
2. Sardana V, Singh S, Sheora P. Influence of sowing dates on the productivity of semi-spreading and bunch type verities of groundnut. Indian Journal of Agricultural Sciences. 2008;78(2):372-4.
3. Sardana V, Kandhola SS. Productivity of semi-spreading and bunch type varieties of groundnut as influenced by sowing dates. Open Accesses. Journal published by ICRISAT. 2007;5:1-3.
4. Subbiah BV, Asija GL. A raped processor of determination of available nitrogen in nitrogen in soil. Current Sciences. 1956;25:259-260.
5. Olsen SR, Cole CV, Watanabe, FS, Dean, LA. Estimation of available phosphorus in soils by extraction with sodium bicarbonate. USDA Circ. No. 939, Washington; 1954.
6. Stanford S, English L. Use of flame photometer in rapid soil tests for K and Ca. Agronomy Journal. 1949;41:446–7.
7. Jackson ML. Soil Che. I Ana. Prentice Hall of India Pvt Ltd, New Delhi; 1973.
8. Leong SK, Ong CK. The influence of temperature and soil water deficit on the development and morphology of peanuts (*Arachis hypogaea* L.). Journal of Experimental Biotechnology. 1983;34: 1551–1561.
9. Nuttonson MY. Wheat-climate relationships and the use of phenology in ascertaining the thermal and photothermal requirements of wheat. Institute of Crop Ecology, Washington; 1955.
10. Gomez KA, Gomaz A. Statistical Procedures for Agricultural Research. John Wiley & Sons, Singapore; 1984.
11. Kingra PK, Kaur P, Effect of dates of sowing on thermal utilization and heat use efficiency of groundnut cultivars in central Punjab. Journal of Agriculture Physics. 2012;12(1):54-62.
12. Gouri V, Reddy, R, Narayansimha SBS, Rao YA. Thermal requirement of *rabi* groundnut in Southern Telangana Zone of Andhra Pradesh. Journal of Agrometeorology. 2005;7(1):09-94.
13. Canavar K, Kaynak, AM. Effect of different planting dates on yield and yield components of peanut (*Arachis hypogaea* L.). Turk Journal of Agriculture. 2008;32(1): 521-528.
14. Hariprasanna KL, Chuni, Radhekrishana T. Relationship between flowering duration and physical-quality traits as well as pod

yield in groundnut (*Arachis hypogaea* L.) Indian Journal of Agricultural Sciences. 2008;78(2):180-2.

15. Gochar R. Effect of cultivars and weed management on late sown groundnut (*Arachis hypogaea* L.) in north western Rajasthan. Agriculture for Sustainable Development. 2013;1(1):45-47.

16. Bala, HMB, Ogunlela, VB, Kuchinda NC, Tanimu B. Response of two groundnut (*Arachis hypogaea* L.) varieties to sowing date and NPK fertilizer rate in a Sami-arid environment: yield and yield attributes. Asian Journal of Crop Sciences. 2013; (3):130-134.

13

Spatial Characterization of Cone Index and Some Nutrients in a Sandy Loam Soil (*Eutric Leptosol*) Using the Multivariate Analysis

David Lomeling[1*], Sebit Mathew Otwari[1] and Yahya Mohammed Khater[1]

[1]Department of Agricultural Sciences, College of Natural Resources and Environmental Studies (CNRES) University of Juba, P.O.Box 82, Juba, South Sudan.

Authors' contributions

This work was carried out in collaboration between all authors. Author DL designed the study, performed the model simulations and statistical analysis and wrote the first draft of the manuscript. Authors SMO and YMK conducted the field and laboratory tests. All authors read and approved the final manuscript.

Editor(s):
(1) Özge Çelik, Department of Molecular Biology and Genetics, Istanbul Kultur University, Turkey.
(2) WenJun Zhang, Sun Yat-sen University, Guangzhou, China and International Academy of Ecology and Environmental Sciences, Hong Kong.
(3) Daniele De Wrachien, State University of Milan, Italy.
Reviewers:
(1) Andrea Trevisan, Department of Cardiologic, Thoracic and Vascular Sciences, University of Padova, Italy.
(2) Sedat Serçe, Nigde University, Turkey.
(3) Anonymous, USA.

ABSTRACT

A multivariate analysis was performed on some soil nutrient and Cone Index (CI) data from the research and demonstration farm of the Dept. of Agricultural Sciences, University of Juba in South Sudan. The main objective of the study was to characterize the spatial distribution of the soil nutrients: N, P, K, Fe and Mn as well as soil penetration resistance CI. A Principal Component Analysis (PCA), Gaussian Mixture Model (GMM) and Hierarchical Cluster Analysis (HCA) were performed on the analyzed samples. The Bayesian Information Criterion (BIC) was used for model selection between the Equal size, Equal shape and Equal orientation (EEE) and Equal size, Equal shape and Variable orientation (EEV) models which defined the size, shape and orientation of the ellipsoid with full covariance matrices. Eigenvalues of the three major principal components F1, F2 and F3 accounted for 75.67% of the total variance of the data. From hierarchical clustering, P was

*Corresponding author: E-mail: dr.david_lomeling@gmx.net;

observed to cluster with Fe, Mn with N which at second level clustered with K then with Cl. The results of the PCA showed that Nitrate-N, Mn and Kwere strongly influenced by Cl and so determining their spatial distribution. This could be associated mainly to earlier anthropogenic activities on the soil. The results of this study also showed spatial relationships between individual soil nutrients with both K and P mutually antagonistic with Nitrate-N, whereas between K and P where mutually synergistic. While P was strongly adsorbed to Fe, this was associated to lithogenic soil materials and therefore interpreted as derived from natural sources of the *Eutric Leptosol.* The goodness-of-fit test using the Kolmogorov-Smirnov (KS) showed that the values of the variables: Cl, K, P and Fe were significant at $p \leq 0.05$ and that the data followed normal distribution, whereas Mn and Nitrate-N were not. The KS test also corroborated the results of strong spatial dependency of each variable at less than 25%. The multivariate GMM adequately described the spatial distribution of all measured variables than the unimodal Gaussian.

Keywords: Dendrogram; gaussian mixture model; hierarchical clustering; kolmogorov-smirnov test; multivariate analysis; spatial distribution.

1. INTRODUCTION

Principal component analysis (PCA) is a multivariate analysis method also known as eigenvector analysis. It has been applied in environmental studies in; assessing heavy metals in urban soils [1], estimating heavy metals in agricultural soils [2], assessing earthworm populations in oil plantation soils [3], on genetic diversity of tomato germplasm [4], on effects of soil parameters and environmental factors on flavonoid contents of medicinal plants [5],soil factors influencing heavy metals in medicinal plants [6], and in combination with geo-statistical tools in assessing fertility of humicrhodichapludox [7], on distribution of trace elements in unsaturated soil profile [8].

Similarly, PCA has been performed to investigate management impacts on soil quality [9], on chemical and microbial properties in histosols as influenced by land-use types [10] and microbial community structure and function [11,12].

PCA is a technique that reduces the number of variables and eliminates the relations among input variables by developing a set of new variables that are linear functions of the original variables. This new variables are denoted as F1, F2, F3 etc. Whereas F1 will give the direction with the greatest variation, F2 orthogonal to F1 will give the direction with the maximum variation left in data. The variables are multiplied by loadings that are vectors of constants generated during PCA and whose values reflect the importance of original variables in the direction of each PC [13]. The resulting PC can be used to project the originally multidimensional data into only a two- or three dimensional space known as a score plot [14].

Hierarchical Cluster Analysis (HCA) is another method applied for geological/hydrological analysis and looks for groups of samples according to their similarities. HCA is a powerful tool for analyzing datasets for expected or unexpected clusters including the presence of outliers. In HCA, each point forms, initially, one cluster, and the preliminary matrix is analyzed. The most similar points are grouped forming one cluster and the process is repeated until all points belong to one cluster [15]. HCA examines distances between samples and datasets. The result obtained could be presented in a two-dimensional plot called dendrogram which illustrates the fusions or divisions made at each successive stage of analysis. Typically, a hierarchical agglomerative cluster analysis of the studied variables is performed and the square of the Euclidean distance taken as a measure of similarity between samples [16]. Samples that are similar will lie close to one another, whereas dissimilar ones will lie distant from each other. Several methods may be applied in deciding similarity or dissimilarity between two groups: (i) single linkage, which uses the minimum distance between points in different groups; (ii) complete linkage, which uses the maximum distance between the furthest points; (iii) mean linkage, which uses the average of all distances between points in the two groups and (iv) centroid linkage, which uses the distances between group centroids (e.g. group means).

The PCA and HCA are both methods for uncovering relationships in large multivariate datasets. However, they are not sufficient for developing a classification rule that can accurately predict the class-membership of unknown sample [17]. Classification or pattern recognition is a method that seeks to develop class-membership based on some specific and measured property within the variables. For cluster classification, the similarity-based classifier e.g. *k-nearest neighbor* (KNN) may be chosen under the premise that distances between points in the measurement space will be inversely related to their degree of similarity.

Another tool for multivariate analysis is the Gaussian Mixture Model (GMM). This is a parametric probability density function (pdf) and is often applied as a model for the probability distribution of continuous measurements of variables in a dataset. As opposed to histogram analysis, which is non-parametric, GMM provides greater flexibility and precision in modeling the underlying statistics and smoothening gaps resulting from sparse sample data by assigning variable to specific membership. Based on the premise of a vector-based variable *x*, a dataset is assumed to have a univariate normal (*Gaussian*) distribution if its probability density function is given by:

$$p(x,\mu,\sigma^2) = \frac{1}{\sqrt{2\pi}\sigma}\exp\left(-\frac{1}{2\sigma^2}(x-\mu)^2\right) \text{(Eqn. 1)}$$

It is also said to have a multivariate normal (Gaussian) distribution if its probability density function is given by:

$$p(x,\mu,\Sigma) = \frac{1}{(2\pi)^{n/2}|\Sigma|^{1/2}}\exp\left(-\frac{1}{2}(x-\mu)^T\Sigma^{-1}(x-\mu)\right) \text{(Eqn. 2)}$$

Where μ is the mean (location) and Σ the covariance matrix of the Gaussian that incorporates the variance σ, and therefore dispersion. Under the GMM, the random variable x is generated from several distinct random and independent processes each generating a specific pdf. A GMM is often a mixture of two or more independent or unimodal pdfs combined into a single bi- or multimodal pdfs. The concept of clustering using the GMM is based upon the fact that individual data points are produced by choosing from one of a set of multivariate Gaussians and the GMM parameters estimated using the iterative Expectation-Maximization (EM) algorithm or Maximum A-Posteriori (MAP). Although the EM algorithm has some limitations (e.g. it is not guaranteed to converge to a global rather than a local maximum of the likelihood), it is generally efficient and effective for the parameters' estimation of GMM [18].

The Gaussian model is then fitted to the data points that is either spherical (circular with off-diagonal correlations equal to zero) symmetry or elliptical (off-diagonal correlation equal to non-zero). To fit a Gaussian model to given data points, the sample mean and variance is computed and the resulting Gaussian model is superimposed as a contour map.

GMM, a parametric probability density function (pdf) was employed and provided the basis for partitioning and fuzzy classification of the data into different clusters. Arguably, this probabilistic method partitions the data by considering that each component represents a cluster through the *k-means* clustering that tends to minimize the *Euclidian distance* or the within-cluster sum of squares (WCSS). A measure of dissimilarity or similarity between observed data sets was used to combine specific clusters (agglomerate) or split (divide) them altogether. This hierarchical clustering is achieved by use of measuring the *Euclidian distance* between pairs of observations and a linkage criterion which specifies the dissimilarity of sets as a function of the pairwise distances of observations in the sets. We used complete linkage as the agglomeration method as suggested by [19] for clustering the same dataset.

The primary goal of any soil scientist is the ability to adequately characterize and describe major soil factors, parameters and processes as well as their spatio-temporal distribution within the context of precise farming. Increasing environmental awareness in trying to reduce the application of especially costly inorganic fertilizers and use of heavy farm implements by farmers has prompted the need for more precise and sustainable farming methods. Introducing precise farming especially in developing countries like South Sudan would not only reduce unnecessary production costs, but promote a more sustainable and environmentally-friendly farming. As South Sudan intends to embark on large scale commercial farming to address food security, such tools as PCA would be utilized in designing precise and sustainable farming methods that would compromise both economic and ecological imperatives.

The objective of this study was to assess how best the PCA in combination with GMM and HCA methods could be used to adequately characterize the spatial distribution of some major soil nutrients such as Nitrate-N, P, K and micro-nutrients: Mn and Fe as well as the soil penetration resistance.

2. MATERIALS AND METHODS

2.1 Study Area

The study was conducted at the demonstration farm of the Department of Agricultural Sciences, College of Natural Resources and Environmental Studies (CNRES), University of Juba, South Sudan. The study area lies within the green belt agro-ecological zone of Central Equatoria State (CES), South Sudan and lies between latitude 4°50'28'' and longitude 31°35'24. The average annual rainfall is about 650 mm during the months from April to October with dry spell in July. The climate of the area is tropical wet during the rainy season with average temperatures at around 27 to 30°C and over 35°C during the dry season of November to March. The soils can be predominantly classified as *Eutric Leptosol* with less associated *Eutric Gleysol* as shown in Table 1.

The experimental area was the 40 m x 80 m demonstration farm or nursery of the Department of Agricultural Sciences, University of Juba and divided in to 32 plots each 10 m x 10 m. Four samples were extruded from each 12 out of the 32 randomly chosen plots for the NO_3-, P, K, Mn and Fe analysis. Similarly, four penetrations at a separation distance of 1 to 2 m per plot to maximum depth of 80 cm were determined using a hand push electronic cone penetrologger (*Eijkelkamp Penetrologger SN*) as reported by [20,21] with a cone type 1.0 cm^2, 60° and a penetration speed of 2 cm/s. Processed soil samples were analyzed for various soil nutrients: nitrate-nitrogen, phosphorus, potassium, manganese and ferric iron using the LaMotte basic Model STH-4 Outfit (Code 5029) and values expressed in kg/ha except for Mn which was expressed in ppm.

2.2 Multivariate Analysis

Datasets consisting of (n=48) from 12 randomly selected plots were analyzed by PCA, HCA and GMM methods. The PCs were used to project new position of variables in space using new matrix which would indicate degree of similarity. Equally, the HCA was applied to cluster two objects by calculating iteratively the dissimilarity between them till a minimum agglomeration criterion: the Bayesian Information Criterion is attained. The Pearson correlation coefficient was used as index of similarity. Meanwhile, the GMM was used to model data using a set of Gaussian distributions. These models were used in the clustering process with each clustered group assigned to each Gaussian. A class model related to the classification of soil nutrients according to their spatial distribution and variability was developed by applying the statistical software package XLSTAT 2014.4.06 (Addinsoft SARL, Paris, France). PCA was used to establish the simplest mathematical model capable of describing the dataset satisfactorily. An overview on the three approaches in shown in Fig. 1.

Table 1. Average values of some selected physical and chemical properties of *Eutric Leptosol*

Soil property	Description
Soil texture classification*	Sandy loam
Drainage class (0-0.5%)*	Moderatelywell
Sand	47.6%
Silt	45.1%
Clay	7.3%
pH (LaMotte STH test method)	7.2
Vol. Water Content	18.4%
Bulk density (gm/cm³)	1.34
Humus content	2.95%

**Source: Harmonized world soil data viewer version 1.2*

2.3 Geo-Statistical Analysis

Spatial variability of soil nutrients as well as penetration resistance measured as cone index were analyzed using geo-statistics. Geo-statistical software GS+™ Version 9 (GAMMA DESIGN SOFTWARE, LLC, PLAINWELL MICHIGAN, USA, 2001) was applied to quantify the isotropic spatial variability and semi-variogram models. Several models: spherical, exponential, Gaussian and linear semi-variogram were considered in selecting the best fitting model based on the values of weighted residual sums of squares, regression coefficient (r^2) and relative spatial structure or dependency measured as the ratio of the nugget to sill.

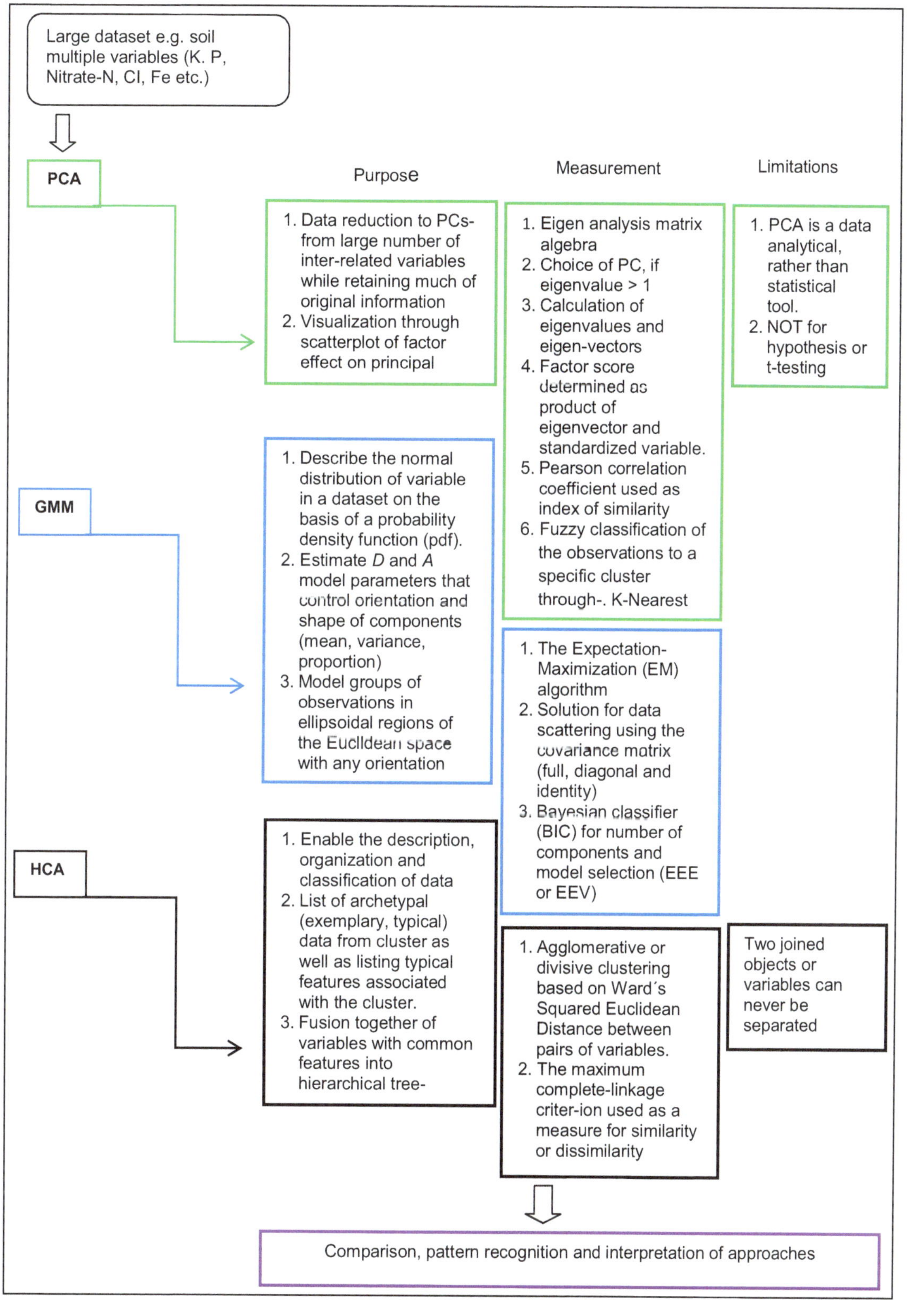

Fig. 1. Diagrammatic representation of the proposed multivariate analysis using the PCA, HCA and GMM approach on soil variables of *Eutric Leptosol*

For the GS+ Version 9, the semi-variance is defined by the following equation:

$$\gamma(h) = \sum_{i=1}^{N(h)} [2(x_i + h) + 2(x_i)]^2 \quad \text{(Eqn. 3)}$$

Where $\gamma(h)$ is the experimental semi-variogram value at distance interval h; N(h) is number of sample value pairs within the distance interval h; and z (x_i+h) is sample value at two points separated by the distance interval h. All pairs of points separated by distance h (lag h) were used to calculate the experimental variogram. Several variogram functions were evaluated to choose the best fit with the data. After several simulation runs, the Gaussian model was found and chosen to be adequate in describing the measured data and was fitted to the empirical semi-variogram as:

$$\gamma(h) = C_0 + C\,(1 - \exp\,(\frac{-h}{A_0})] \quad \text{(Eqn. 4)}$$

Where $\gamma(h)$ = semi-variance for interval distance class h, h= lag interval, C_0 = nugget variance ≥ 0, C = structural variance ≥ C_0, and A_0 = range parameter. In the Gaussian model, both the effective range A = $3^{0.5}$. A_0 and sill ($C+C_0$ that lies within 5% of the asymptote) are less discernible owing to the gradual and asymptotic rise of $\gamma(h)$.

3. RESULTS

3.1 Principal Component Analysis

Application of the PCA three principal components were extracted that cumulatively explained for 75.67% of the total data variability with an eigenvalue of greater than 1 (Table 2). The other three components had an eigenvalue less than 1 and were not considered. In our study, the first two components accounted for 56.65% of total data variability as represented the single largest variability contained in the data set [22].

The first component F1 correlated with 3 of the 6 soil nutrients analyzed. The nutrients Nitrate-N, P and Fe correlated positively with the F1, while it negatively correlated with Mn and Cl and neither with K (Fig. 2).

Table 2. Principal component analysis

	F1	F2	F3	F4	F5	F6
Eigenvalue	1.881	1.517	1.142	0.671	0.484	0.305
Variability (%)	31.356	25.291	19.025	11.187	8.063	5.078
Cumulative %	31.356	56.647	75.672	86.859	94.922	100.000

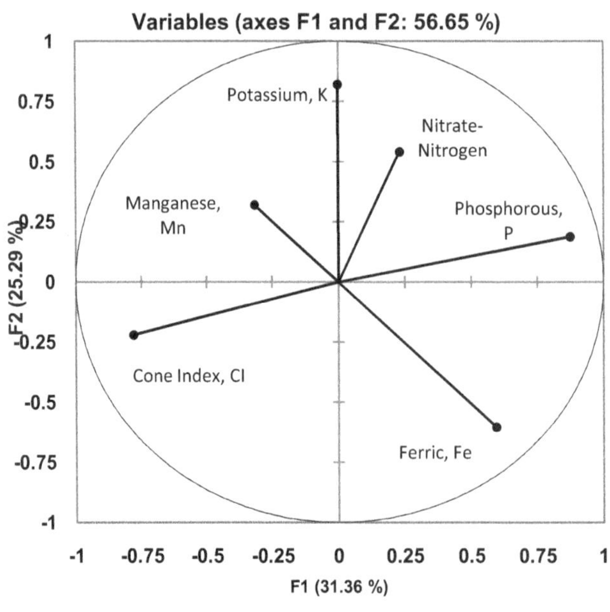

Fig. 2. Loading plots of principal components analysis on soil nutrients: Nitrate-N, P, K, Fe, Mn as well as soil penetration resistance measured as Cone Index, CI

Similarly, the F1 *versus* F2 loading plot showed that, variable K was not in any way correlated with CI, however, the variable CI was negatively correlated with both Nitrate-N and P and positively correlated with Mn. Variables with significant loadings on F1 Nitrate-N (factor loading= 0.231), P (factor loading = 0.875), Fe (factor loading = 0.594), while Mn (factor loading= -0317) and CI (factor loading=-0.781) had negative loadings. Interestingly, variable CI had negative loadings on both F1 and F2, whereas Nitrate-N, P, K and Mn had positive loadings on F2 and Fe negative loading on F2.

Biplot generated by the PCA in Fig. 3 showed the spatial distribution of the different variables in the different plots. For example, the amounts of P, NO_3-N, K and Fe in kg/ha were higher in plots 1, 3, 4, 5, 6 and 12 than in the rest plots. Similarly, CI and Mn was higher in plots 7, 8, 9, 10 and 11 than in the rest plots. The CI as first principal component F1 significantly influenced the spatial distribution of the soil nutrients.

Ward's method of hierarchical agglomerative clustering was used to characterize soil parameter CI and soil nutrients N P K, Fe and Mn. The method is based on minimum variance in which groups are formed so that the pooled within-group sum of square is minimized. Fig. 4 shows the hierarchical agglomerative clusters. The first clusters (Phosphorous, P and Ferric, Fe) and (Manganese. Mn and Nitrate-Nitrogen, N) are fused at 0.3412 and 0.2542 respectively. The second cluster (Mn, N and Potassium, K) are fused at around 0.1042. The third cluster (Mn, N, K and CI) is fused slightly above the truncation level of 0.0542.

For all the measured soil variables, there was only one class of spatial dependency as shown in Table 3. Spatial dependency was classified on the basis on nugget/sill ratio. Spatial class ratio between 0 to 25% were considered strongly spatially dependent; those between 25 to 75% moderately spatially dependent; and greater than 75% as weakly spatially dependent [23]. The structural variance of all estimated parameters showed a narrow range of variability within the plots between 8.6 and 15.5% thereby indicating a strong spatial dependency.

The results of the Pearson's correlation matrix have been presented in Table 4. The correlation matrix identified the relationship among the soil nutrients and CI with values between +1 or -1 revealing either a positive or negative relation between the variables. In general, the CI showed a negative correlation with most soil nutrients: NO_3-Nitrogen, P, K and Fe (*p<0.0001*) and a weak positive correlation with Mn. However, between the individual soil nutrients, P showed a weak positive correlation with K, Nitrate-Nitrogen Fe. Similarly, Mn showed low negative correlation with bothP and Fe (*p<0.0001*) as well as with K and Fe (*p<0.0001*).

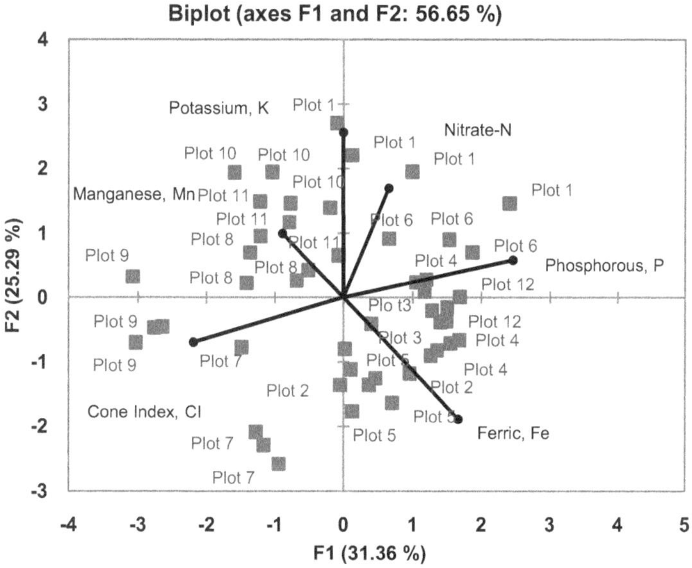

Fig. 3. Biplot generated by the principal component analysis for some soil nutrients and CI distributed within the different plots of the sandy loam soil (*Eutric Leptosol*)

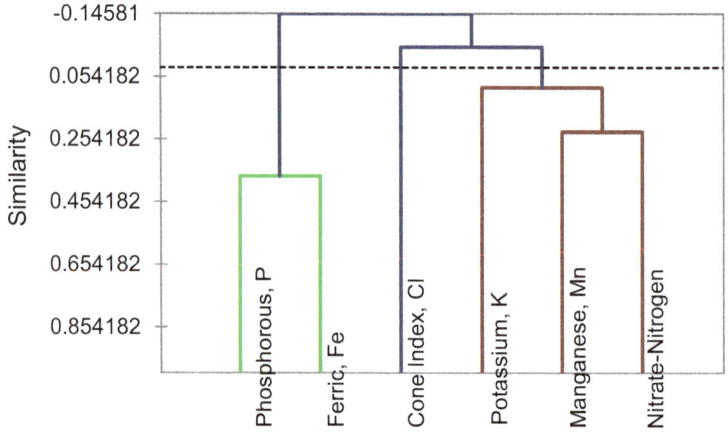

Fig. 4. Dendrogram generated by the agglomerative Hierarchical Cluster Analysis (HCA) for some soil nutrients and CI of a sandy loam soil (*Eutric Leptosol*)

Table 3. Geo-statistical parameters for the different soil nutrients including the Cone Index of a sandy loam soil (*Eutric Leptosol*) (Adapted from Lomeling [21])

Parameter	NO_3^--N	P	K	Mn	Fe	CI
Model	**Gaussian**	**Gaussian**	**Gaussian**	**Gaussian**	**Gaussian**	**Gaussian**
Nugget variance, C_0	104.24	2402.50	1200.00	0.065	0.089	0.0378
Sill, $C+C_0$	827.00	27930.00	7758.00	0.674	0.774	0.282
Range (m), A_0	14.85	8.50	16.93	98.76	1.95	8.80
R^2	0.709	0.865	0.793	0.532	0.304	0.232
RSSC_0/$C+C_0$	904998	4.04E+08	2.14E+07	0.0425	0.136	0.261
	0.126 (12.6%)	0.086 (8.6%)	0.155 (15.5%)	0.096 (9.6%)	0.115 (11.5%)	0.134 (13.4%)

0-0.25 or 0-25% strong dependency; 0.25-0.75 or 25-75% moderate dependency; >0.75 or >75% weak dependency

Table 4. Pearson correlation coefficient matrix (r) of CI and some soil nutrients

Variables	Cone Index, CI	Manganese, Mn	Phosphorous, P	Potassium, K	Ferric, Fe	Nitrate-Nitrogen
Cone Index, CI	1					
Manganese, Mn	0.138	1				
Phosphorous, P	-0.566	-0.172	1			
Potassium, K	-0.139	0.032	0.155	1		
Ferric, Fe	-0.208	-0.127	0.373	-0.455	1	
Nitrate-Nitrogen	-0.117	0.232	0.231	0.148	-0.025	1

Table 5. Factor loadings of the first 5 components on some soil nutrients and CI of a sandy loam soil (*Eutric Leptosol*)

	F1	F2	F3	F4	F5
Cone Index, CI	-0.781	-0.221	0.130	0.333	0.443
Manganese, Mn	-0.317	0.320	0.717	-0.524	0.082
Phosphorous, P	0.875	0.188	-0.002	-0.020	0.265
Potassium, K	-0.002	0.821	-0.395	-0.066	0.329
Ferric, Fe	0.594	-0.603	0.326	-0.020	0.290
Nitrate-Nitrogen	0.231	0.542	0.591	0.530	-0.135

The results of factor loadings are presented in Table 5. The first three principal components F1, F2 and F3 had eigenvalues greater than 1 and accounted for 75.67% of total variation of the data with values at 31.36%, 25.29% and 19.03% respectively. F1 showed high loadings of P, Fe and low Nitrate-Nitrogen with positive effect and CI, Mn and K with negative effect. F2 showed high loadings of K, Nitrate-Nitrogen and moderate loadings of Mn and P with positive effects whereas it showed negative effects with CI and Fe. F3 showed high loadings of Mn, Nitrate-Nitrogen and moderate to low loadings of Fe and CI with positive effects with negative effects of both P and K.

We employed a Gaussian mixture model to analyze the dataset and estimate the probability density functions. For illustrative purposes, we chose one example for CI and Mn to demonstrate the performances of both single and multimodal Gaussian distribution of the GMM as in Fig. 5. Whereas Fig. 5(a) showed a unimodal Gaussian pdf with mean value at about 1.81 MPa, Fig. 5(b) showed a GMM multivariate distribution with four components. Due to the closeness of the components 2 and 3 in Fig. 5(b), only three were chosen: 1, 2 and 4 with peak CI values at 1.2, 1.9 and 3 MPa respectively. The mixture model showed an average CI value of about 2 MPa. The GMM analysis was corroborated by scatter plots showing four clusters as in Fig. 6. The ellipsoid approximated the standard deviation of the data distribution within each cluster with the size, shape and orientation of each cluster indicating the means, co-variance matrices and correlations of the variables. With the full covariance matrix, the mixture fits the four cluster shapes well with the pdf covering the space containing most of the observations.

Four 2-D probability density function ellipsoids of the GMM of CI on Mn are shown in Fig. 6 with wide range of CI values between 0.6 to 3.5 and centered means: μ_1 at 1.91, μ_2 at 1.24, $\mu_3$at 3.00 and μ_4 at 1.89. A closer look at Fig. 6 shows that for CI values between 1.5 to 2.5generated a wider Mn spatial variation with two clusters centered at levels 4.5 and 5.3 ppm suggesting that this CI value had a comparatively greater influence on Mn distribution ranging between 4.5 to 5.5 ppm than CI values lesser than 1.5 or greater than 3.0.

Fig. 7 shows both the unimodal and multimodal Gaussian pdfs of Mn distribution. The single Gaussian pdf showed a mean value at about 4.5 ppm whereas the GMM showed 2 peak values at about 4.2 and 5 ppm for components 1 and 2 respectively. As aforementioned, the CI mean values between 1.5 to 2.5 had significant influence on Mn distribution.

With inferences from Fig. 6, we attempted to highlight the significance of CI on Mn by focusing mainly on two clusters. Fig. 8 shows 2 two-dimensional Gaussians of CI on Mn. CI values with center means between 1.5 and 2.0 that generated two Mn clusters with average values at 4 and 5 ppm respectively. With the Normalized Entropy Criterion (NEC) at 0.206 and therefore less than 1, there was clustering structure in the data with many observations centered around 5 ppm value. The observations were spread across a wider range of CI values up to 3.5 than those around 4 ppm value. The results showed the positive correlation between CI and Mn suggesting that high CI tended to favor increased Mn and *vice-versa* [20]. The illustrations in Figs. 5 and 7 for both CI and Mn showed that the weighted mixture of the components using the GMM provided good and accurate approximation for the spatial data variation than the single Gaussian pdfs.

Fig. 9 shows a group of 5 clusters (*representing 5 probability density functions*) of Gaussian distribution between the F3 and F4 with the Normalized Entropy Criterion (NEC) as basis for model selection at 0.009 (i.e. less than 1). This showed that within the complete dataset five center locations for both point P and K were evident with lowest P concentrations at 64 k/ha and for K at around 232 kg/ha respectively. The fifth cluster and highest concentrations for P and K were at around 512 kg/ha and 488 kg/ha respectively suggesting a more synergistic relationship between both elements. Between 95-99% of soil P is insoluble P and unavailable for plants [24,25]. Although the full covariance matrix as opposed to either diagonal or identity covariance matrix was precise and covered most of the observations within the ellipsoid, not all were captured owing to the large variability and inability of the GMM to minimize the Euclidean distance between adjacent observations.

Fig. 10(a) showed the F3 loading on F6 with four cluster classes (*representing four probability density functions*) with equal volumes shapes but different orientations. The NEC was 0.067 with EEV as best model. Although not quiet discernible, a more, accurate and clearer antagonistic relationship especially between classes 2 and 4 was noticeable. At lower P between 60-250 kg/ha, the Nitrate-N was around 70 kg/ha and reduced to about 30 kg/ha with increased P at between 350-550 kg/ha. Similarly, Fig. 10(b) showed clustering of data with NEC at 0.007 and EEV as best model. There was less discernible relationship between F4 and F6 showing both antagonistic and synergistic features, however, with a predominating antagonistic relationship. In class 1, with F4 at around 280 kg/ha, the F6 was at around 60 kg/ha. In class 3 and 4, F4 was at 380-520 kg/ha and 440-520 kg/ha respectively. There was generally a negative correlation between F4 and F6.

The estimated parameters of each GMM component for the CI and some soil nutrients are summarized in Table 6. Note that in all cases, the GMM components have more or less similar proportions or weights. However, large standard deviation values especially of P, K and Nitrate-N and the few observations poorly captured by the full covariance matrix within the thin-shaped and stretched ellipsoids suggest wide spatial variability and apparently poor clustering of data and *vice-versa*. Generally, the results of our study suggest that any measured soil variable, irrespective of its spatial distribution, can still be approximated well using GMMs.

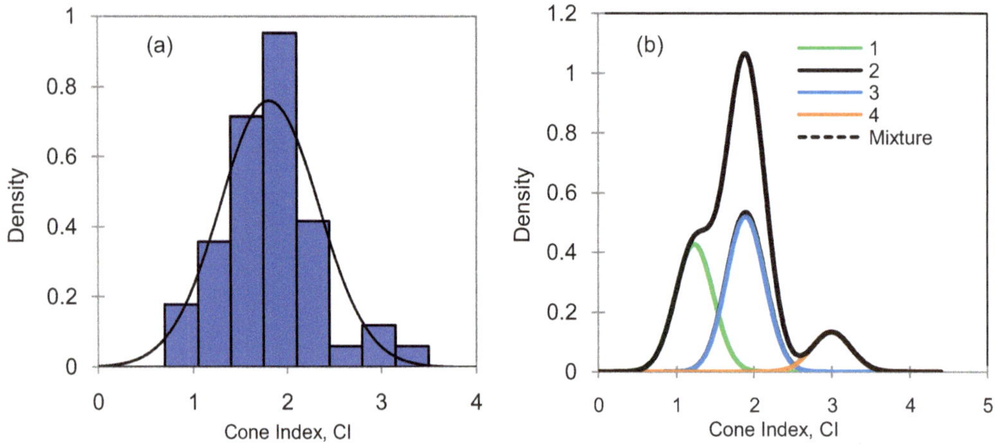

Fig. 5. Single Gaussian pdf approximation (a) and the GMM multivariate normal distribution of CI (b) showing the four components

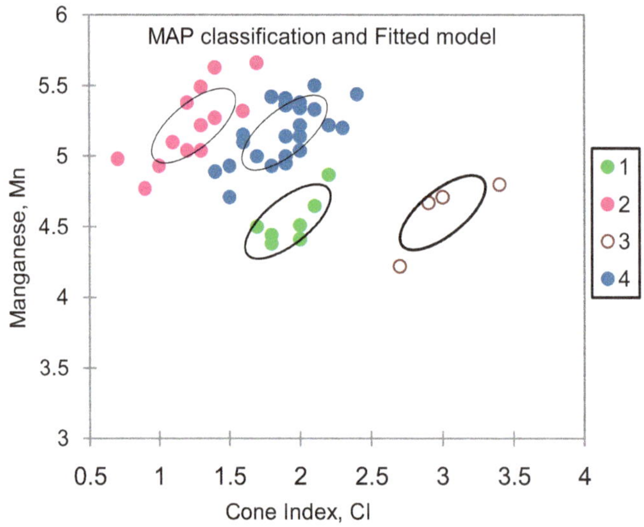

Fig. 6. Scatter plots of 4 two-dimensional Gaussians of CI on Mn with full covariance matrices. Best model was the EEE (Equal volume, Equal shape and Equal orientation)

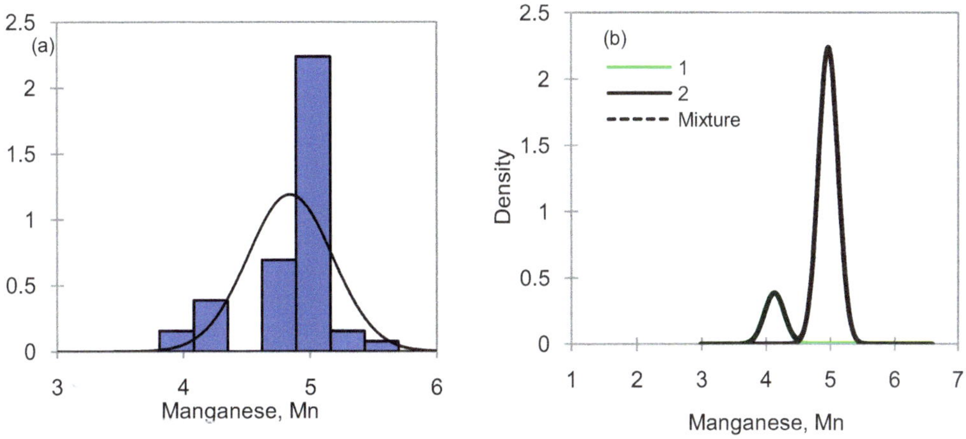

Fig. 7. Single Gaussian pdf approximation (a) and the GMM multivariate distribution of Mn (b) showing the two components

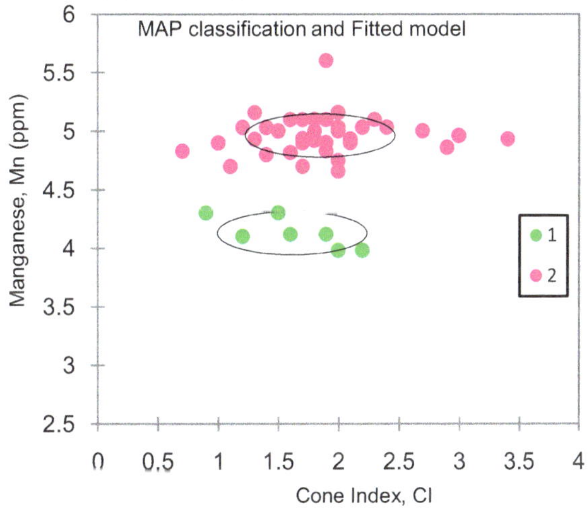

Fig. 8. Scatter plots of 2 two-dimensional Gaussians of CI on Mn with full covariance matrices. Best model was the EEE (Equal volume, Equal shape and Equal orientation)

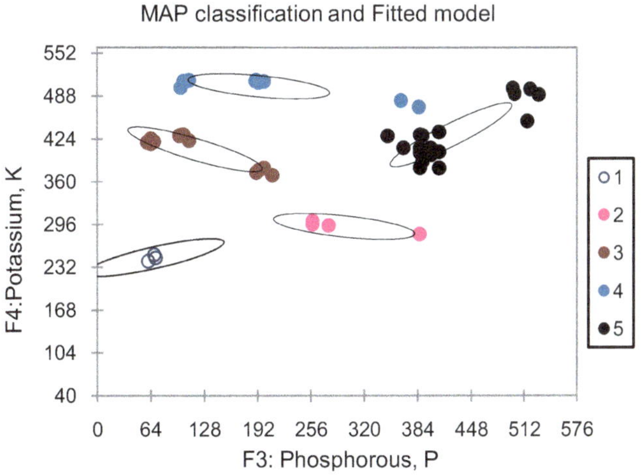

Fig. 9. Scatter plots of 5 two-dimensional Gaussians of phosphorous on potassium with full covariance matrices. Best model was the EEV (Equal volume, Equal shape and Variable orientation)

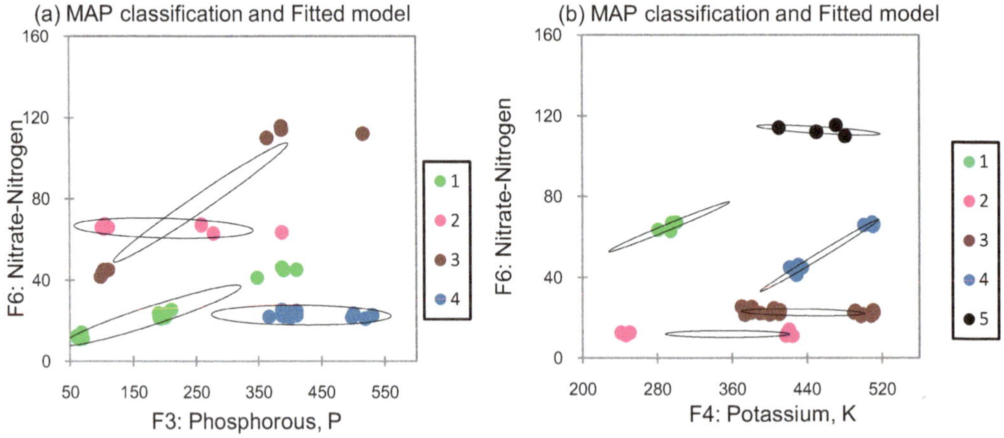

Fig. 10. Scatter plots of 4 two-dimensional Gaussians of phosphorous on Nitrate-Nitrogen (a) (b) and a 5 two-dimensional Gaussian of potassium on Nitrate-Nitrogen with full covariance matrices. Best model for both cases was the EEV

Table 6. Results of GMM parameters of Cone index and some soil nutrients of a sandy loam soil

Parameter				
Cone index, CI	**Class/cluster**	**Proportion**	**Meanvector**	**Standard deviation**
	1	0.168	1.91	
	2	0.275	1.25	
	3	0.083	3.00	
	4	0.474	1.89	0.526*
Manganese, Mn	1	0.146	4.13	
	2	0.854	4.96	0.335*
Phosphorous, P	1	0.063	69.03	
	2	0.083	295.89	
	3	0.271	117.31	
	4	0.208	193.77	
	5	0.376	423.93	154.422*
Potassium, K	1	0.063	246.52	
	2	0.083	292.66	
	3	0.271	408.61	
	4	0.208	502.24	
	5	0.376	429.83	74.679*
Ferric, Fe	1	0.168	4.54	
	2	0.276	5.22	
	3	0.083	4.60	
	4	0.474	5.17	0.353*
Nitrate-Nitrogen	1	0.167	11.89	
	2	0.417	23.61	
	3	0.167	44.04	
	4	0.167	65.60	
	5	0.083	112.98	28.845*

** Standard deviation for all observations of measured variable*

Contour maps were generated by the interpolation method (Inverse Distance Weighting, IDW) as in Fig. 11. Although the plots between 1, 3 and 9 remained fallow during the last five years, CI values were high between 2.26 to 2.94 MPa due to the presence of tree roots and termite anthills that registered high values during measurement. CI values tended to decrease in an easterly direction to values as low as between 0.73 – 1.07 MPa. Generally, the CI values in all plots were below the critical value of 3 MPa on sandy loam soil as observed by [26].

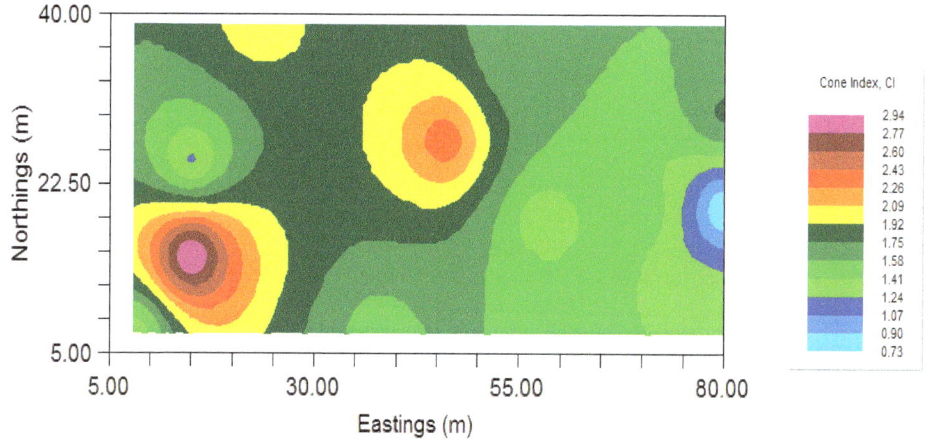

Fig. 11. IDW contour maps showing the distribution of the Cone Index CI in a *Eutric Leptosol* (Adapted from Lomeling [21])

Table 7. Statistical parameters of some soil nutrients and CI of a *Eutric Leptosol*

Variable	Mean	Variance	Skewness (Pearson)	Kurtosis (Pearson)	Kolmogorov-Smirnov Test	Chi-Test
CI	1.840	0.270	0.587	0.866	0.368*	0.348*
K	416.307	5576.883	-0.708	-0.132	0.443*	<0.0001
Mn	4.842	0.112	-1.173	1.139	0.008	<0.0001
P	260.021	23846.11	0.136	-1.465	0.121*	<0.0001
Nitrate-N	39.087	832.009	1.317	0.906	0.002	<0.0001
Fe	5.027	0124	-0.353	-0.721	0.912*	0.277*

* *significant at $p \leq 0.05$*

Table 7 shows some statistical parameters that were tested using the Kolmogorov-Smirnov (KS) and Chi-tests for unimodal distribution of CI and soil nutrients. The test using the (KS) showed that the values of the variables: CI, K, P and Fe were significant at $p \leq 0.05$ and that the data followed normal distribution, whereas Mn and Nitrate-N did not. Conversely, the Chi-test showed that all values except of the CI and Fe variables were not significant at $p \leq 0.05$ and so did not follow normal distribution. No reasons could be given for this contradiction in the fit-of-goodness of both test procedures. Despite these anomalies, it can be concluded that the KS test provided better fits for the spatial distribution of CI and soil nutrients.

3.2 Nutrient Distribution and Spatial Dependency

It was also reported that, soil nutrients generally tended to increase with the sand fraction whereas this was the reverse trend for silt and clay fractions. Sand particles have relatively higher friction coefficients than silt and clay particles, which would explain the increasing trends of CI with sand fraction [20].

From the dendrogram in Fig. 4 it may be seen that the first four variables (P, Mn, Fe and Nitrate-N) are clustered together at first level with strong spatial dependency at 8.6%, 9.6%, 11.5% and 12.6% respectively. Meanwhile, K was clustered to Nitrate-N and Mn at second level with strong spatial dependency at 15.5%. Results of our study showed that lower spatial dependency values implied higher similarities between the compared variables.

4. DISCUSSION

The results of PCA and cluster analysis showed that the spatial distribution of soil nutrients was most strongly correlated with soil CI within the different plots. The results further showed that CI was the most important major factor that affected the spatial distribution of the soil nutrients. These results are in agreement with those reported by [20] on *Eutric Leptosol*. Increasing CI in the different plots was a major aspect that influenced the spatial distribution of soil nutrients. With an increase in the CI, the available $N0_3$-N, P, K and Fe decreased correspondingly as shown by the negative Pearson´s correlation. For example, the amounts of P, $N0_3$-N, K and Fe in kg/ha were higher in plots 1, 3, 4, 5, 6 and 12 than in the rest plots. Similarly, CI and Mn was higher in plots 7, 8, 9, 10 and 11 than in the rest plots. The CI as first principal component F1 significantly influenced the spatial distribution of the soil nutrients.

Although the first three factors F1, F2 and F3 explained most of the variance, components F4 to F6 were also chosen to explain any interactions and loading effects of each individual variable. The idea was to assess, if there were any antagonistic or rather synergistic effects of any of the variables on the other. The study also tried to assess, if indeed anthropological influences had any significant roles on spatial distribution of some soil nutrients, or rather if spatial distribution was random and only dependent on lithogenic influences. Associating the respective variables to each component, (CI to F1, F2 to Mn, F3 to P, F4 to K, F5 to Fe and F6 to Nitrate-N) and incorporating geo-statistical parameters obtained from Gaussian model (Table 3), the short-range component (at 8.8 m) of F1 positively correlated with F2 (at range 98.76 m). Conversely, F1 negatively correlated with F5 (at range 1.95 m). Component F3 positively correlated with F4 with ranges at 8.5 and 16.93 m respectively.

The plots on the research and demonstration farm showed heterogeneous distribution and spatial variations of soil nutrients and cone index. The results also indicated that not all soil nutrients and cone index values were homogenously distributed in all plots. Plots in the same quadrant with similar predominating soil nutrients could practically be managed alike. It also implied that, similar plot management practices would be tailored on each quadrant within the context of precision farming. Kriged maps reported by [21] showed more CI, Nitrate-N and P on the left-hand side gradually decreasing in the easterly direction, whereas Mn and Fe tended to increase in the opposite direction. Implicitly, this did not suggest any negative correlations between CI and both Nitrate-N and P, on the contrary, high CI negatively correlated with both Nitrate-N and P as reported by [20].

The CI values were generally high (2.94 MPa) at the left-hand side of the farm and gradually decreased in an easterly direction to as low as 0.73 MPa at the extreme right edge of the farm. One reason for this is probably the differentiated tillage activities on the farm. There was hardly any tillage activity on the left hand side of the farm in the last 3 years as opposed to much soil tillage on the right hand side which must have significantly altered the structural and aggregative nature of the soil.

The interaction CI and Mn (or F1 and F2) found in the present study could be described as synergistic. Increase in CI would increase inter-particle contacts with correspondingly reduced pore space. This would tend to favor anaerobic soil conditions thereby increasing Mn availability. In the present study, we also found out that on average, increasing P concentration in soil did not accentuate K availability in soils. Although the *Eutric Leptosol* has a low organic carbon (0.72% by weight) and low $CaCO_3$ (0.2% weight), there appeared to be specific preference for Ca^{2+} as opposed to K^+ ions in soils in forming both in- and organic complexes with phosphate ions. Such mutually synergistic relationship would suggest that increasing concentration of P ions would enhance further adsorption to Ca^{2+} and so increasing the concentration of K^+. Conversely, a mutually antagonistic relationship was also found between P and Nitrate-N whereby there was specific and preferential adsorption of the phosphate than Nitrate ions [27]. Furthermore, our work showed that Nitrate-N was clustered with Mn. This mutually synergistic relationship would be the result of high water table and poor drainability of the *Eutric Leptosol* following the heavy rains between April and November. In the absence of cultivated crops (*that would utilize the Nitrate-N*), much of the Nitrate-N was stored within the rhizophere. Equally, the high CI values tended to reduce soil drainability thereby enhancing the development of anaerobic conditions which favored reduction processes and therefore Mn increase [20]. Based on these groups, the clustering of these nutrients in the studied soil could be ascribed to two different components: P and Fe that have a common lithogenic origin and the CI, K, Mn and Nitrate-N to have a common anthropogenic influence.

The K^+ is required for cells to maintain the osmotic balance and regulating stomatal opening and closure during photosynthesis [26,28]. It is also an essential co-factor for many enzymes. Similarly, P concentration in plant cells is closely associated with photosynthetic rate [29] and; therefore, decrease of P in leaves will reduce the plant growth. By implications, soils low in K^+ would not only reduce stomatal activity, but also low available P would lead to reduced photosynthesis and hence reduced plant growth rates. Analysis of the dendrogram confirmed that the CI is an independent variable that tended to influence spatial variations of K and Nitrate-N suggesting that both nutrients reduced with increased CI whereas Mn increased with increasing CI and *vice-versa.*

P showed a significant and positive relationship with Fe (r = 0.37, $p < 0.0001$). The significant relationship between these nutrients could be as a result of adsorption resulting from degradation of manure or organic material and mineralized Fe-oxides. Especially under anaerobic conditions influenced by both high soil compaction (high CI values) and high water table, this would favor reduction processes and enhancement of *Vivianite* $Fe_3(PO_4)_2.8H_2O$ [30]. Clustering and the mutually synergistic relationship between P and Fe was to be expected as much of the P ions ($H_2PO_4^-$) were easily adsorbed to Fe ions (Fe^{2+}) especially under poor drainable conditions (*conditioned by high CI values or high soil penetration resistance*) where Fe(III)-oxide was reduced to Fe^{2+}.

The set of 6 original soil properties as in Fig. 4 with low similarity measure at 0.054 (*truncation or cut-off threshold level*), created three classes that explained 75.67 % of the data variability. Increase of the cut-off level would not only imply increase in the number of classes but also represent percentage increase in similarities suggesting that comparison was better under a large number of classes. For example, increasing similarity at cut-off threshold level between 0.2542 and 0.4518 would increase the number of classes to 5 therefore enhancing better comparison of the variables.

In our study, the Euclidian distance was utilized as a basis for classifying the clusters that demonstrated both mutual antagonistic and synergistic relationships between the different principal components. However, the Euclidean distance, suffers from the so-called "*scaling effect*" [17] due to the inadvertent weighting of the variables in the analysis that can occur due to differences in magnitude among the measurement variables. For example, the measured concentration of two variables: P and Nitrate-N is soils with varying ranges. The measured amount of P in the studied soil varied between 61 and 511 kg/ha, whereas that of Nitrate-N between 11 and 116 kg/ha approximately one-fifth of P. A five-fold increase in the P concentration will have a greater effect on Euclidean distance than a similar increase in the Nitrate-N concentration. However, effect of variable scaling on the Euclidean distance by the different measured variables may be mitigated by standardizing the measurement variables, so that each variable has a mean of zero and a standard deviation of 1.

One of the major problems in hierarchical clustering is setting a threshold value or truncation level that would define similarity within a cluster. This value is often subjective [17]. By varying the truncation or cut-off level, similarity within a cluster may increase or decrease. This is conditioned by practical considerations as to whether or not one needs a high degree of similarity within a given cluster. If the similarity value is substantially larger, the random variation and dissimilarity within the data is probably negligible and *vice-versa.* Our study also showed that clustering of the different variables gave rise to correspondingly different number of classes, e.g. between F1 and F2 gave 3 cluster classes, whereas between F4 and F6 gave 5 cluster classes. Practically, smaller number of classes e.g. 2 to 3 would suggest lesser variability with a more homogenous spatial distribution than a larger number of cluster classes that would show more variability and heterogeneity.

5. CONCLUSION

The applications of PCA, HCA and GMM were used to grouping multi-nutrient and Cone Index data on sandy loam soil (*Eutric Leptosol*) from the University of Juba demonstration farm, in South Sudan. The first principal component F1, explained 31.36%, the second principal component F2, accounting for 25.29% and F3 19.06%. GMM analysis for the rest components F3 to F6 showed that these components representing individual variables had either antagonistic or synergistic relationships, which explained their spatial distribution and variations. The results of PCA made it possible for the initial six variables to be reduced to three factors representing 75.67% of the total variance. From hierarchical clustering, CI was observed to be clustered with K, Mn and Nitrate-N whereas P was clustered with Fe. It can be inferred from these results that, the spatial distribution of K, Mn and Nitrate-N was predominantly due to anthropological influences as conditioned by CI and less lithogenic as determined by nature of *Eutric Leptosol*.

The PCA is a standard tool in modern data analysis owing to its simplicity, non-parametric method for extracting relevant information from large and confusing data sets. The PC is a less complicated approach that can provide a roadmap for reducing complex dataset and as well disclose less discernible structures. However, PCA is not a statistical method from the viewpoint that there is no probability distribution specified for the observations. Therefore, it is important that only in combination with other pattern recognition tools does it best serve to explain underlying relationships between the different variables within a data set.

The PCA, HCA and GMM methods used herein have been found to be most useful in identifying the structure and inherent inter-linkages within the variables contained in the dataset. The approaches described in these methods rely heavily on mathematical algorithms whose results are presented in graphical form as clusters, scatter plots and dendrograms for easy patter recognition and interpretation. For two independent variables or components, a mixture of Gaussian probability density functions is best modeled under the assumption that the distribution is bi- or multimodal as it is poorly approximated using a single Gaussian.

In conclusion, the PCA, HCA and GMM helped reveal some underlying relationships between soil nutrients and soil parameter CI. They were shown to be useful methods for studying the spatial distribution and variations of soil nutrients and variables either due to intrinsic factors (*lithogenic*) or anthropological influences. More research on wider range of soil mechanical-physical as well as chemical parameters would provide more knowledge on their interactions and inter-linkages.

ACKNOWLEDGMENTS

This research was partly supported by the Rebuilding Higher Education in Agriculture (RHEA) Project funded by USAID (HED072-9742-SDN-11-01) at the College of Natural Resources and Environmental Studies (CNRES), University of Juba through the purchase of the geo-statistical software GS+™ Version 9. We equally thank the Norwegian funded project NORAD for the purchase of penetrologger and soil testing kit. Appreciation is also extended to Sebit Mathew Otwari and Yahya Mohammed Khater for conducting field trials, analyzing soil physical and chemical properties.

COMPETING INTERESTS

Authors have declared that no competing interests exist.

REFERENCES

1. Cleto MS de Moura, et al. Analysis and assessment of heavy metals in urban surface soils of Teresina, Piauí State, Brazil: a study based on multivariate analysis. ComunicataScientiae. 2010;1(2):120-127.
2. Abollino OM, et al. Heavy metals in agricultural soils from Piedmont, Italy. Distribution speciationandchemometricdata reatment.Chemosphere. 2002;49:545–557.
3. Sabrina T, et al. Earthworm populations and cast properties in the soils of oil palm plantations. Malaysian J. of Soil Sci. 2009;13:29-42.
4. Osei MK, et al. Genetic diversity of tomato germ-plasm in Ghana using morphological characters. Int. J. of Plant & Soil Sci.2014;3(3):220-231.
5. Reimberg MC, et al. Multivariate analysis of the effects of soil parameters and environ-mental factors on the flavonoid content of leaves of *Passiflora incarnate* L., Passifloraceae. Bras. J. of Pharma. 2008; 19(4):853-859.
6. Dayang SN, Fauziah CI. Soil factors influencing heavy metal concentrations in medicinal plants. Pertanika J. Trop. Agric. Sci. 2013;36(2):161-178.
7. Silva SA, Lima JSS. Multivariate analysis and geo-statistics of the fertility of a humicrhodic hapludox under coffee cultivation. R. Bras. Ci. Solo. 2011;36:467-474.
8. Bam EK, et al. Multivariate cluster analysis of some major and trace elements distribution in an unsaturated zone profile, Densu river basin, Ghana. African J. of Environ. Sci. and Tech. 2011;5(3):155-167.
9. Wander MM, Bollero GA. Soil quality assessment of tillage impacts in Illinois. Soil Sci. Soc. Am. J. 1999;63:961–971.
10. Ye RZ, Wright AL. Multivariate analysis of chemical and microbial properties in histosols as influenced by land-use types. Soil & Till. Res. 2010;110:94–100.

11. Allison VJ, et al. Changes in enzyme activities and soil microbial community composition along carbon and nutrient gradients at the Franz Josef chrono sequence, New Zealand. Soil Biol. Biochem. 2007;39:1170–1781.
12. Cookson WR, et al. Controls on soil nitrogen cycling and microbial community composition across land use and incubation temperature. Soil Biol. Biochem. 2007;39:744–756.
13. Smejkalova D, et al. Multivariate analysis of CPMAS ^{13}C-NMR spectra of soils and Humic matter as a tool to evaluate organic carbon quality in natural systems. European J. of Soil Sci. 2008;59:496–504.
14. Brereton RG. Appendices, in Chemometrics: Data Analysis for the Laboratory and Chemical Plant, John Wiley & Sons, Ltd, Chichester, UK; 2003. DOI: 10.1002/0470863242.app.
15. Davis JC. Statistics and data analysis in geology, 3rd ed. John Wiley and Sons, Inc. New York; 2002.
16. Everitt B, et al. Cluster Analysis, 5th ed., John Wiley & Sons, Ltd, Chichester, UK; 2011.
17. Lavine BK, Mirjankar N. Clustering and classification of analytical data. John Wiley & Sons Ltd, Chichester, UK; 2000.
18. Hu B, et al. Method for Extraction of Remote Sensing Information Based on Gaussian Mixture Model. Remote Sensing for Land & Res. 2012;24(4):41-47.
19. Clarkson D, et al. S+ Functional data analysis user guide, Springer, New York; 2005.
20. Lomeling D, Abakr AA. Variability of cone index on seedling emergence rate and growth establishment of cowpea in a sandy loam soil (*Eutric Leptosol*). Int. J. of Sci. Basic and Applied Res. 2014;14(1):34-48.
21. Lomeling D. Correlating the spatial distribution of some macro- and micronutrients with cone index in a sandy loam soil (*Eutric Leptosol*). Int. J. of Agri Science. 2014;4(2):89-101.
22. Johnson RA, Wichern DW. Applied multivariate statistical analysis, 5th ed. New Jersey, Prentice Hall; 2002.
23. Cambardella CA, et al. Field-scale variability of soil properties in central Iowa soils. Soil Sci. Soc. of Amer. J. 1994;58:1501-1511.
24. Sridevi M, et al. Isolation and characterization of phosphate- solubilizing bacterial species from different crop fields of Slem, Tamil Nadu, India. Int. J. ofNutri., Pharma., Neurol. Diseases. 2013;(3):1.
25. Gyaneshwar P, et al. Role of soil microorganisms in improving P nutrition of plants. Plant Soil. 2002;245:83–93.
26. Håkansson I, Lipiec J. A review of the usefulness of relative bulk density values in studies of structure and compaction. Soil Till.and Res. 2000;53:71-85.
27. Yeo A. Molecular biology of salt tolerance in the context of whole-plant physiology. J. Exp. Botany.1998;49(323):915-929.
28. Chow W, et al. Growth and photosynthetic responses of spinach to salinity: implications of K+ nutrition for salt tolerance. Funct. Plant Biol. 1990;17(5): 563 578.
29. Overlach S, et al. Phosphate translocator of isolated guard-cell chloroplasts from *Pisum sativum* L. transports glucose-6-phosphate. Plant Physiol. 1993;101(4): 1201-1207.
30. Schachtschabel, et al. Lehrbuch der Bodenkunde. 12th ed. Ferdinand Enke Verlag, Stuttgart Germany; 1989.

14

Effects of High Night Temperatures on Cotton Leaf Gas Exchange and ATP Levels at Flowering

Dimitra A. Loka[1*] and Derrick M. Oosterhuis[1]

[1]Department of Crop, Soil and Environmental Sciences, University of Arkansas, Altheimer Laboratory, Fayetteville, AR, USA.

Authors' contributions

This work was carried out in collaboration between both authors. Author DAL designed and executed the study, performed the statistical analysis and wrote the manuscript. Author DMO managed the corrections to the manuscript. Both authors read and approved the final manuscript.

Editor(s):
(1) Sławomir Borek, Department of Plant Physiology, Adam Mickiewicz University, Poland.
Reviewers:
(1) Inês Cechin, Department of Biological Sciences, Faculty of Science, Sao Paulo State University, Unesp, Brazil.
(2) Anonymous, Czech Republic.

ABSTRACT

Aims: To monitor the effects of high night temperatures on leaf photosynthesis and respiration, stomatal conductance and adenosine triphosphate (ATP) levels of cotton during its reproductive stage.

Study Design: A two-factor factorial, the two factors being temperature and time (weeks), with 40 replications in each of the temperature treatment.

Place and Duration of Study: Altheimer Laboratory, Department of Crop, Soil and Environmental Sciences, University of Arkansas, between September 2013 and June 2014.

Methodology: Growth chamber experiments were conducted using cotton (*Gossypium hisrsutum* L.) cultivar ST5288B2F with the treatments consisting of normal day/night temperatures (32/24°C) and high night temperatures (32/30°C) for two weeks at flowering. Measurements of leaf photosynthesis, respiration, stomatal conductance and ATP levels were conducted in the end of the first and the second week after imposition of stress.

Results: Leaf photosynthetic rates and stomatal conductance rates remained unaltered under higher night temperatures during both weeks of the experiment. In contrast, a significant increase in leaf respiration rates was observed at the end of the second week of the experiment with plants grown under conditions of high night temperatures increasing their respiration rates by 30%

**Corresponding author: E-mail: dloka@uark.edu;*

compared to those grown under normal temperatures. Conversely, leaf ATP levels were significantly decreased under conditions of elevated night temperatures.
Conclusion: It was concluded that higher than optimum temperatures during flowering had no significant effect on cotton leaf photosynthesis and stomatal conductance in contrast to leaf respiration and ATP levels that were significantly decreased.

Keywords: Cotton; photosynthesis; respiration; stomatal conductance; night temperatures; ATP levels.

1. INTRODUCTION

Heat stress constitutes a major threat to crop production worldwide and significant increases in global temperatures are anticipated by the end of the century [1]. Extensive research has been dedicated on the negative effects increased temperatures have on plant growth and development but interestingly enough, night temperatures are expected to increase faster and to a greater extent than day temperatures as a result of increased cloudiness and consequently decreases in heat radiant loss [2]. Cotton (*Gossypium hirsutum* L.) may be a perennial that originates from hot and arid areas but with its physiological optimum at 28°C [3] heat stress becomes a constraint more often than not during the growing season. Considerable research has been conducted on the effects of higher than optimum temperatures during daytime on cotton recording significant inhibition of photosynthetic rates [4-6] and increases in respiration [7] that are accompanied by reductions in pollination and fertilization [8] leading ultimately to compromised reproductive efficiency and decreased yields [9]. Significantly less attention has been given on the subject of elevated night temperatures and their effects on cotton physiology and growth. Previous research has reported that night temperatures between 28 and 42.5°C caused an in increase in the Q_{10} of respiration rates by 1.86 [10] while Reddy et al. [11] indicated that high night temperatures inhibit photosynthesis. However, in their experiments, day and night temperatures were raised proportionally to each other and therefore it was not possible to determine the effect of night temperature alone. Loka and Oosterhuis [12] on experiments with higher than optimum night temperatures during cotton's vegetative stage of growth observed significant increases in respiration rates. Consequently depletion in leaf carbohydrates content and significant reductions in leaf energy levels were reported [12] ultimately resulting in yield reduction [13].

The reproductive stage appears to be more heat susceptible to heat stress than the vegetative stage [14] and research in other crops has indicated that high night temperatures occurring during the reproductive stage result in significant yield reductions due to increases in male sterility and floral abscission [15,16], delays in flowering [17] and decreases in pollen viability and grain filling [18]. Variable responses have been reported on photosynthesis [19-22] while respiration have been shown to increase under conditions of high night temperature stress [18, 23,24].

As mentioned above, little or no attention has been given to the effects of increasing night temperatures during the reproductive stage on leaf gas exchange and energy levels. Since night temperatures are projected to increase faster and to a longer extent compared to day temperatures it is important that research should focus on their effects. It was hypothesized that high night temperatures will have a negative effect on cotton photosynthesis, respiration, and stomatal conductance that would result in significant reductions of energy levels. The objective of these studies was to evaluate and quantify the effect of high night temperatures on cotton photosynthesis, respiration, stomatal conductance and subsequent ATP content.

2. MATERIALS AND METHODS

Growth chamber studies were conducted and repeated at the Altheimer Laboratory, University of Arkansas. Cotton (*Gossypium hirsutum* L.) cultivar ST5288B2F was planted in 2L pots containing horticulture mix (SunGro Distribution Inc., Bellevue, WA). Pots were arranged in a growth chamber (Conviron PGW36) that was equipped with incandescent and fluorescent lamps and set for a 14 h photoperiod with a photosynthetic flux density (PPFD) of 800-850 μmol/m^{2}s, relative humidity of 60% and CO_2 concentration was 340-370 ppm. Half-strength Hoagland's nutrient solution was applied daily to all pots in order to maintain adequate nutrients and water. Plants were grown until first flower stage (approximately eight weeks after planting) under normal day/night temperatures of 32/24°C (maximum during the day, minimum during the night, respectively) simulating a normal diurnal variation. At the first flower stage, pots were divided in two groups and one group was transferred into a second growth chamber, with similar conditions of photon flux density, humidity and photoperiod as the first chamber, but with night temperatures raised to 30°C for 6 h at the start of the dark period (18:00-24:00) and a gradual decrease to 24°C, for an overall duration of two weeks, while the control plants remained under normal conditions (32/24°C). Measurements of photosynthesis, respiration, stomatal conductance and ATP levels were conducted at the end of the first and second week from a total of 40 plants (10 replications for each week-group). All measurements were taken with the plants inside the growth chambers.

2.1 Photosynthesis and Respiration Measurements

A Li-Cor Model 6200 portable photosynthesis system (LiCor Inc., Lincoln, NE) was used to determine photosynthetic and respiratory rates from the fourth uppermost main-stem fully expanded leaf (n=10). Photosynthesis measurements were taken at 13:00 at the end of the first and second week. Respiratory rates were taken at 22:00 at the end of the first and second week. The results were expressed as $\mu mol/m^2s$.

2.2 Stomatal Conductance Measurements

Stomatal conductance measurements were taken at the end of each week from the fourth uppermost main-stem fully expanded leaf from 11:00 until 12:00 using a Decagon SC-1 Porometer (Decagon Inc. Pullman, WA). Three measurements of various areas of the leaf were taken and then averaged and the results were expressed as $mmol/m^2s$.

2.3 ATP Content Measurements

ATP content was measured according to Loka and Oosterhuis [12]. Six leaf disks per subtending leaf were excised using a cork borer (d=1cm) at the end of the dark period. The leaf disks were placed immediately in tubes containing 10 ml TRIS solution and the tubes were put in a boiling waterbath for 10 min, for the extraction of ATP. The resulting aliquot was then stored in 1.5 ml microcentrifuge tubes at -80°C. The substrate-enzyme complex of luciferin-luciferase (ATP bioluminescent assay kit, Sigma Chemical Company, St. Louis, MO) that converts the chemical energy associated with ATP into light was used and the light produced (proportional to the ATP content of the sample) was determined with a 20/20n Luminometer (Turner Biosystems Inc. Sunnyvale, CA) and the help of a standard curve. The concentrations of the standards were 0, 0.00001, 0.0001, 0.001, 0.01, 0.04 µg ATP/ml.

2.4 Statistical Analysis

The experimental design was a two-factor factorial, the two factors being temperature and time (weeks), with 40 replications in each of the temperature treatment. The trends of the two growth chamber studies were similar and the data were pooled. No significant interaction was observed between the two factors across all measurements and the effects of temperature were analyzed for each week using a Student's *t*-test. Means were considered significantly different at $P \leq 0.05$.

3. RESULTS

3.1 Photosynthesis and Respiration

Photosynthetic rates were not significantly affected by high night temperatures (Fig. 1). No significant interaction was observed between the main factors (time and temperature, $P = 0.89$ for photosynthesis and $P = 0.45$ for respiration) and the effect of temperature was analyzed using Student's *t*-test for each week. Plants grown under high night temperatures increased their photosynthetic rates by 5% and 6.5% the first and the second week respectively, after initiation of treatment compared to the control. In contrast to photosynthesis, respiration rates significantly increased at the end of the second week of high night temperatures with heat-stressed plants having a 30% higher respiration rates compared to the control (Fig. 2).

3.2 Stomatal Conductance

Stomatal conductance measurements indicated that high night temperatures had no significant effect on the function of stomates. Again, no significant interaction was observed between the main factors (time and temperature, $P = 0.49$) and the effect of temperature was analyzed using Student's *t*-test for each week. A non-statistically significant increase of 6% and 2.5% was recorded the first and the second week respectively (Fig. 3), compared to the control.

3.3 ATP Levels

Contrary to both photosynthesis and stomatal conductance functions, leaf ATP levels were considerably affected by high night temperatures (Fig. 4). No significant interaction was observed between the main factors (time and temperature, P = 0.35) and the effect of temperature was analyzed using Student's t-test for each week. Heat-stressed plants had 14% and 18% lower ATP levels at the end of the first and the second week respectively, after initiation of stress compared to control.

4. DISCUSSION

Temperature plays a major role regulating plant growth and development and ultimately yield since it determines the function of several physiological and metabolic processes. Photosynthesis, a function closely related with crop productivity is considered to be directly affected by temperature, being one of the most heat-sensitive processes in plants. Leaf photosynthetic rates are known to decrease under conditions of high day temperature stress

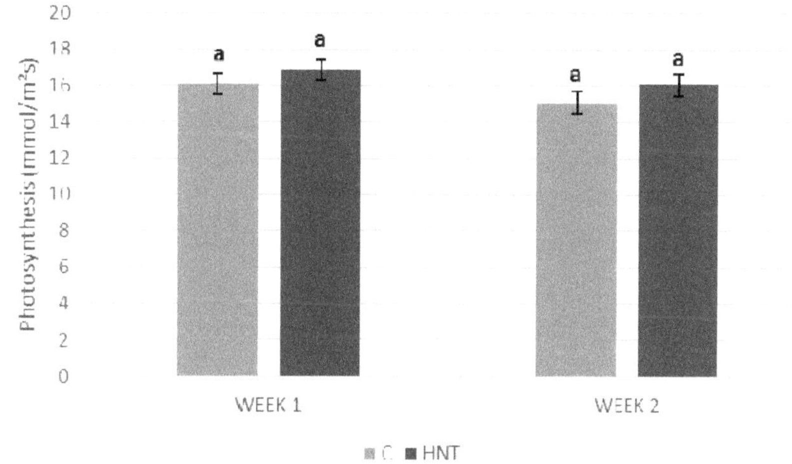

Fig. 1. Effect of high night temperatures on leaf photosynthesis rates one and two weeks after the night temperatures were raised

Pairs of columns of columns within each time interval with the same letter are not significantly different (P ≤ 0.05). C = 24°C night temperature, HNT = 30°C night temperature. Error bars indicate ±1 standard error

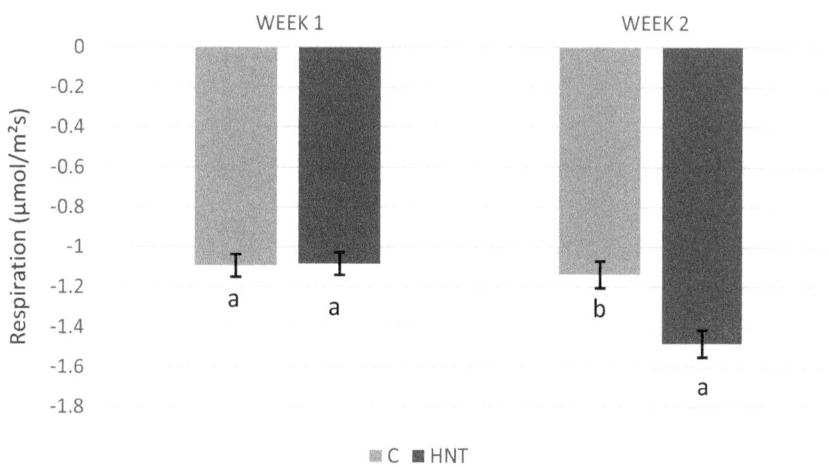

Fig. 2. Effect of high night temperatures on leaf respiration rates one and two weeks after the night temperatures were raised

Pairs of columns of columns within each time interval with the same letter are not significantly different (P ≤ 0.05). C = 24°C night temperature, HNT = 30°C night temperature. Error bars indicate ±1 standard error

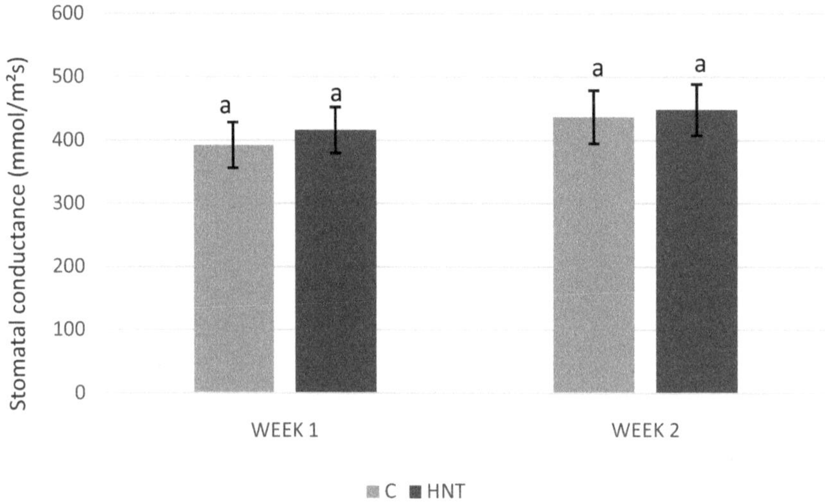

Fig. 3. Effect of high night temperatures on leaf stomatal conductance rates one and two weeks after the night temperatures were raised

Pairs of columns of columns within each time interval with the same letter are not significantly different (P ≤ 0.05). C = 24°C night temperature, HNT = 30°C night temperature. Error bars indicate ±1 standard error

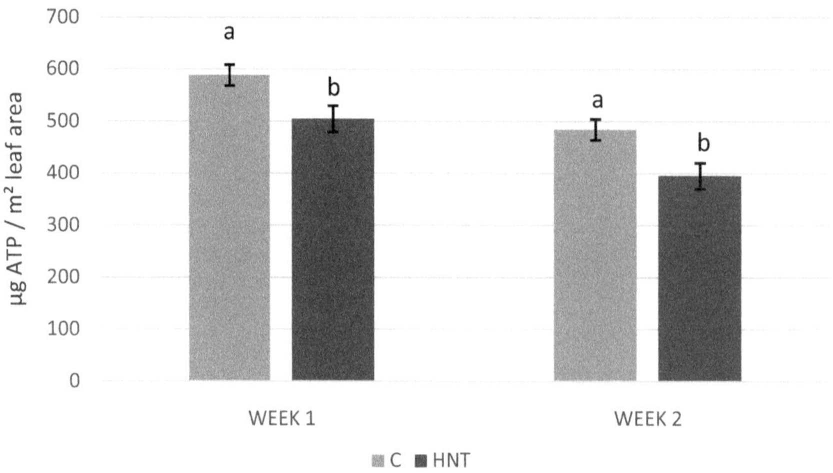

Fig. 4. Effect of high night temperatures on leaf ATP content one and two weeks after the night temperatures were raised

Pairs of columns of columns within each time interval with the same letter are not significantly different (P ≤ 0.05). C = 24°C night temperature, HNT = 30°C night temperature. Error bars indicate ±1 standard error

[4,6,25]; however, controversial reports exist on the photosynthetic response under high night temperature stress. Kanno et al. [22] and Turnbull et al. [19] with experiments on rice (*Oryza* sativa) and cottonwood (*Populus deltoids*) respectively, indicated that leaf photosynthetic rates were increased under higher than optimum night temperatures, while the opposite was reported by Prasad et al. [26] and Djanaguiraman et al. [24] on wheat (*Triticum aestivum*) and soybean (*Glycine max*) respectively. In our study leaf photosynthetic rates remained unaffected from the high night temperature regime during the first and the second week of our experiment which is in accordance with Mohammed and Tarpley [18] and Prasad et al. [26], where no change in the leaf photosynthetic rates of rice and sorghum (*Sorghum bicolor*) respectively, was observed. In studies with cotton, Reddy et al. [11] reported that photosynthesis decreased after heat stress;

however, in their experiment day and night temperatures were increased.

Carbon dioxide supply for photosynthesis is dependent on stomatal function and in our study higher than optimum night temperatures had no significant effect on leaf stomatal function. Contrary to our results, Prasad et al. [26] and Djanaguiraman et al. [24] reported significant decreases in leaf stomatal conductance rates under conditions of high night temperature stress which were accordingly to their photosynthetic rates. They suggested that this could be due to indirect effects of high night temperatures on leaf senescence.

Respiration, is another key metabolic process controlling crop growth. Dark respiration (in contrast to photorespiration and photosynthesis) occurs during day and night with the energy being produced divided between growth and maintenance components [27]. Similarly to photosynthesis, respiration is greatly controlled by temperature with Salvucci and Crafts-Brandner [10] reporting high night temperatures between 28 and 42.5°C causing an increase in the Q_{10} of respiration rates by 1.86 in cotton plants. As our results have shown, high night temperatures of 30C increased leaf respiration rates during the second week of our experiments but not the first. Loka and Oosterhuis [12] also reported significant increases in respiration rates under high night temperature conditions during cotton's vegetative stage indicating that reproductive and vegetative stage are equally sensitive to high night temperatures. Research in other crops such as sorghum [28], soybean [24,29] and rice [18] also noticed that higher than optimum night temperatures resulted in increasing rates of respiration.

The significantly lower leaf ATP levels observed in plants grown at higher than optimal night temperatures compared to the levels of those grown at normal temperatures are attributed, we assume, to the increased respiration rates that resulted in significant decreases in leaf ATP levels. Similar to our results, Djanaguiraman et al. [24] reported significant decreases in leaf ATP levels of soybean plants grown under high night temperatures conditions and they speculated that this was due to increases in thylakoid membrane's permeability which resulted in leakage of protons. In support of our assumption, Loka and Oosterhuis [12] reported substantially lower levels of leaf ATP in cotton under increased night temperatures during cotton's vegetative stage. Furthermore, Lawrence and Holaday [30] observed lower ATP values in expanding cotton leaves for plants grown and 30/28C compared to those for plants grown at 30/19°C leading us to assume that leaf ATP levels are susceptible to increased night temperatures at all growth stages of cotton.

5. CONCLUSION

In conclusion, contrary to our expectations and hypothesis, high night temperatures had no significant effect on leaf photosynthesis and stomatal conductance with plants grown at higher temperatures having almost similar photosynthetic and stomatal conductance rates with those grown under normal conditions. However, leaf respiration rates were significantly increased under high night temperatures which consequently resulted in considerable decreases in leaf ATP pools leading us to conclude that both functions, respiration and ATP production during cotton's reproductive stage are as sensitive, to higher than optimum night temperatures, as during its vegetative stage.

COMPETING INTERESTS

Authors have declared that no competing interests exist.

REFERENCES

1. IPCC (Intergovernmontal Panel on Climate Change). Climate Change 2007: Impacts, Adaptation and Vulnerability. In: Contribution of Working Group II to Fourth Assesment Report of the Intergovernmental Panel on Climate Change. M.L. Parry, O.F. Cnaziani, J.P. Palutikof, P. J. van der Linden, and C.E. Hanson (eds.) Cambridge University Press, Cambridge, UK; 2007.
2. Alward RD, Detling JK, Milchunas DG. Grassland vegetation changes and nocturnal global warming. Science. 1999; 283(5399):229-231.5399.
3. Conaty WC, Burke JJ, Mahan JR, Neilsen JE, Sutton BG. Determining the optimum plant temperature of cotton physiology and yield to improve plant-based irrigation scheduling. Crop Sci. 2012;52(4):1828-1836.
4. Bibi AC, Oosterhuis DM, Gonias EG. Photosynthesis, quantum yield of photosystem II and membrane leakage as

affected by high temperatures in cotton genotypes. J. Cotton Sci. 2008;12(2):150–159.

5. Burke JJ, Hatfield JL, Mahan JR. Crop specific thermal kinetic windows in relation to wheat and cotton biomass production. Agron. J. 1988;80(4):553–556.
6. Snider JL, Oosterhuis DM, Kawakami EM. Diurnal pollen tube growth rate is slowed by high temperature in field-grown *Gossypium hirsutum* pistils. J. Plant Physiol. 2011;168(5):441-448.
7. Krieg DR. Feedback control and stress effects on photosynthesis. In: Proceedings of the Beltwide Cotton Conference. Natl. Cotton Counc. Am., Memphis, TN. 1986; 227–243.
8. Snider JL, Oosterhuis DM, Skulman BW, Kawakami EM. Heat stress induced limitations to reproductive success in *Gossypium hirsutum*. Physiol. Plant. 2009; 137(2):125–138.
9. Wahid A, Gelani S, Ashraf M, Foolad MR. Heat tolerance in plants: An overview. Environ. Exp. Bot. 2007;61(3):199-223
10. Salvucci ME, Crafts-Brandner SJ. Inhibition of photosynthesis by heat stress. The activation state of rubisco as a limiting factor in photosynthesis. Physiol. Plant. 2004;120(2):179–186.
11. Reddy KR, Hodges HF, McKinion JM. Modeling temperature effects on cotton internode and leaf growth. Crop Sci. 1997; 37(2):503–509.
12. Loka DA, Oosterhuis DM. Effect of high night temperatures on cotton respiration, ATP levels and carbohydrate content. Environ. Exp. Bot. 2010; 68(3):258-263.
13. Echer FR, Oosterhuis DM, Loka DA, Rosolem CA. High night temperatures during the floral bud stage increase the abscission of reproductive structures in cotton. J. Agron. Crop Sci. 2014; 200(3):191-198.
14. Hall AE. Breeding for heat tolerance. Plant Breed. Rev. 1992;10:129-168.
15. Warrag MOA, Hall AE. Reproductive responses of cowpea to heat stress. II. Responses to night air temperature. Field Crops Res. 1984; 8:17-33.
16. Guinn G. Abscission of cotton floral buds as influenced by factors affecting photosynthesis and respiration. Crop Sci. 1974;14(2):291-293.
17. Zinn KE, Tunc-Ozdemir M, Harper JF. Temperature stress and plant sexual reproduction: Uncovering the weakest links. J. Exp. Bot. 2010; 61(7):1959-1968.
18. Mohammed AR, Tarpley L. Impact of high nighttime temperature on respiration, membrane stability, antioxidant capacity and yield of rice plants. Crop Sci. 2009; 49(1):313-322.
19. Turnbull MH, Murthy R, Griffin KL. The relative impacts of daytime and night-time warming on photosynthetic capacity in *Populus deltoides*. Plant Cell Environ. 2002; 25(12): 1729-1737.
20. Gibson LR, Mullen RE. Influence of day and night temperature on soybean seed yield. Crop Sci. 1996;36(1):98-104.
21. Frantz MJ, Cometti NN, Bugbee B. Night temperature has minimal effect on respiration and growth of rapidly growing plants. Ann. Bot. 2004;94(1):155–166.
22. Kanno K, Mae T, Makino A. High night temperature stimulates photosynthesis, biomass production and growth during the vegetative stage of rice plants. Soil Sci. Plant Nutr. 2009;55(1):124-131.
23. Manunta P, Kirkham MB. Respiration and growth of sorghum and sunflower under predicted increased night temperatures. J. Agron. Crop Sci. 1996;176(4):267–274.
24. Djanaguiraman M, Prasad PVV, Schapaugh WT. High day- or nighttime temperature alters leaf assimilation, reproductive success and phosphatidic acid of pollen grain in soybean [*Glycine max* (L.) Merr.]. Crop Sci. 2013; 53(4):1594-1604.
25. Crafts-Brandner SJ, Salvucci ME. Rubisco activase constrains the photosynthetic potential of leaves at high temperature and CO_2. Proc. Natl. Acad. Sci. USA. 2000; 97(25):13430-13435.
26. Prasad PVV, Pisipati SR, Ristic Z, Bukovnik U, Fritz AK. Impact of nighttime temperature on physiology and growth of spring wheat. Cop Sci. 2008;48(5):2372-2380.
27. Amthor JS. The McCree-de Witt –Penning de Vries – Thornley respiration paradigms: 30 years later. Ann. Bot. 2000;86(1):1-20.
28. Prasad PVV, Djanaguiraman M. High night temperature decreases leaf photosynthesis and pollen function in grain

sorghum. Funct. Plant Biol. 2011;38(12):993-1003.

29. Zheng SH, Nakamoto H, Yoshikawa K, Furuya T, Fukuyama M. Influence of high night temperature on flowering and pod setting in soybean. Plant Prod. Sci. 2002; 5(4):215–218.

30. Lawrence C, Holaday AS. Effects of mild night chilling on respiration of expanding cotton leaves. Plant Sci. 2000;157(2):233–244.

15

Evaluating the Impacts of Land Degradation on the Quality of Soils and Their Variations between Different Clusters in Mosiro Irrigation Scheme, Narok County, Kenya

Edward Mare Muya[1*], Violet Kirigua[2], Victor Wasike[2], Leonard Simiyu Nafuma[3], Esther Macharia[3], Elias Gitonga Thuranira[1], Peter Onyango Owenga[1], Haron Goro[4], Benard Kariuki Waruru[1], Alla Lusi Chek[1], Jeremiah Kimigo[1] and Peter Mbindio Maingi[1]

[1]Kenya Agricultural and Livestock Research Organization (KARLO), Food Crop Research Institute, P.O.Box 14733, Nairobi, Kenya.
[2]Kenya Agricultural and Livestock Research Headquarters, P.O.Box 57811-00800, Nairobi.
[3]KARLO-Njoro, P.O. Njoro, Kenya.
[4]Small Scale Horticultural Development Project, Ministry of Agriculture, Livestock and Fisheries, P.O.Box 30028-00200, Nairobi, Kenya.

Authors' contributions

This work was carried out in collaboration between all authors. Authors EMM and ALC designed the study, came up with the research protocol, coordinated the data collection processes and wrote the first draft of the manuscript. Authors VK, PMM, BKW and HG described and mapped out the study area, evaluated and rated different custers and assisted in writing the methodology. Authors EGT, JK and POO reviewed the experimental design, tabulated and synthesized the data and carried out statistical analysis. Authors VW, LSN and EM were responsible for quality control to ensure the flow of research materials from the introduction to conclusions and tallied the whole all the paper contents with what was expressed in the summary to make sure that there was consistency in the paper. All authors read and approved the final manuscript.

Editor(s):
(1) Daniele De Wrachien, State University of Milan, Italy.
Reviewers:
(1) Anonymous, USA.
(2) Anonymous, Tunisia.
(3) Jayath P. Kirthisinghe, Department of Crop Science, University of Peradeniya, Sri Lanka.

**Corresponding author: E-mail: edwardmuya@yahoo.com;*

ABSTRACT

Aims: To study the impact of land degradation on soil quality
Study Design: Comparative analysis of soil quality indicators in year 2002 and 2013 in Mosiro Irrigation Scheme.
Methodology: Changes in soil indicators determined in 2002 and 2013 were analyzed using ANOVA at 95% confidence level through Genstat Computer Software. The same indicators were applied to identify and characterize different clusters whose degree of variations were analysed using the same statistical method.
Results: The results showed that all the soil quality attributes changed in the range of 21.4 and 79.1%. The greatest change was recorded in potassium which decreased by 79.1%, followed by phosphorous (60% decrease).
The increase in sodium by 47% had a negative implication in terms of its increased potential to cause soil structural deterioration, while the increase of soil pH from 6.74 to 8.18 implied increased tendency of soil to fix most of the nutrients rendering them unavailable to plants. The soil organic carbon, nitrogen, phosphorous and potassium decreased by over 20%, which is much higher than the permissible threshold of 5% in ten years. The variations in soil characteristics between the five clusters identified were found to be significant for: cation exchange capacity ($P=0.002$), magnesium ($P=0.003$) and bulk density ($P=0.01$). There was no significant difference for calcium ($P=0.147$). All the textural characteristics of soil vary significantly with the highest being clay content ($P=0.001$), followed by silt ($P=0.0018$) and sand ($P=0.008$). For the micro-nutrients, the variation of manganese ($P<0.001$), zinc ($P=0.003$), copper ($P=0.008$) and iron ($P=0.031$) between different clusters was found to be significant.
Conclusion: Because the variations between different clusters in terms of both physical and chemical soil quality attributes was found to be significant except for calcium, each cluster must have different irrigation schedules and fertilizer inputs (except for calcium). This should form an important consideration in designing the irrigation water supply scheduling for all the clusters. Similarly, the prescription of the quantity of the micro-nutrients to apply should be clusters-specific. This is because the blanket recommendations across the clusters, where variations in terms of the levels of micro-nutrients are significant, would result into lower fertilizer use efficiency.

Keywords: Land degradation; soil quality, clusters and productivity.

1. INTRODUCTION

Delineation of soils into clusters which vary in productivity to an extent that they would respond differently to management inputs is a preamble to precision agriculture, aimed at improved soil quality and water use efficiency [1]. Soil is a key natural resource and its quality is as a result of integrated effects of all soil forming factors and management [2]. As a result of human activities on land over time, the capacity of a given soil to produce goods and services is often decreased at much more rate than the formation of topsoil [3]. A cluster of soils or production systems are said to be biophysically sustainable if the compounded sufficiency of land quality attributes (productivity index) does not deteriorate over a realistic time horizon [4]. Indexing and monitoring soil productivity under crop and soil specific production system is the means to judge whether or not the system is sustainable in biophysical sense [5]. According to Muchangi et al. [6], agricultural production from most irrigation schemes in Kenya has either reduced significantly over time or is far much below the level expected per unit of water supplied, and with advancement of agriculture under irrigation, soils are being degraded, thereby causing tremendous decline in their quality and productivity. Maintaining soil quality at a desirable level is a complex issue due to involvement of the interactive actions of the soil forming factors on the parent materials to form soil properties that determine its quality and productivity [7,8]. The results of these interactions are the vertical, spatial and temporal variations in soil properties or soil quality indicators within and between clusters of soils under different landforms. The assessment and evaluation of soil quality indicators and their variations are required, not only to determine the type and quantity of agricultural inputs to apply, but also soil and water management strategies for sustained ecosystem services under irrigated agriculture. In this regards, there is an urgent need to adopt appropriate soil and plant management practices to reduce land degradation and maintain soil quality at a desirable level, based on the evaluation of homogeneity and heterogeneity of the delineated clusters of soils under different landforms or slopes [9]. Low irrigation and fertilizer use efficiency as well as declining quality and productivity of soil in most of the schemes are exacerbated by the blanket fertilizer recommendations that ignore natural and biophysical diversity of production systems [10]. For example, fertilizer recommendations in Kenya by Kenya Agricultural Research Institute were based on trials without detailed characterization and clustering of production systems, based on relevant soil quality indicators [11]. Soil quality assessment, indexing and clustering of production systems under different landforms and slopes is a new paradigm in soil science research that is applied in vertical, spatial and temporal evaluation of land degradation, aggregation and management effects on soil quality and productivity. The degree of land aggregation that caters for soil structural formation and stabilization are influenced by the factors of soil formation including geology (parent material), climate, topography, time and human activities. The degree of variations in the resulting soil properties depends on the scale. At farm level, the effects of climate and geology are generally insignificant, while the effects of topography and human activities on soils under different clusters over specified time becomes an important issue to examine at site specific scale. In this context, the objective of this research was to examine soil conditions within different topographical facets to evaluate the homogeneity of soils and the impacts of land degradation on their quality as a basis of formulating the cluster-specific interventions for improved nutrient and water use efficiency in Mosiro Irrigation Scheme. The homogeneity of different clusters of biophysical domains was assessed, based on the physical parameters and chemical characteristics which influence the availability of water and plant nutrients, while the impacts of land degradation on soil productivity was examined by indexing the dynamic and management dependant soil quality indicators.

2. MATERIALS AND METHODS

Mosiro Irrigation Scheme is situated South-West of Nairobi in Narok County, about 12 km South of Mosiro trading centre, and about 82 km from Ntulele Market Centre. The intersection of the grid line 36° 04' E and 1° 28' S marks approximately the centre of the scheme, at an elevation of approximately 1265 m above the sea level. The area lying to the North of the scheme is the main source of sediment, causing increased land degradation and siltation of irrigation channels [12]. Therefore, water use efficiency and biophysical sustainability of the scheme depends on the extent to which the excess flows are checked and soil aggregate stabilization is maintained [12]. The Physiography of the scheme comprises two major landforms, namely flat to gently undulating erosion plain and undulating old river alluvial plain. The soils of erosion plain are developed on volcanic tuff while those of river alluvial plain are derived from alluvium and volcanic ash mixtures respectively [13]. The study area was divided into five sampling areas, with slopes of 0-0.5%; 0.6-1.5%; 1.6-2.5%; 2.6-4.0% and 4.1-4.5%. Systematic soil investigation and mapping was done on transects across all the sampling areas from the upper erosion plain into the bottomlands of the river alluvial plain in the same area soil survey and mapping were carried in the year 2002. Clustering of the production systems was done, based on the visual assessment of observable soil parameters (soil colour, depth, texture, consistence and structure) across the five sampling areas. In each of the five sampling area, five representative soil profiles were identified for detailed description of vertical and spatial characteristics of soils. For the analysis of the impacts of land degradation on soil quality and productivity, composite soil samples were collected using a river auger at the depth of 0-20 cm around each of the representative soil profiles. The soil quality indicators (Table 1) selected for this purpose are those whose quantitative relationships with maize yield have been developed by Aune and Lal [14].

For the assessment of homogeneity of different clusters, calcium (Ca), magnesium (Mg), bulk density (BD) and particle size distribution (texture) were used. They were selected because they form an important measure of soil productivity as is explained by the merits indicated in Table 2 [15].

Prior to laboratory analysis of the selected soil quality attributes, the soil samples were air dried and sieved through a 2-mm sieve. Soil pH was measured in 1:2.5 soil to water mixture, using the relevant electrodes according to Hinga et al. [16]. Organic carbon was oxidized with concentrated H_2SO_4 and K_2CrO_7 and determined calorimetrically. Total N was determined using the method provided by Okaleb et al. [17]. Cation exchange capacity (CEC) and exchangeable cations were extracted using 1N ammonium acetate at pH 7.0, followed by flame photometry for the determination Na, K, Mg and Ca, using flow analyzer [17]. Soil texture was determined using hydrometer method [18]. Indexing of soil productivity was done, using semi-quantitative land evaluation methods [19]. In this case, a range of numerical values of the selected soil quality indicators were rated and assigned fractions in percentage, being guided by the critical limits of the indicators. The critical limit of an indicator is defined as the numerical value of the soil property where crop yield is 80% of the maximum yield [14].

Table 1. Indicators for assessing soil productivity

Soil quality indicators	Description
Soil pH	Controls several factors that influence soil's functions, toxicity of soil's internal environment and availability of micro-nutrients.
Soil organic carbon	Influences both nutrient availability and soil aggregate formation.
Nitrogen, phosphorous and potassium	They are macronutrients that determine crop growth.

Table 2. Indicators for evaluating the homogeneity of different clusters

Soil quality indicators	Merits for selection
Calcium and magnesium	The levels of these elements and their ratios determine the availability of potassium (one of the macro-nutrients).
Bulk density	It is one of the most important soil structure attributes that controls the circulation of air, water and nutrients between the soil media and plants, hence a principle determinant of soil health.
Particle size distribution	Determines the retention and availability of soil moisture to plants.

Source: Amacher et al. [15]

Productivity index (PI) was determined using parametric methods of land suitability assessment provided by Driessen and Konijn [19]. This method involves: assigning ranges of numerical values and percentage fractions to each soil property selected as key soil quality indicators, ranking (Table 3) and combining all the single factor valuations in one mathematical equation that produces a numerical expression of the system performance or a relative index of performance (compounding) as follows:

$$PI=(SQ1/100) \times (SQ2/100) \times (SQ3/100) \times (SQn/100)$$

Where:

PI=Productivity index in % and SQ1, SQ2, SQ3, SQn are percentage ratings of soil quality indicator number 1, 2, and number n.

To assess the homogeneity of the identified clusters the variations in the selected soil quality attributes and PI within and between the clusters were evaluated by subjecting the data obtained from laboratory determinations (in the 2002 and 2012) to analysis of variance (ANOVA) at 95% confidence level where those soil quality indicators with significant levels were separated using Genstat Computer Software.

3. RESULTS AND DISCUSSIONS

3.1 Clusters Identified and Their Characteristics

Five clusters were identified and described as shown in Table 4.

3.2 The Impacts of land Degradation on Soil Quality and Productivity

All the soil quality attributes changed in the range of 21.4 and 79.1% between the year 2002 and 2012 (Fig. 1). According to Arshad and Martin [20], changes in the soil attributes could be used as measure of the impacts of land degradation on soil quality. The increase in soil pH from 6.74 to 8.18 implied an increased tendency of soils to fix phosphorous and render it unavailable to plants [21]. At soil pH above 8.0, the increased alkalinity triggers the release of aluminum, manganese and molybdenum to a toxic level that impairs the root uptake of micro-nutrients such as copper, zinc and iron [22]. Soil organic carbon, nitrogen, phosphorous and potassium decreased by over 20% which is much higher than the permissible threshold decrease of 5% in ten years [19,23]. The greatest percentage change was recorded in potassium which decreased by 79.1%, followed closely by phosphorous (60% decrease).

The increase in ESP by over 47% had a negative implication in terms of increased deterioration of soil structure by the increased accumulation of sodium [13].

3.3 Characteristics and Homogeneity of Different Clusters

The mean values of soil quality indicators for different clusters are presented in Table 5. The soil characteristics with the highest variation between the clusters is the clay (P<0.003) followed by CEC (P=0.002). Significant variations occur also in other characteristics, except calcium and sodium.

3.4 Variations in Soil Textural Characteristics and Their Management Implications

As is indicated in Fig. 2, the textural characteristics of soils, namely sand, silt and clay content vary significantly between different clusters, and this has an important implication in irrigation water management. Muya [24] showed significant relationships between these attributes and available soil moisture holding capacity, which may be applied in designing the irrigation water scheduling for different clusters. For example, cluster five has the highest sand content, implying the need for lighter and more frequent irrigation, while cluster one has the highest clay content, meaning more and less frequent water application than any other clusters.

3.5 Variations in Soil Micro-nutrients and Their Management Implications

The variations of the micro-nutrients between different clusters were found to be significant for all the micro-nutrients except sodium (Table 6). The variation of manganese (P<0.001), zinc (P=0.003), copper (P=0.008) and iron (P=0.031) between different clusters was found to be significant. This had an important implication on the prescription of the quantity of these micro-nutrients for different clusters during implementation of the envisaged management strategies. According to White and Zasoski [26], precision agriculture, aimed at improved nutrient

Table 3. Ratings of soil quality indicators

Soil quality indicator	Ranges of numerical values	Assigned values in %	Ratings	Remarks
Soil pH	4.8-5.5 or 5.6-6.8 or	100	1	80% of the maximum yield of
	4.8-5.5 or 6.9-7.5	80	2	maize obtained from pH of
	4.0-4.7 or 7.6-8.7	60	3	5.1 (Aune and Lal, 1997) [14]
	3.5-4.5 or 8.7-10.0	40	4	
	<3.5 or >10.0	20	5	
Exchangeable sodium percentage	<2.0	100	1	The permissible environmental threshold is 6 while maize yield is 80% (Waruru et al., 2002).[13]
	2.1-10.0	80	2	
	10.1-20.0	60	3	
	20.1-35.0	40	4	
	>35.0	20	5	
Bulk density (g/cc)	<1.2	100	1	Bulk density changes according to the degree of erosion and values of 1.0-1.4 gave sufficiency of 100%, (Pierce et al., 1983) [25]
	1.3-2-1.5	100	2	
	>1.5	75	3	
Potassium (m.e.%)	>0.5	100	1	80% of the maximum yield obtained by the value 0.7 (Aune and Lal, 1997) [14]
	0-2-0.5	80	2	
	0.1-0.2	60	3	
	<0.1	40	4	
Phosphorous (ppm)	>60	100	1	7.6 ppm gave 80% of the maximum yield of maize (Aune and Lal, 1997) [14]
	21-60	90	2	
	10-20	80	3	
	<20	70	4	

use efficiency should be based on the understanding of the variations in the levels of micro-nutrients as a basis of prescribing the soil and site specific types and quantity of micro-nutrients required as is normally done for macro-nutrients.

Table 4. Identified clusters and their characteristics

Clusters		Description
Cluster no	**Physiographic unit**	
C1	Upper level structural plains	Very deep clay, in places, has crusting, causing surface water stagnation, especially, on bare ground, showing high susceptibility to disruptive external forces including animal trapping and raindrop impacts.
C2	Lower level structural plains	Shallow to moderately deep sandy clay loam to clay, in places gravelly, pale yellow to reddish brown clay over murrum.
C3	Old alluvial plain	Moderately structured and deep clay loam to clay with relatively high carbonate concentration, in places shallow and highly calcareous
C4	Highly degraded, gently sloping structural plains	Dominantly shallow sandy loam to loam soils, on relatively steep slopes towards the low-lying area with low water uptake and retention capacity, being extremely calcareous, saline, sodic and highly degraded with relatively low productive capacity.
C5	On alluvial flood plains	Extremely deep stratified sandy clay loam to loam.

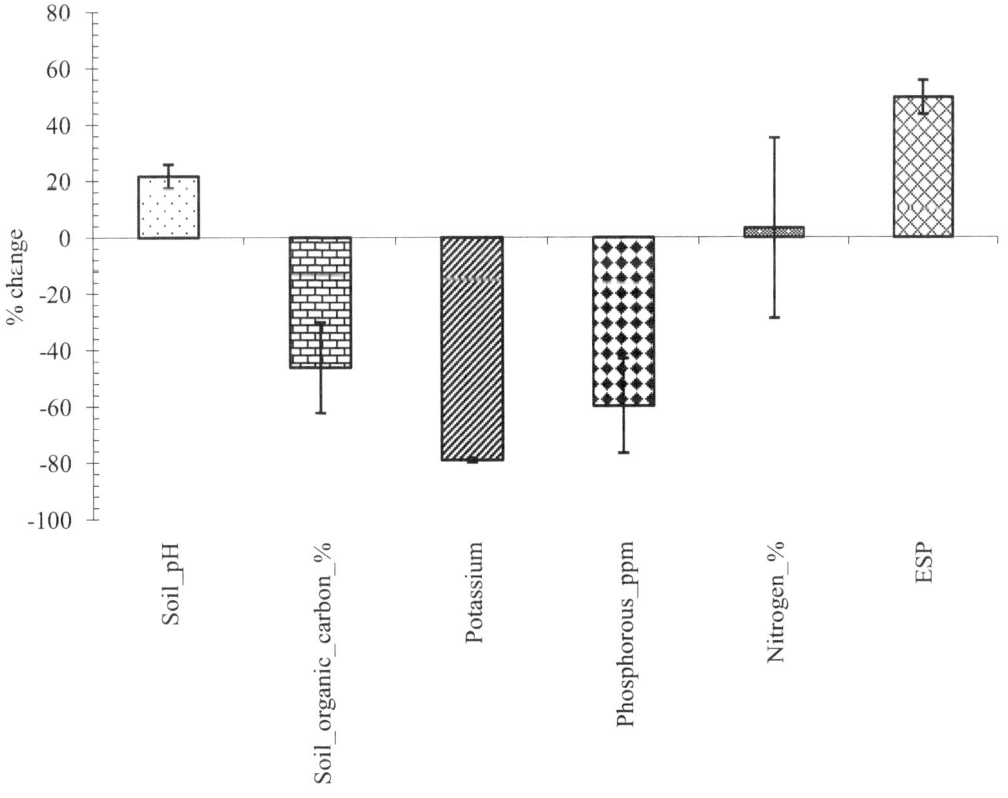

Table 5. Physical and chemical characteristics of different clusters

Clusters	Physical and chemical characteristics of soils						
	Bulk density	Textural characteristics			Ca me%	Mg me%	CEC me%
	(g/cc)	% Sand	% Silt	% Clay	52.7	4.1a	23.6a
C1	1.34b	17.3c	19.3c	63.3a	52.8	2.7b	22.1a
C2	1.27c	42.0ab	28.7bc	29.3bc	63.2	3.1ab	24.5a
C3	1.28c	36.0b	28.0bc	36.0b	33.5	0.7c	26.8a
C4	1.56a	54.0a	31.3ab	14.7d	36.0	2.4b	13.1b
C5	1.18d	36.0b	41.3a	22.7cd	26.6	1.3b	
P	<0.01	0.008	0.018	0.001			

Key: g/cc=grams/cubic centimetre; Ca=Calcium, Mg=Magnesium; CEC=Cation exchange capacity; ME=Miliequivalent

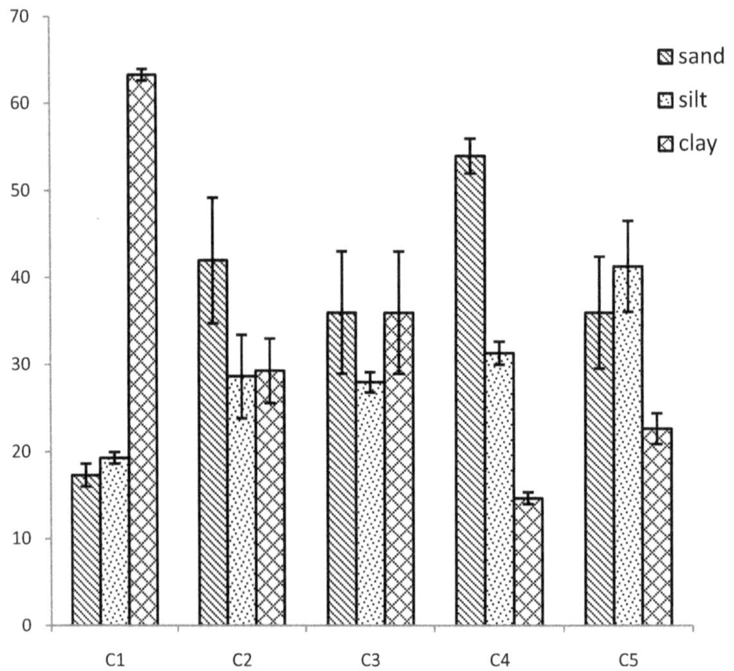

Fig. 2. Variations in % sand, silt and clay contents between different clusters

Table 6. Variations in Micro-nutrients between different clusters

Clusters	Micro-nutrients				
	Mn me%	Cu ppm	Fe ppm	Zn ppm	Na me%
C1	4.0a	0.5c	65.4a	7.1b	0.7
C2	0.9b	0.6bc	23.8b	4.6b	1.1
C3	0.8b	0.8ab	30.7b	6.6b	0.5
C4	0.1d	0.5c	11.8b	0.4c	0.7
C5	0.4c	0.9a	39.9ab	4.7a	0.8
P	<0.001	0.008	0.031	0.003	0.155

Key: Mn=Manganese; Cu=Copper; Zn=Zinc; Na=Sodium; me=Miliequivalent; ppm=parts per million

4. CONCLUSION

The study demonstrated the impacts of land degradation on soil quality in terms of the change in soil quality attributes between the year 2002 and 2013. The change in all the soil quality attributes examined was found to be in the range of 21.4 to 79.1%. The increase in soil pH from

6.74 to 8.18 implied an increased tendency of soil colloids to fix nutrients and render them unavailable to plants, which could explain the decreased availability of nutrients. The decrease in phosphorous, potassium, nitrogen and soil organic carbon by over 20% was found to be much higher than 5%, which is the value permissible (threshold) within 10 years.

The five clusters identified were described as: C1: very deep clay with crusts that caused water stagnation on the surface; C2: shallow to moderately deep sandy clay loam to clay in places gravelly over murrum; C3: moderately structured and deep clay loam to clay with high carbonate concentration; C4: Shallow and highly degraded sandy loam to loam; and C5: extremely deep, stratified sandy clay loam to clay loam. The variations in soil characteristics between the five clusters were found to be significant for cation exchange capacity (P=0.002), magnesium (P=0.003) and bulk density (P=0.01). There was no significant difference for calcium (P=0.147). All the textural characteristics of soil varied significantly with the highest being clay content (P=0.001), followed by silt (P=0.0018) and sand (P=0.008). This meant that the design of irrigation scheduling for different clusters should be based on their textural differences, which have an important bearing on their water uptake and retention capacities.

The variation of manganese (P<0.001), zinc (P=0.003), copper (P=0.008) and iron (P=0.031) was found to be significant, meaning that the quantities of the fertilizers applied to supply these nutrients should be different for different clusters as opposed to the current blanket recommendations that ignore these differences. This had an important implication on the prescription of the quantity of these micro-nutrients for different clusters during implementation of the envisaged management strategies.

ACKNOWLEDGEMENT

Funding from African Development Bank to Small Scale Horticultural Development Project and timely facilitation for the implementation of the project activities by the Project Coordination Unit in the Ministry of Agriculture, Livestock and Fisheries and Kenya Agricultural and Livestock Research Organization are highly appreciated.

COMPETING INTERESTS

Authors have declared that no competing interests exist.

REFERENCES

1. Muya EM, Maingi PM, Chek AL, Gachini GN, Otipa M, Kahiga P, Njeru E, Kilonzo E, Anyango JJ, Macharia E, Yegon JV, Kirigua V, Lusike L, Atonya B. Analysis of soil quality, irrigation system and cluster-specific interventions for improved production of high value vegetables. A Technical Report; 2011.
2. Sharma KL, Mandal UK, Srinivas K, Vittal, KPR, Mandal B, Grace JK, Ramesh V. Long-term management effects on crop yields and soil quality in dryland Alfisol. Soil and Tillage Research. 2004;83:246-259.
3. Rijberman P. Effects of erosion on land and agricultural productivity. A review paper; 1984.
4. Driessen PM. Biophysical sustainability of land use systems. ITC Journal. 1997;243-247.
5. Hurni H. Concepts of sustainable land management. ITC Journal. 1997;210-214.
6. Muchangi P, Sijali I, Wendot H, Lemperiere P. Irrigation systems, their operation and maintenance: Training Source Book; 2005.
7. Prasad RN, Biswas PP. Soil Resources of India. In: Fifty Years of Natural Resource Management Research, Natural Resource Management Division, India Council for Agricultural Research, New Delhi. 1999;13-30.
8. Arshad MA and Gilley JE. Determinants of soil quality and health. In: Soil quality and soil erosion (R. Lal, ed.) CRC Press. Soil and Water Conservation Society, Iowa; 1999.
9. Smith JL, Elliott LF. Tillage and residue management effects on soil organic matter dynamics in semi arid regions. Adv. Soil Sci. 1990;13:69-85.
10. Muya EM, Njeru E, Maingi PM, Chek AL, Otipa M. Irrigation suitability assessment for improved soil quality and water use efficiency. A technical Report; 2012.
11. Sanginga N, Woomer PL. (eds). Integrated Soil Fertility Management in Africa: Principles, Practices and Development Processes. Tropical Soil Biology and Fertility Institute of the International Centre

for Tropical Agriculture (TSBF-CIAT); 2009.
12. Muya EM, Nafula LS, Macharia E, Owenga P. Land suitability assessment for improved crop production and environmental sustainability in Mosiro Irrigation Scheme, Narok County: A Technical Report. 2013;3.
13. Waruru BK, Muya EM, Gicheru PT, Wanjogu SN. The soils and their suitability for surface irrigation in the proposed Mosiro Irrigation Scheme, Mau Division, Narok District. Detailed Soil Survey Report No D74; 2002.
14. Aune JB, Lal R. Agricultural productivity in the tropics and critical limits of properties of Oxisols, Ultisols and Alfisols. Trop. Agric. (Trinidad). 1997;74(2):96-102.
15. Amacher MC, Katherine PO, Charles HP. Soil vital signs: a new soil quality index for assessing forest soil health. USDA, department of agriculture and forest service. A Handbook. 2007;1.
16. Anderson JM, Ingram JSL. Tropical Soil Biology and Fertility: A handbook of Methods, CAB International, Wallingford [4] Doran JW, Jones, AJ. 1993;1.
17. Okaleb JR, Gathua KW, Woomer PL. Laboratory methods of soil analysis: A working manual (second edition); 2002.
18. Hinga G., Muchena FN, Njihia CM. Physical and chemical methods of soil analysis. A Handbook. 1980;1.
19. Driessen PM, Konijn NT. Land use systems analysis. Lecture Notes, Wageningen Agricultural University; 1992.
20. Arshad MA, Martin S. Identifying critical limits of key soil quality indicators in agro-ecosystems. Agricultural Ecosystems and Environment. 2002;88:153-160.
21. Silveria M. Effects of soil acidity on nutrient use efficiency. A technical report, Soil and Water; 2012.
22. Whiting D, Reeder J. Soil pH and its effects on nutrient availability. Colorado Master Gardener Programme and extension document; 2009.
23. Steer A. Making development sustainable. Adv. Geo. Ecol. 1998;31:857-865.
24. Muya EM. An investigation into numerical relationships between soil water contents at several suctions and other soil properties. MSc Thesis. Cranfield University, U.K; 1996.
25. Pierce FJ, Larson WE, Downdy RHG. The relationships between land quality and crop performance (unpublished information); 1983.
26. White JG, zasoski RJ. Mapping soil micronutrients. Field Crop Research. 1999;60(1-2):11-26. Elsevier.

16

Effect of Organic and Chemical Fertilizers on Yield and Yield Components of Peppermint (*Mentha piperita* L.)

Z. Sheykholeslami[1], M. Qasempour Almdari[1*], S. Qanbari[1] and M. Akbarzadeh[2]

[1]Department of Agronomy, Qaemshahr Branch, Islamic Azad University, Qaemshahr, Iran.
[2]Department of Botany, Mazandaran Agricultural and Natural Resources Research Center, Sari, Iran.

Authors' contributions

This work was carried out in collaboration between all authors. All authors read and approved the final manuscript.

Editor(s):
(1) Vladimir K. Mukhomorov, Agrophysical Institute, 14 Grazhdanskiy Ave., Russia.
Reviewers:
(1) Changbing Yu, Oil Crops Research Institute, Chinese Academic of Agricultural Science, China.
(2) Raúl Leonel Grijalva Contreras, Instituto Nacional de Investigaciones Forestales Agrícolas y Pecuarias. Mexico.

ABSTRACT

The current study examined the effect of chemical and organic fertilizers on yield components and essence content of peppermint (*Mentha piperita* L.). The experiment was carried out in randomized complete design with 7 treatments and 4 replications in Sari Agricultural Research Center, Iran, in 2013. Treatments were: 1-control, 2-sheep manure, 3-vermicompost, and basic chemical fertilizers at different contents as 4-N60,P50,K60 kg ha^{-1} which equals 0.24, 0.2 and 0.24 g per pot, respectively), 5-N60,P80,K60, 6-N90,P50,K80 and 7-N90,P80,K80. There were 28 pots that 10 kg soil was added to each of those and chemical fertilizer (at determined rates), sheep manure and vermicompost fertilizers (both at the content of 20 ton ha^{-1} equal 80 g per pot) were also added in per selected pot as treatment. The results showed that the effect of all treatments was very significant on the all agronomic properties and yield of peppermint. Some agronomic and quantitative traits of peppermint like plant height, number of chain, stem wet and dry weight, leaves wet weight and total wet and dry weight were obtained at the maximum values when basic chemical fertilizers was applied at the rate of 90-80-80. The maximum number of stems, number of flowers, number of suckers, leaves dry weight and essence content were also attained by sheep manure. Thus, to prevent soil, plant and finally environment pollution by using chemical fertilizers

**Corresponding author: E-mail: ghasempour3m@yahoo.com;*

and also to achieve the maximum leaves dry weight and essence content of peppermint, application of sheep manure as organic fertilizers is recommended.

Keywords: Vermicompost; sheep manure; chemical fertilizer; yield; essence; peppermint.

1. INTRODUCTION

Medicinal plants are included most sources of secondary metabolites which provides effective material of some drugs. Biosynthesis of secondary metabolites will be controlled by genetics and also climate factors [1]. An effective material of peppermint plant is included essence as essential oil (1-2%), Tanon, Flavonoid, Colien and bitter materials. Medicine effects of this plant is concerned to chemical compounds of essence in leaves which is more than 20 types such as Menthol (40-60%), Menthofuran, Menthone, Piperitone, Pulygone and Cineole. So, the factors which affect on the essence content of the plant; must be attended as growth restricted factors like warm or cool stresses, nutrient deficiency into the soil, water deficiency and the other factors. [1] Investigation of the plant nutrition is one of the most important programs in the management of crop plant cultivation especially medicinal plants. An adequate nutrient and water supply into the soil is needed to healthy growth and produce more effective materials for medicine plant especially aromatic and essence-bearing plants like peppermint. The use of chemical fertilizers in the arable soils of the world is imbalanced and does not take into account the actual necessary need of the crop plant or the nature of the soil [2]. Among the essential elements macro and also micronutrients such as N, P, K, S, Fe and Zn are necessary to suitable growth and increase essence content and wet or dry yield of medicine plants. To provide these essential elements, not only chemical fertilizer but also the use of manure fertilizers is recommended which are included macro and also micro nutrients [2]. Patrui and Tabara [3] reported that the manure fertilizer (as organic) increased the yield of peppermint compared to the chemical fertilizer (as inorganic). Dalvand et al. [4] and Mehr-afarin et al. [5] have investigated the effect of bio and manure fertilizers on morphological traits and essence content of peppermint and resulted that these fertilizers increased dry yield and the percentage of essence which is due to existence of more microelements at the compounds of these organic fertilizers. This fertilizer will be decomposed and released microelements gradually during the harvesting time that will be more taken up by plant and will be increased essence content of the leaves. Daramol [6] expressed that biomass yield and dry weight of peppermint increased by using compost and manure fertilizer at the content of 1-2 kg and 250-500 g per pot, respectively. But, Mahmod et al. [7] reported that nitrogen fertilizer as urea and phosphorous as Triple super phosphate increased dry matter and essence yield of this plant. Niyakan et al. [8] reported that wet and dry yield and also the content of essence as qualitative trait were increased using nitrogen and potassium elements at more content compared to the less content. The reason of the effect of nitrogen on wet and dry weight is existence of it in the chemical compound of some molecules like Protein, amino acid and nucleic acid [9,10]. Regarding to the effects of chemical and manure fertilizers on the qualitative and quantitative characteristics of medicinal plants, the present research is devoted more to the comparative study of chemical and manure fertilizers on yield, yield components and essence content of peppermint plant.

2. MATERIALS AND METHODS

The present experiment was done in Sari Agricultural Farm, Dodangeh region (Long. 52°15' E. to Lat. 36°15' N), in northern part of Iran in 2013, where the general altitude is about 700 m above the sea level. The mean precipitation and daily mean temperature during growth season of peppermint plant in the studied region is 94.5 mm and 31.19°C, respectively. The experiment was carried out as randomized complete design with 7 treatments and 4 replications. Treatment were as control, vermicompost, composed sheep manure and basic chemical fertilizers which are N60,P50,K60; N60,P80,K60; N90,P50,K80 and N90,P80,K80. Initially, the studied soil and sheep and vermicompost (from plant sources) fertilizers were tested at laboratory to determine the element contents and other their chemical and physical properties which are shown in Tables 1A and 1B.

All studied plants as seedling were cultivated into the pots (size 30*40cm^2) on May 2013 which located on the farm (outdoor condition). Each pot was filled 10 kg soil and chemical fertilizer (were chosen as treatments) and sheep and vermicompost fertilizers (both at the content of 20 ton ha^{-1} equals 80 g per pot) were also added in per selected pot as treatment. Then, all the pots were irrigated two times per week and were also weeded out during growth period. First and second harvesting times were done at August and November, 2013. Essential oil content was determined by hydro distillation method by submitting aerial part of dried plants (100 g) in modified a British-type Clevenger's apparatus system model 7890 A. After 3 hours distillation was stopped so essential oil ratio was measured by using dry yield (biomass yield) of peppermint. The measured characteristics in the current study were plant height, the total number of stem, the number of flowers, the number of suckers, the number of chain, leaves wet weight, stem wet weight, total wet weight (wet yield), leaves dry weight, stem dry weight, total dry weight (dry yield) and the content of essence. Data were analyzed following the analysis of variance technique (ANOVA) by using Microsoft-MSTAT-C and the mean differences were adjudged by Duncans' multiple range tests (DMRT) at the level of 5%, probability [11].

3. RESULTS AND DISCUSSION

The results showed that the effect of all used treatments was very significant on all the measured characteristics, statistically (Table 2). According to analysis of variation of data, the essence content and also all agronomic characteristics of peppermint were very significantly influenced by both organic and chemical fertilizers. Coefficient of variation of all the measured characteristics is less than 10% at the present study which is showed the experiment was carefully done at suitable conditions.

As it can be seen at Table 3, the maximum values of the number of flowers, the number of stem, the number of suckers, leaves dry weight and essence content of peppermint were attained when organic fertilizers (sheep manure) was applied. Using sheep manure fertilizers, the number of chain per pot 212%, leaves wet weight 29%, stem wet weight 52%, wet yield 39%, leaves dry weight 84%, stem dry weight 97%, dry yield 90% and essence content 70% increased compared to the control pot. The maximum values of plant height, the number of chains, leaves and stem wet weight, total wet weight, stem dry weight and total dry weight were also attained when chemical fertilizers applied at the ratio of 90-80-80 and then 90-50-80. Comparison between the control pot and the use of organic and chemical fertilizers showed that using sheep manure fertilizer dry yield and essence content increased 90 and 70% compared to the control pot, respectively. Essence content increased 23% when sheep manure was applied compared to chemical fertilizer at the ratio of 90-80-80. Table 4 shows that all the agronomic characteristics, quantitative yield and essence content of peppermint were influenced by both organic and chemical fertilizers which had significant difference compared to control pot. But, organic fertilizers group (sheep and vermicompost fertilizers) affected significantly on the number of flowers, the number of stem, the number of suckers, leaves dry weight and essence content of peppermint and increased those traits compared to chemical fertilizers group, statistically. Other measured traits were more influenced by chemical fertilizers group.

The results of this study showed that not only organic fertilizers but also basic chemical fertilizers such as nitrogen, phosphorus and potassium increased plant height, wet and dry weight of stem and leaves, total wet and dry weight and the content of leaves essence of peppermint significantly compared to control pot. The obtained results are in agreement with the observations by Mahmod et al. [7]. They investigated the effect of nitrogen fertilizer on the growth and essence yield of peppermint and announced that dry matter and essence yield was significantly increased when nitrogen fertilizer applied. Not only the nitrogen element does not exist into the essence compound, but also it is involved in the chemical compound of some molecules like Protein, amino acid and nucleic acid [9]. The use of this element causes increasing extraction gland of essence at the leaves of peppermint [12]. It is because of producing and consuming simple glucoses, increasing vegetative growth and developing leaves surface [13].

Table 1A. Elements content and chemical and physical properties of the studied soil

Properties	Depth (cm)	Base saturation (%)	E.C (ds m^{-1})	pH	Organic matter (%)	Organic carbon(%)	P (mg kg^{-1})	K (mg kg^{-1})	N (%)	Ca (%)	Mg (%)	Fe (mg kg^{-1})	Mn (mg kg^{-1})	Zn (mg kg^{-1})	Cu (mg kg^{-1})	Soil texture
	0-30	40	3.5	7.4	2.2	1.28	14.3	137	0.128	-	0.04	15.1	11.3	2.6	0.7	Sand-Loam*

*,Clay 10%, Sand 60% and Silt 30%

Table 1B. Elements content and chemical and physical properties of the sheep and compost fertilizers

Properties	E.C (ds m^{-1})	pH	Organic matter (%)	Organic carbon(%)	P (mg kg^{-1})	K (mg kg^{-1})	N (%)	Ca (%)	Mg (%)	Fe (mg kg^{-1})	Mn (mg kg^{-1})	Zn(mg kg^{-1})	Cu (mg kg^{-1})
Vermicompost	5.25	7.42	24.42	14.2	0.44	0.28	1.1	2.32	0.41	12.5	428	43	21
Sheep	25.3	7.56	48.5	28.2	0.46	2.79	2.1	1.32	0.36	1110	290	25	19

Table 2. Square mean of the effect of organic and chemical fertilizers on yield and yield components of peppermint

S.O.V	Df	Plant height (cm)	Number of stem	Number of chain	Number of flowers	Number of suckers	Leaves wet weight (g pot^{-1})	Stem wet weight (g pot^{-1})	Wet yield (kg ha^{-1})	Leaves dry weight (gpot^{-1})	Stem dry weight (gpot^{-1})	Dry yield (kgha^{-1})	Essence content (g 100g^{1})
Treatment	6	95.25**	1.5767**	63.774**	2.3748**	1.5767**	62.187**	113.1**	2747722**	0.67**	2.2014**	33161**	0.715**
Error	21	0.6548	0.0129	0.3724	0.0196	0.0129	0.7395	0.5692	16507.6	0.0129	0.0571	723.21	0.039
C.V(%)	-	1.50	5.039	5.72	8.15	9.07	8.69	7.63	7.22	4.82	10.20	6.36	7.840

**: high significant at the level of %1

Table 3. Mean comparison of the effect of chemical and organic fertilizers on yield and yield components and essence content of peppermint

Treatments	Plant height(cm)	Number of stem	Number of chain	Number of flowers	Number of suckers	Leaves wet weight (g pot^{-1})	Stem wet weight (g pot^{-1})	Wet yield(kg ha^{-1})	Leaves dry weight (g pot^{-1})	Stem dry weight (g pot^{-1})	Dry yield (kg ha^{-1})	Essence content (g 100g^{-1})
Control pot	45.8^{f}	2.75^{c}	3.08^{f}	1.65^{c}	1.75^{c}	4.58^{g}	3.25^{g}	704^{g}	1.48^{f}	1.13^{g}	234^{g}	1.85^{g}
Vermicompost	55.3^{c}	2.93^{ab}	8.85^{de}	2.43^{b}	1.93^{ab}	6.68^{e}	4.9^{ef}	1042^{e}	2.6^{ab}	2.1^{cdef}	423^{cde}	2.2^{f}
Sheep manure	51.3^{e}	2.98^{a}	9.63^{d}	3.1^{a}	1.98^{a}	5.95^{ef}	4.95^{e}	981^{ef}	2.73^{a}	2.23^{cd}	446^{bc}	3.15^{a}
Mean manure	53.3	2.95	9.24	2.76	1.95	6.31	4.92	1011.5	2.66	2.16	434.5	2.67
N60-P80-K60	51.3^{e}	1.8^{e}	13.6^{b}	1.35^{de}	0.8^{e}	12.8^{abc}	12.1^{d}	2241^{bcd}	2.5^{bc}	2.15^{cde}	419^{cdef}	2.65^{bc}
N60-P50-K60	54.8^{cd}	1.45^{g}	11.2^{c}	1.03^{fg}	0.45^{g}	13.3^{ab}	13.5^{bc}	2410^{bc}	2.43^{cd}	2.38^{c}	432^{cd}	2.5^{cde}
N90-P50-K80	57^{b}	1.73^{ef}	13.6^{b}	1.05^{f}	0.73^{ef}	12.6^{abcd}	14.3^{b}	2414^{b}	2.43^{cd}	2.9^{b}	479^{b}	2.85^{b}
N90-P80-K80	61^{a}	2.13^{d}	14.7^{a}	1.43^{d}	1.13^{d}	13.4^{a}	16.2^{a}	2657^{a}	2.3^{de}	3.53^{a}	524^{a}	2.55^{bcd}
Mean chemical	56.02	1.77	13.27	1.21	0.77	13.02	14.02	2430.5	2.41	2.74	463.5	2.63

A given means per each column with the same letters, have not significant difference, statistically ($p<0.05$).

Table 4. Comparison of average of yield and yield components of peppermint by the effect of organic and chemical fertilizers

Properties	Plant height (cm)	Number of stem per pot	Number of chain per pot	Number of flowers per pot	Number of suckers per pot	Leaves wet weight (g pot^{-1})	Stem wet weight (g pot^{-1})	Wet yield (kg ha^{-1})	Leaves dry weight (g pot^{-1})	Stem dry weight (g pot^{-1})	Dry yield (kg ha^{-1})	Essence content (g $100g^{-1}$)
Control	45.8^{f}	2.75^{c}	3.08^{f}	1.65^{c}	1.75^{c}	4.58^{g}	3.25^{g}	704^{g}	1.48^{f}	1.13^{g}	234^{g}	1.85^{g}
Organic mean	53.3	2.95	9.24	2.76	1.95	6.31	4.92	1011.5	2.66	2.16	434.5	2.67
Chemical mean	56.02	1.77	13.27	1.21	0 77	13.02	14.02	2430.5	2.41	2.74	463.5	2.63

The effect of sheep manure fertilizer on the some agronomic characteristic like the number of flowers, the number of stem, the number of suckers, leaves dry weight and essence content of peppermint was significant compared to vermicompost fertilizer, statistically. The values of electrical conductivity, total nitrogen, organic carbon, organic matter, total potassium and iron concentration existing sheep fertilizer are 5, 2, 2, 2, 10 and 88 times more than vermicompost fertilizer, respectively (Table 1). It seems that high nutrient values of sheep fertilizer could be increased vegetative growth content, leaves surface developing, photosynthesis content, some agronomic properties and finally the content of leaves essence. Existence of some macronutrients like nitrogen and potassium and also micronutrients especially iron into the sheep manure fertilizer increased essence content as qualitative trait which is in agreement with some observations by Dalvand et al. [4] and Mehrafarin et al. [5]. The effect of chemical fertilizer as 90-80-80 and then 90-50-80 on the some measured properties like plant height, the number of chains, leaves, stem and total wet weight, stem and total dry weight are more than other chemical fertilizer treatments, statistically; which is because of high nitrogen and potassium content at these types of chemical fertilizers. It can be concluded that quantitative properties of peppermint will be increased by increasing N and K elements up to 90 and 80 kg ha^{-1}, respectively. The same results were attained by Niyakan et al. [8]. They reported that some quantitative characteristics like total wet and dry weight and also the content of essence of peppermint increased significantly using high N and K content compared to the low content ones. Zhao [10] expressed that the reason of the effect of nitrogen on increasing wet and dry weight of peppermint is due to exist of this element into the big molecules composition like protein, amino acid and nucleic acids. Thus it may be resulted that there is an interaction effects (positive correlation) between N and K elements into the plant body. Therefore, increasing quantitative and qualitative traits of peppermint by both chemical and sheep fertilizers can be raised due to high content of N and K elements in both these fertilizers. The results of the present study are in agreement with the observations by Clark and Menary [14], Singh et al. [15], Fernander [16] and Gerder et al. [17] and Arabasi and Bayram [18] who reported that essence content increases using nitrogen fertilizer.

The content of some measured quantitative traits is more using chemical fertilizers than organic fertilizers. Chemical fertilizers especially nitrogen will be rapidly decomposed into the soil by micro-organisms and will be changed to absorbable inorganic materials (mineralization period). These absorbable nutrients will be taken up by plants as soon as possible and will increase vegetative growth of plant. While, organic fertilizers will be decomposed into the soil by micro-organism gradually and released their nutrients slowly [9]. Thus, the content and content of nitrogen uptake by the plant at organic fertilizers is less than chemical fertilizers. Accordingly, organic fertilizers may be affected on the plant growth at the end of growth stages or at the second or third harvesting times. On the other hand, the content of some measured quantitative traits is also more using organic fertilizers than chemical fertilizers. It is because of high microelement concentration in organic fertilizers which can be increased the number of stem, the number of suckers, the number of flowers, leaves dry weight and the content of essence. This finding is in agreement with those reported by Tahami et al. [19]. They reported that the maximum sub-stem per plant was attained when sheep manure fertilizer applied compared to chemical, vermicompost and chicken fertilizers.

4. CONCLUSION

Some agronomic and quantitative traits of peppermint like plant height, number of chain, stem wet and dry weight, leaves wet weight and total wet and dry weight were obtained at the maximum values when basic chemical fertilizers was applied at the rate of 90-80-80. The maximum number of stems, number of flowers, number of suckers, leaves dry weight and essence content were also obtained by sheep manure. Thus, to prevent soil, plant and finally environment pollution by using chemical fertilizers and also to achieve the maximum leaves dry weight and essence content of peppermint, application of sheep manure as organic fertilizers is recommended.

COMPETING INTERESTS

Authors have declared that no competing interests exist.

REFERENCES

1. Omid Beygi R. Production and processing of medical plants. 2nd valium. 5th ed. Qods razavi publication. Mash-had, iran (in persian); 2000.
2. Qasempour Alamdari M. The effect of macro and micronutrients on the growth and yield of rice (*Oryza sative* L.). Ph.D thesis, Pune University, India; 2004.
3. Patrui O, Tabara V. Research on the quality yield of herbal mint (*Mentha piperita* L.) under the influence of mineral and organic fertilization in climate condition n S.C.D.A.LOVRIN, Bulctinul AGIR nr; 2011.
4. Dalvand Y, Sadeqi A, Asadi Gh, Karimi A. Effect of *Streptomyse* and different content of vermicompost on yield and essence of peppermint (*Mentha piperita*) under farm condition. Iranian 7th congress of horticulture sciences; 2011. Available:http://www.civilica.com/Paper-BAGHBANI07-BAGHBANI07_890.html
5. Mchrafarin A, Naqdi Badi H, Pour Hadi M, Hadavi A, Qavami N, Kadkhoda Z. Response of photochemistry and agronomy of peppermint to use of bio-fertilizer and urea fertilizer. Medical Plant Journal. 2010;2(1):63-74
6. Daramol OS. Effects of organic fertilizers on growth and yield of MINT (*Mentha piperita*). Department of Horticulture Trans. By-Mycol, Soci. 2011;55:158-61.
7. Mahmod S, AsafarY, Al-Hassan M. Effect of nitrogen and phosphorus fertilizers on growth and oil yield of indigenous Mint (*Mentha longifolia* L.). Biotechnology. 2009;3:380-384.
8. Niyakan M, Khavarinezhad R, Rezaee M. Effect of different proportion of basic fertilizers N, P and K on wet and dry weight and leaves surface and essence content of peppermint (Mentha piperita). Iranian Aromatic and Medical Plants Research Journal. 2004;20(2):131-148.
9. Kolata E, Beresniiewicz A, Krezel J, Nowosielshi L, Slow O. Slow release fertlizers on organic carriers as the source of N for vegetab cops production in the open field. Acta- Horticaulhtre. 1992;339:241-249.
10. Zhao J. The effect of nitrogen fertilization on spearmint. Journal of Essential Oil Research. 2006;18:452-455.
11. Gomez KA, Gomez AA. Statistical procedures for agricultural research (2nd ed). Gohn wiley sons, New York, Chichester, Bribane, Toronto, Singapor; 1984.
12. Marotti M, Piccaglia R, Crout W, Craufutd K, Deans S. Effect of planting time and mineral fertilization on peppermint (*Mentha piperita* L.) essential oil composition and its biological activity. Flavor and Fragrance Journal. 2004;9(3):125-129.
13. Brown B. Mint soil fertility research in the PNW. Western Nutrient management Conf. 2003;5(3):54-60.
14. Clark RJ, Menary R. The effect of irrigation and nitrogen on yield and composition of peppermint oil (*Mentha piperita* L.). Applied-Plant Science. 1999;62(2):68-71.
15. Singh VP, Chatterjee BN, Singh P. Response of mint species to nitrogen fertilization. Journal of Agricultural Science. 2003;113(2):267-271.
16. Fernander CH. Nitrogen and water management for medicinal and aromatic plants. Acta Horticulture. 2006;132(2):203-215.
17. Gerder HV, Vangelder H, Mucciarelli N. Influenco of nitrogen fertilizer application level on oil production and quality in *Mentha spp*. Applied Plant Science. 1993;92(2):68-71.
18. Arabasi D, Bayram E. The effect of nitrogen fertilization and different plant densities on some agronomic and technologic characteristic of (*Ocimum basilicum* L.). Essential Oil Research. 2005;17:203-205.
19. Tahami Zarandi S, Rezvani P, Jahani M. Compression of the effect of chemical and organic fertilizers on yield and essence percentage of *Ocimum basilicum* L. Agro-Ecology Journal. 2010;2(1):63-74.

17

An Assessment of Demographic Characteristics of Women Involvement on the Improved Methods of Groundnut Processing in Three (3) Local Government Areas of Niger State, Nigeria

U. Mohammed [1*] and R. S. Olaleye[2]

[1]Niger State Ministry of Agriculture and Rural Development, Minna, Nigeria.
[2]Department of Agricultural Economic and Extension Technology, Federal University of Technology Minna, Niger State, Nigeria.

Authors' contributions

This work was carried out in collaboration between all authors. Author UM designed the study, wrote the protocol and wrote the first draft of the manuscript. Author RSO reviewed the experimental design and all drafts of the manuscript. Authors UM and RSO performed the statistical analysis. All authors read and approved the final manuscript.

Editor(s):
(1) Moreira Martine Ramon Felipe, Department of the Universidade de Santiago de Compostela, Spain.
Reviewers:
(1) Anonymous, USA.
(2) Anonymous, China.
(3) Anonymous, Ghana.

ABSTRACT

There is gender division of women in farm labour in Niger State. The gender division of women in farm labour assigns women more works in the processing of groundnut as Agricultural food products and yet, women have no access to improved methods of groundnut processing and depend mainly on the traditional methods the purpose of this study therefore were to determine the demographic characteristics of women involvement on the improved methods of groundnut processing in there (3) Local Government Areas of Niger State. Two specific objectives that guided the study are. 1. Examine the demographic characteristics of women involvement on the improved methods of groundnut processing in the study areas.

Identify the constraints faced by women processors in the study area.

A multi-stage sampling procedure were employed in the selection of the respondents. The first

**Corresponding author: Email: usmanmohammed.usmanou@gmail.com;*

state was purposive selection of Agricultural zone one of the State AMDA because of the high concentration of the respondents in the zone. This was followed by random selection of there (3) Local Government Areas (LGAs) from the zone which is equivalent to three AMDA Extension blocks. The third stage was random selection of four (4) extension cells from each of the Extension blocks, from an existing list of registered women groundnut processors Association (314) with sate AMDA, a total population samples size of 180 respondents was randomly selected. A structured questionnaire was used to collect data pertinent to the study with the help of Enumerators. Data were descriptively analyzed. Demographically, majority of the respondents were between the age distribution of 21-50 years and married, majority of the respondents in the study area had their household size of between 6-10 people. About 50% of the respondents had no access to formal education and 57% had 1-6 years in schooling. The result also revealed that about 32% of the respondents had 15-20 years of groundnut processing experience and about 55% of the respondents had their major occupation has full time processor. The recommendation made include need for the government and interested non government organization (NGOs) to provide credit facilities for respondents to enable them involved into improved methods of groundnut processing technologies.

Keywords: Assessment; demographic; women; involvement; methods; processing.

1. INTRODUCTION

In today's digital age, the latest technologies always reach the rural poor last in spite of recent advances in communication and information technologies, Technical Centre for Agricultural and Rural Cooperation [1]. Rural women are much less likely to have access to new agricultural technologies because they are generally less educated and possess less economic and political power relative to their men folk. In Nigeria, reports indicated that women play more important roles in agricultural processing compared to men [2]. While records further show that the agricultural labour force is made up of about 60-80 percent women depending on the region, and two thirds of the foods crops is processed by the women [3], Despite these facts there still exists wide spread assumption that men and not women make the key management decisions [4]. As a result of this, most extension activities targeted at women emphasize their domestic roles with topics such as child care and family nutrition while excluding activities involving agricultural processing.

Basically, women are involved in the supply of labour, processing of food crops and livestock processing and transportation farm produce for effective storage and marketing [5,6]. Despite the high level of their involvement in agricultural processing activities such as groundnut processing, they are inadequately recognized and undervalued [7,8,9]

Women's role in groundnut processing is not a new phenomena. Their role is fundamental to agricultural growth/ development especially via Agricultural food processing. They account for 100 percent for those who process agricultural food stuff [10] In Nigeria, women take part actively in the groundnut processing activities in addition to their domestic household responsibilities [11]. Women are said to be involved in over 95 percent of groundnut milling activists in "Zone A" agro ecological area (Bida of Niger State).

2. PROBLEM STATEMENT

The declaration of 1975-1985 as decades for women by member countries of the United Nations (UN) marked the beginning of recognition of gender issues in development circle worldwide. This resulted in active discussion among researchers policy makers, educationists and developmental partners on roles of women vis- a vis those of men [12]. This led to the creation of Women In Agriculture (WIA) programme within the existing state Agricultural Development Projects (ADPs) and the conversion of women home economists to female Agricultural processors, all in an attempt to correct the gender related deficiencies and recognition of the role play in extension and processing of agricultural commodities. There is gender division of women in farm labour in Nigeria, [13]. The gender division of women in farm labour assigns women more works in the processing of groundnut as agricultural food products and yet, women have no access to improved methods of groundnut processing and depend mainly on the traditional methods, due to the constraints that are responsible for it, such as poor electricity supply, lack of credit facilities, high purchasing price of technologies increase in price of petrol, poor processing equipments poor quality of groundnut and absence of sustainable policy for groundnut processing activities. All this constraints affect women processors involvement of improved groundnut processing technologies.

3. PURPOSE AND OBJECTIVES

The purpose of the study was to determine the demographic characteristics of women involvement on the improved methods of groundnut processing in three (3) Local Government Areas of Niger State.

Specific objectives were to:

1. Examine the demographic characteristics of women involvement on the improved methods of groundnut processing in the study areas.
2. Identify the constraints faced by women processors in the study area.

4. METHODOLOGY

4.1 Study Area

The study was conducted in *Mokwa, Lavun and edati* LGAs of Niger state. Niger state occupies also north-central part of Nigeria, and shares common boundaries with North by *Zanfara* state, to the north east by *kebbi* state and north east and federal capital territory(FCT) bordered the state to north east and south east. The state also shares a common international boundary with the republic of Benin along *Newbussa, Aguara* and *Wushishi* local government Areas. The state is located in the north central savannah of the vegetation zones between latitudes 8° and 11°N and longitudes 3°E and 7°E of the prime meridian [14].

Population of Niger state is the largest in Nigeria with a total land mass representing about 9.3 percent of the total land mass of the country. The population of the state was 3,950,429 in 2006 [15]. The bureau of statistics has maintained an approximate population growth rate of 2.5% geometrically. Based on that, the projected population in 2012 was estimated to be 4,883,036. The state occupies on area of about 58,676.2 square kilometers [15]. The state is made up to 25 LGAs namely *Agaie, Agwara, Bida, Borgu, Bosso chanchaga, Edati, Gurara, Katch, Kontagora, Lapai, Lavun, Magama, Mariga, Mashegu, Mokwa,Munya, Paikoro, Rafi, Shiroro, Suleja, Tafa,* and *Wushishi* local government Areas.

Niger state is divided in to three Agricultural zones namely I, II and III for better Administration of agricultural activities each of these Agricultural zones has their headquarters at *Bida, Kuta* and Kontagora respectively. The Climate and Ecological conditions of the state is favored with mean annual rain fall of 782-1250 mm and temperature is about 82°C or 27.7°C [16]. The state has abundant wild vegetation of Shea trees and dominated by small scale farmers. The major crops cultivated are millet, rice, maize, guinea corn, beans, cassava, groundnuts and sweet potatoes. Majority of the famers keep livestock like poultry, goat and sheep. Other engaged in crafts such as sculptures, weaving and blacksmith [16]. About 85% of the population of the state are farmers, while 15% are involved in other vocation such as white collar jobs, businesses, crafts and arts [16].

4.2 Method of Data Collection and Sampling Procedure

The data used were obtained mainly from primary sources, through the use of structure questionnaires. With the help of enumerators. The sampling method adopted for the study were multi-stage sampling procedure techniques. The first stage was purposive selections of Agricultural zone one of the state AMDA. This was followed by random selection of three (3) Local Government Areas (LGAs) from the zone which is equivalent to three AMDA extension blocks.

The third stage was random selection of four (4) extension cells from each of the extension blocks, giving a total sampling size of 180 women processors used in the study area.

5. ANALYTICAL TECHNIQUE

Analysis of the data was done using descriptive statistics which include measure of central tendency such as means, percentage, table and frequency distribution.

$Y = f(X_1\ X_2\ X_3 \ldots\ldots\ldots\ldots\ldots\ldots\ldots\ldots + X_7$ Explicit

Explicit form (Linear)

$$\frac{Ypi}{I-pi} = bo + b_1X_1+b_2X_2+b_3X_3+b_4X_4+b_5X_5+b_6X_7 + e$$

Where:

Ypi = (Probability of high involvement)
I-pi =(Probability of low involvement)
Logy =dischotomous variable high = 1, otherwise = 0
X_1 = Age (years)
X_2 =Education level (years)
X_3 =Marital Status (Number)
X_4 =Household size (Number)
X_5 =Years of processing experience (Years)
X_6 = Year of Schooling (Years)
X7 =Major Occupation

6. RESULTS AND DISCUSSION

6.1 Demographic Characteristics

The data in (Table 1) revealed that there are significant differences between the demographical characteristics of women involvement and improved groundnut processing methods. The Demographic characteristics of the respondents in the study area were examined and described with respect to their Age, marital status, Household size, educational level, year of schooling, year of processing experiences and major occupation.

6.2 Age

The age distribution of the respondents in the study area were between 21-50 years (81.1%). This implies that they were young and energetic within the productive age which could increase their groundnut processing activities. This was in agreement with finding of [17] in indigenous resources managements among communities in North West Zone of Nigeria which revealed that the women processors are of the middle age category of 40-50 years of age. This implies that, the productive age groups of women processor in the study area were actively involved in groundnut processing activities and still have energy to cope with rigors of groundnut processing.

6.3 Marital Status

About 67% of the respondents in the study area were married. The implication is that respondents that were married are more involved in the improved groundnut processing technologies in the study area.

This was in line with study by [18]. Which revealed that marital status of women processors play a significant role, in groundnut processing activities where improved technologies are involved for instance married women processors with large family size may have large involvement and readily supply of labour, which reduces the cost of hired.

6.4 Household Sizes

About 48% of the respondents has household size of between 6 and 10 people. The implication is that the higher the number of household size the lower the cost of labour this was in agreement with [19] which revealed that In agricultural food processing, household size determines the availability of labour. The additional labour to be hired depends on the amount of family labour available and reduce the cost of hiring more hand for such activities.

6.5 Educational Level

About 50% of respondents in the study area have no access to formal education. The implication of this result implies that the low level of formal education is obviously a disadvantage to the respondents, due to the fact that improved technologies would have to be communicated to these respondents in native dialects. This view was in line with findings of [19]. Which revealed that education enables every individual to gain knowledge and skills and this increase their power of understanding.

6.6 Years of Schooling

About 57% of the respondents had 1-6 years in school which implies that majority of the respondents had primary school qualification. The implication of this result is that 1-6 years may not be enough for the respondent to have more knowledge about improved technologies.

6.7 Years of Processing Experience

About 32% of the respondents had 15-20 years of processing experience. This result was in line with finding of [20]. The study revealed that high experience in agricultural processing can raise productivity.

The mean years of experience of the respondents were 15.57 years. The high mean years of processing experience suggests that the respondents would not have many problems in involving and effective use of improved technologies.

Table 1. Demographic characteristics

Demographic information	Frequency	Percentage	Mean
Age (years)			
21-30 years	19	10.6	
31-40 years	42	23.3	42.99
41-50 years	85	47.2	
Above 50	34	18.9	
Marital status			
Married	121	67.2	
Single	18	10.0	
Divorce	18	10.0	
Widowed	23	12.8	
Household size(Number)			
1-5	56	31.1	
6-10	86	47.8	6.76
11-15	35	19.4	
16-20	3	1.7	
Educational Level			
Primary	19	10.6	
Post Primary	9	5.0	
Secondary	13	7.2	
Post Secondary	7	3.9	2.67
Adult Education	4	2.2	
Qur'anic Education	38	21.1	
None	90	50.0	
Years of schooling			
1-6	102	56.7	
7-12	75	41.7	
13-18	3	1.7	
Years of processing experiences			
1-5	12	6.7	
6-10	42	23.3	
11-15	33	18.3	15.57
15-20	57	31.7	
above 20	36	20.0	
Major occupation			
Processors	99	55.0	
Self employed	39	21.7	
Company employed	5	2.8	
Processor	37	20.6	

Source: Field survey, 2013

6.8 Major Occupation

About 55% of the respondents in the study area are full time processors as their major occupation. The implication of this result is that respondents will involved and adopt improved technologies. This view was in line with the study by [12], which revealed that major occupation determine the level of involvement and adoption of improved technologies.

7. RESPONDENTS CONSTRAINTS

The data in (Table 2) revealed that there are significant differences between constraints faced by women processors and improved groundnut processing methods.

About 92% of the respondents in the study area reported that poor electricity constraints were among of their problems faced. about 80% of the respondents also indicated that lack of credit facilities were also their problems such as loan, condition attached to loan disbursement, interest rate, collateral and several trips to the bank before loan is granted. This finding was in agreement with [21] which revealed that women face a number of barriers to obtained credit from lending institutions because most of them have no collateral. The implication of this is that, without loan to facilitate the involvement and adoption of improved groundnut technologies, the respondents will not expand the scope of processing women processor will continue to queue for long at the extraction point. Another problem experienced by the respondents were increase in price of Petrol.

About 95% the respondents reported that petrol engine is used for processing when there is no electricity supply, and the women tend to spend more as the cost of kneading is high which is not economical for the respondents. The constraints increase in price of petrol has negative influence on involvement and adoption. It can be concluded here that, if prices of petroleum products are not reduced some of the women processors might revert to the use of the traditional methods of groundnut processing.

About 94% of the respondents reported that they still faced with constraints of poor processing equipment because these technologies are expensive to acquired and then emphasized in traditional methods of groundnut processing.

Table 2. Respondents constraints

Constraints	Frequency	Percentage %
Poor electricity	165	91.7
Lack of credit Facility	144	80.0
High purchasing Price of Technologies	170	94.4
Increase in price of petrol	171	95.0
Risk associated with technologies	165	90.0
Price flotuation	164	91.1
Poor processing Equipment	170	94.4
Poor quantity of G/Nut	167	92.8
Poor capacity Building	162	90.0
Lack of viable commercial practice	163	90.6
Need for organization producer	164	91.1
Absence of Sustainable policy	144	80.0

*Source: Field survey, 2013; * Multiple Responses*

8. CONCLUSION AND RECOMMENDATIONS

The study has provided information on demographic characteristics of women involvement on the improved methods of groundnut processing in three Local Government Areas of Niger State, Nigeria. Majority of the respondents do not have access to formal education and this can affect disseminations of information (printed materials) on any new technologies to the respondents. The constraints such as poor electricity high cost of petrol supply, lack of credit facilities, high purchasing price of technologies poor and lack of operational facilities for processing groundnuts affect the level of women involvement on the improved of groundnut processing technologies in the study area. Base on the findings in the study the following recommendations are made-

1. In view of the high cost of petroleum products and irregular electricity supply, which are required for groundnut processing activities. It was recommended that provision of solar powered electricity by interested NGOs to address the irregular power supply should be made a priority intervention in the study area.
2. It was found that majority of the respondents in the study area lack access to credit facilities. It is recommended that credit facilities should be provided by the government and interested non Government organization to increase the scope of groundnut processing business and improved their level of living.
3. The study revealed that respondents faced with high purchasing price of technologies. Constraints it is recommended that improved groundnut technologies should be made available for the respondents at the subsidies rate. This can be done though appeal to interested Non Governmental Organizations (NGOs) private volunteers and other organized bodies to assist government in complementing the present high purchasing price of technologies to the respondents.

COMPETING INTERESTS

Authors have declared that no competing interests exist.

REFERENCES

1 Centre for Agricultural Cooperation (CTA). The roles of a women in agricultural food processing evidence from women right place in groundnut processing, *spore bi-monthly bulletin* (CTA) Netherlands. 2003;44

2 Hassan H, Usman Y, Ajayi O. Factors influencing the level of women participation in agricultural processing activities. Paper Presented At The Conference of Agricultural and green revolution in Africa (AGRA) Togo. 2012;1-98.

3 World Bank. Oil Processing Intermediate Technology Publication. London No 4: 2003;1-26.

4 Christiana HG, Ken LB, Abe GC, Prefer TH, Max LKJ, Donna NK, et al. Gender and soil fertility in Africa. Soil science society of America special publication 2007;51:220.

5 Shannon SB. Female agricultural workers wages, female wages differentials, and agricultural growth in a Developing Country. Indian Economic Development and cultural changes. India. 2005;40:46-98.

6 Okonko DO, Emmanuel JE. Involvement of improved technologies for groundnut production by farmers in Kebbi State of Nigeria. Samaru Journal of Agricultural Research. 2013;(8):23-31.

7 Ajayi AR. Training needs of village Extension workers in Osun State Agricultural Development Projects. Nigerian Journal of Agriculture and Teacher's Education. 2005;(2):1-10.

8 Henn JK. Feeding the cities and feeding the peasants. What roles for Africa's Women Farmers; in: A Sessay and A. Odebiyi (eds). *Women in society and Development* 11 (12) Ibadan: Dokun publishing House. 2005;5-12.

9 Nwachukwu I, Jibowo AA. Evaluation of the utilization of rice processing technology among women farmers in the Central Zone, Nigeria. Journal of Agricultural Extension. 2000;(4):63.

10 United Nations. United Nations Surveying Nigerian Women Processors Discussion Paper. at Nico-Nogan 13th March; 2004.

11 Abba A, Ndanitsa MA, Umar IS, Ibrahim M, Abubakar B. Pilot report for the preparation of One Local Government one Product (OLOP) Programme in Niger State. JICA-OLOP Technical cooperation in Nigeria. 2011;1-89.

12 Yahaya MK. Gender and communication variable in agricultural information dissemination in two agro-ecological zones of Nigeria. Research Monograph University of Ibadan. 2007;68.

13 Nkoh NE, Domenico CM. Enhancing the roles of women in food processing. A case study of igbo women in Nigeria. World Development Publications Elservier Science Ltd. Great Britain. 2005;23(10):1731-1744.

14 Niger State Bureau of Statistics year book 2013;4(17):8-19.

15 National Population Commission. 2006;9(13):87-117.

16 Tsado PB. Futures and prospects of Women Farmers in Niger State Agricultural Zone 1 *Paper Presentation* at Women in Agricultural Programme (WIA) Conference at Niger State 1st March. 2013;1-34.

17 Musa MW. Indegenous resource management system among communities in North West Zone of Nigeria and their relevance for poverty reduction. Ph.D Thesis Submitted to the Department & Agricultural Economics And Rural Sociology, Ahmadu Bello University, Zaria; 2006, ("in press").

18 Maigida DN. Challenges affecting the involvement of women groundnut processor: in Women Agricultural Programme (WIA) in Kano State. Journal of Agricultural Extension. 2012;2(7):43-51.

19 Adepoju SO, Umar AG, Agun JO. An appraisal of women in Kano Agricultural and Rural Development Authority (KNARDA) Extension Delivery programmes. Journal of Agricultural Extension. 2007;10:22.

20 Johnson DT. The business of processing, a guide to processing business. Macmillian Education ltd, London, England. 2009;174-190.

21 Saito KA. Agricultural extension for women farmer in Nigeria. World Bank Discussion Paper, Washington DC. USA. 2009;103-57.

18

Comparative Evaluation of Organic and Inorganic Manure on Sweet Pepper Performance in Two Ecological Zones of Nigeria

Fabiyi Esther Funsho[1], Ademiluyi Benson Oluwafemi[2*] and Abiodun Joseph[3]

[1]Department of Agricultural Economics and Extension, Landmark University, Omu-Aran, Nigeria.
[2]Department of Plant Science, Ekiti State University, Ado Ekiti, Nigeria.
[3]Department of Crop and Soil Science, Landmark University, Omu-Aran, Nigeria.

Authors' contributions

This work was carried out in collaboration between all Authors. Authors FEF and ABO designed the study. Author ABO wrote the protocol and wrote the first draft of the manuscript. Author AJ reviewed the experimental design and all drafts of the manuscript. All Authors managed the analyses of the study. Author ABO performed the statistical analysis. All authors read and approved the final manuscript.

Editor(s):
(1) Velu Rasiah, Department of Environment and Resource Management, Australia and School of Science, Information Technology & Engineering, University of Ballarat, Australia.
(2) Daniele De Wrachien, State University of Milan, Italy.
Reviewers:
(1) Anonymous, Pakistan.
(2) Anonymous, Israel.

ABSTRACT

Organic (poultry manure) and inorganic (Urea) fertilizers were comparatively studied with the aim of improving the performance of sweet pepper, Capsicum annum (called 'Tatase' in Nigeria) in two ecological zones of Nigeria (Ado Ekiti, Ekiti state and Omu-Aran, Kwara State). Poultry manure applied at the rate of 8ton ha^{-1} gave higher fruit yield than urea at the rate of 0.2ton ha^{-1} twice equally at two weeks after transplanting (WAT) and at fruiting. Although higher leaf number and branches were produced in the plots that received inorganic fertilizer, the yield produced was lower. It is indicative from this study that pepper can be successfully produced organically using poultry manure in the two ecological zones under study. It also suggests that our soils can be protected from the level of acidity associated with synthetic fertilizer use. Pepper fruits produced from such will be purely organic and safer for human consumption.

**Corresponding author: E-mail: femnikben@yahoo.com;*

Keywords: Sweet pepper; organic manure; inorganic manure; yield.

1. INTRODUCTION

The provision of quality food starts with planting, production and preparation of food. Organic farming is one of the fastest growing sectors of agriculture worldwide. Its main objective is to create a balance between the interconnected systems of soil organisms, plants, animals and humans [1]. Plants need nutrients in specific proportions to be present in the soil [2] and fertilizer use is a key factor in soil fertility management and yield increase in crop production [3].

Pepper is one of the most important vegetable crops grown in Nigeria [4], where it is commonly used as condiments [5]. Pepper had even been reported to prevent and treat certain ailments like cold and fever, and also improve other food intake [6,7]. Among the various reported benefits resulting from the use of organic fertilizer in crop production includes; enhancement of soil productivity, increase in the soil organic carbon content, soil micro-organism activities, improvement of soil structure, increase in nutrient status balance of the soil and enhancement of crop yield [6,8,9].

Although the applications of organic and inorganic fertilizers have been reported to improve the growth and development of pepper [10], but superior performance of pepper and other plants under poultry manure had been reported than under the inorganic fertilizer application [11,12,4].

The trend in the world today is to produce food in the organic way for quality food that will promote good health of people. The present study was therefore carried out to evaluate the comparative effect of poultry manure and inorganic fertilizer on the growth and yield of sweet pepper in two ecological zones of Nigeria.

2. MATERIALS AND METHODS

The experiments were carried out at the experimental site of the of the Department of Plant Science, Ekiti State University, Ado Ekiti, Ekiti state, Nigeria (7°40′ N, 5°15′ E) and in Landmark University Teaching and Research Farm, Omu-Aran, Kwara State, Nigeria (8°25′ N, 4°40′ E). Both locations have a bimodal rainfall pattern with an annual mean of 1,400mm and 1232mm respectively. Ado Ekiti is in the rain forest zone of south-western Nigeria while Omu-Aran is in the derived savana of North-southern Nigeria. The experiments were carried out in the cropping season of 2012 between April and August in Ado Ekiti, and between February and July in Omu-Aran in 2014. The experiment in Omu-Aran started with irrigation between February and April and later rain fed for the remaining period.

The land was cleared manually with cutlass and ridges made with Nigerian hoe. The soil at the experimental site in Ado Ekiti was analysed to be a sandy clay loam with Soil Organic Matter content-4.3%, 0.48%N, 5.95mgkg^{-1}P, 0.08Cmol K, and a pH of 6.65. Each ridge was 4m long with 1m between ridges. Four ridges constituted a plot and the experiment was laid out in a randomized complete block. Each treatment was replicated four times. The treatments were: Urea; poultry manure; and control. Sweet pepper, *Capsicum annum (*commonly called Tatase in Nigeria) seedlings were transplanted at four weeks after sowing [13] at the spacing of 0.6 mx1 m. The poultry manure was applied on the ridge three weeks before transplanting at the rate of 8tha^{-1} [14]. Urea was applied at the rate of 200kgha^{-1} twice equally at two weeks after transplanting (WAT) and at fruiting. Weeds were manually removed by hand pulling.

The growth measurements include: plant height; stem girth; number of green branches at 8 WAT; number of leaves at 8 WAT while the yield parameters determined include: number of fruits per plant; fruit girth; and fruit weight (tha^{-1}). Assessments were made from ten randomly selected plants per plot and averaged for analysis.

Collected data were statistically analysed using the analysis of variance and means compared using the fisher's least significant difference (LSD).

3. RESULTS

Effects of organic and inorganic fertilizer on pepper height and stem girth are presented in Table 1. Pepper heights observed in poultry manure and Urea were not significantly different but higher than the one obtained in the control experiment. Stem girth obtained in the poultry manure applied plots were highest in both years except in 2014 when it was identical with those of urea applied plots. The control plots gave the lowest stem girth.

Table 2 shows the effects of organic and inorganic fertilizer on number of pepper branches and leaves. The highest number of branches and leaves per plant were obtained in the urea applied plots but not significantly different from the poultry manure applied plots. The control experiment consistently resulted to the lowest number of branches and number of leaves per plant.

The highest number of fruits per plant was obtained in the poultry manure applied plots in both trials while the least was observed in the control plots. Urea produced higher number of fruits than the control but lower than the poultry manure plots (Table 3). Fruit girth obtained from poultry manure plots and those of urea were identical but higher than the ones obtained from the control plots.

Table 4 shows the effects of organic and inorganic fertilizer on pepper yield. The highest fruit yield of pepper in both years was recorded in the plots that received poultry manure while the lowest was observed in the control plots. Poultry manure was significantly higher than urea in terms of fruit yield production.

Table 1. Effects of organic and inorganic fertilizer on pepper height and stem girth

Treatments	Plant height (cm)		Stem girth (mm)	
	2012	2014	2012	2014
Control	34.6b	33.4b	10.2c	12.0b
Poultry manure	45.3a	42.9a	18.3a	17.1a
Urea	43.4a	45.2a	16.1b	15.4a
LSD	5.55	6.81	1.64	2.28

Means with the same alphabet(s) within columns are not significantly different (P=.05)

Table 2. Effects of organic and inorganic fertilizer on number of branches and number of leaves of pepper

Treatments	Number of branches $plant^{-1}$		Number of leaves $plant^{-1}$	
	2012	2014	2012	2014
Control	9b	5b	109.1b	96.5b
Poultry manure	23a	21a	135a	129.5a
Urea	24a	23a	137a	129.0a
LSD	6.45	3.71	5.73	2.64

Means with the same alphabet(s) within columns are not significantly different (P=.05)

Table 3. Effects of organic and inorganic fertilizer on number of fruits and girth of fruits of pepper

Treatments	Number of fruits $plant^{-1}$		Fruit girth (mm)	
	2012	2014	2012	2014
Control	5c	7c	34.5b	31.7a
Poultry manure	21a	17a	40.1a	39.9a
Urea	15b	13b	40.2a	36.5a
LSD	3.51	2.76	4.63	NS

Means with the same alphabet(s) within columns are not significantly different (P=.05)

Table 4. Effects of organic and inorganic fertilizer on pepper yield (g per plant)

Treatments	Pepper Yield ($t\ ha^{-1}$)	
	2012	2014
Control	7.40c	6.02c
Poultry manure	22.71a	20.09a
Urea	15.12b	13.17b
LSD	3.96	3.64

Means with the same alphabet(s) within columns are not significantly different (P=.05)

4. DISCUSSION

The results obtained from this study clearly showed that the organic manure (poultry manure) used in the two locations of the experiment substantially increased the performance of pepper in terms of growth and yield.

The highest growth parameters and yield obtained from the plots that received poultry manure application probably reflected high release of essential nutrients especially nitrogen and Phosphorous from this organic fertilizer (poultry manure). Reports from other researchers have shown that organic-amended soils have twice the level of N as conventional soils [15-17]. Poultry manure had been reported to possess essential nutrient elements which are associated with high photosynthetic activity capable of promoting root and vegetative growth of plants [18,4]. Higher but insignificant number of leaves and branches obtained from Urea applied plots than poultry manure in this work may suggest over supply of nitrogen which increased vegetative growth without a corresponding increase in yield. Similar vegetative growth in pepper had been reported when NPK fertilizers increased the number of leaves but lower yield than Tithonia diversifolia leaf biomass [19]. Similar report had shown that hairy vetch/rye mineralization allowed for a slower release of N and more efficient absorption by sweet corn [20].

5. CONCLUSION

The present work had clearly shown that pepper can be successfully produced organically using poultry manure in the two ecological zones under study. Since poultry manure is readily available as wastes from poultry farmers, it will be useful for farmers at very low cost and thus reduce the high cost associated with inorganic fertilizers. Our soils will also be protected from the level of acidity associated with synthetic fertilizer use. The pepper fruits produced from such will be purely organic and safer for human consumption.

COMPETING INTERESTS

Authors have declared that no competing interests exist.

REFERENCE

1. Karanatsidis G, Berova M. Effect of Organic-N fertilizer on growth and some physiological parameters in pepper plants (*Capsicum annum* L.) Biotechnol. & Biotechnol. EQ. 2009;254-257.
2. Fagbayide. Growth and nutrient concentration of pepper fruits as affected by phosphorus fertilizer. Proceeding, 15th annual conference. Horticultural society of Nigeria. 1997;40-42.
3. Akinrinde EA. Issues of optimum nutrient supply for sustainable crop production in Tropical Developing Countries. Pakistan Journal of Nutrition. 2006;5(4):377-387. Asian Network for Scientific Information.
4. Aliyu L. The effect of organic and mineral fertilizer on growth yield and composition of pepper (*Capsicum annum* L.) Biol Agric Hort. 2000;18:29-36.
5. Alabi DA. Effect of fertilizer phosphorus and poultry droppings treatments on growth and nutrient components of pepper (*Capsicum annum* L.). African Journal of Biotechnology. 2006;5(8):671-677.
6. Udoh DJ, Ndon BA, Asuquo PE, Ndaeyo NU. Crop Production Techniques for the Tropics. Concept Publication, Lagos, Nigeria. 2005;446.
7. Grubben GJ. Tropical Vegetables and their Genetic Resources. IPRGR, Rome, Italy. 1977;197.
8. FAO. Fertilizers and their uses. 2000;26-39.
9. Ademiluyi BO, Omotoso SO. Comparative evaluation of *Tithonia diversifolia* and NPK fertilizer for soil improvement in maize (*Zea mays*) production in Ado Ekiti, South-western Nigeria. Research Journal of Agronomy. 2008;2(1):8-11.
10. Malik AA, Chattoo MA, Sheemar G, Rashid R. Growth, yield and fruit quality of sweet pepper hybrid SH-SP-5(*Capsicum annuum* L.) as affected by integration of inorganic fertilizers and organic manures (FYM), Journal of Agricultural Technology. 2011;7(4):1037-1048.
11. Husseib TO. Effect of poultry manure on the growth of tomato. Proceeding of 15th Annual Conference of Horticultural Society of Nigeria (HORTSON), April, 8-11 soil analysis Manual series Nigeria. 1997;43-45.
12. De-Lannoy G, Romain HR. Crop Production in Tropical Africa. London, Mc. Graw Hill Publishers Ltd. New York. 2001;234-245.
13. Ademiluyi BO. Studies on Transplanting Age and Pepper (*Capsicum annum* L) Performance in a South-western Nigeria

Location. Bulletin of pure and Applied Sciences. 2012;31B(2):53-57.

14. Ikeh AO, Ndaeyo NU, Uduak IG, Iwo GA, Ugbe LA, Udoh EI, Effiong GS. Growth and yield Responses of Pepper (*Capsicum frutescens* L.) to Varied Poultry Manure Rates in Uyo, Southeastern Nigeria. ARPN Journal of Agricultural and Biological Science. 2012;7(8):735-742.
15. Burger M, Jackson LE. Microbial immobilization of ammonium and nitrate in relation to ammonification and nitrification rates in organic and conventional cropping systems. Soil Biol. Biochem. 2003;35:29–36.
16. Delate K, Cambardella C, Mckern A. Effects of Organic Fertilization and Cover Crops on an Organic Pepper System. Hort Technology. 2008;18:215–226.
17. Dauda S, Aliyu NL, Chiezey UF. Effect of variety, seedling age and poultry manure on growth and yield of garden egg (*Solamun gilo* L.) Aliju Acad Forum. 2005;988-995.
18. John GC, Almazan LP, Paria J. Effects of Nitrogen fertilizer on the intrinsic rate of the rusty plum aphid. Environmental Entomology. 2004;34(4):938-943.
19. Ademiluyi BO. Comparative evaluation of NPK fertilizer and *Tithonia diversifolia* biomass in sweet pepper (*Capsicum annum*) Production in Ado Ekiti, Nigeria. Journal of Life Sciences. 2013;7(3):288-291.
20. Carrera LM, Abdul-Baki AA, Teasdale JR. Cover crop management and weed suppression in no-tillage sweet corn production. Hort Science 2004;39:1262–1266.

19

Molecular Characterization of *Solanum* Species Using EST-SSRs and Analysis of Their Zinc and Iron Contents

L. A. Oppong[1*], M. D. Quain[1], A. Oppong[2], H. A. Doku[1], A. Agyemang[3], and K. Offei Bonsu[4]

[1]Molecular biology section, Biotechnology, Seed and Food Science Division, CSIR-Crops Research Institute, P.O.Box 3785, Fumesua-Kumasi, Ghana.
[2]Plant Health Division, CSIR-Crops Research Institute, P.O.Box 3785 Fumesua-Kumasi, Ghana.
[3]CSSS Division, CSIR-Crops Research Institute, P.O.Box 3785, Fumesua-Kumasi, Ghana.
[4]Horticulture Improvement Division, CSIR-Crops Research Institute, P.O.Box 3785, Kwadaso - Kumasi, Ghana.

Authors' contributions

This work was carried out in collaboration between all authors. Authors LAO and MDQ designed the study, wrote the protocol, and managed the literature searches. Authors AO, HAD and AA performed statistical analysis and wrote the first draft of the manuscript. Author EOB provided germplasm and other technical support. All authors read and approved the final manuscript.

Editor(s):
(1) Özge Çelik, Department of Molecular Biology and Genetics, Istanbul Kultur University, Turkey.
(2) Daniele De Wrachien, State University of Milan, Italy.
Reviewers:
(1) Anonymous, Università Mediterranea di Reggio Calabria – Italy.
(2) Anonymous, Luxembourg.
(3) Anonymous, Arba Minch University, Ethiopia.

ABSTRACT

Aims: The objectives of study were to establish genetic diversity among the populations of three *Solanum* sp: *Solanum torvum*, *Solanum aethiopicum* and *Solanum anguivi* and identify species with the highest concentration of iron and zinc which can be recommended for improved maternal and child nutrition. It also aimed to identify polymorphic markers useful in *Solanum* species diversity studies or screening in Ghana.
Place and Duration of Study: CSIR-Crops Research Institute, Molecular Biology Lab, Kumasi. The study was carried between March 2011 and June, 2012.

**Corresponding author: Email: abroklin1984@yahoo.com;*

Methodology: Investigations were carried out on 30 accessions of three *Solanum* species assembled from five geographical regions of Ghana. EST-SSR and AAS were employed for the estimation of genetic diversity and assessment of mineral concentrations respectively. Data generated were analysed using POPGENE version 1_32, Genstat software version 9 and Microsoft excel software (Windows 2007).
Results: All the markers showed allelic polymorphism in all the accessions studied. Overall genetic diversity was considerably high (I =1.0032, He = 0.4942, Nei's He = 0.4859) and fixation index statistics (Fst) shows that 10% of the variation exists among the population. Clustering of the accessions did not exactly coincide to the populations of the different *Solanum* sp under study. Outcrossing rate (t = 0.8154) and Gene flow (Nm = 2.2071) of populations of the accessions were extremely high. Assessment of their mineral contents suggests that *S. torvum* and *S. aethiopicum* species are rich in Fe and Zn respectively.
Conclusion: EST-SSR markers and mineral analysis revealed genetic diversity among the different accessions. However the population studies suggest gradual homogeneity of these accessions due the high gene flow and outcrossing rate over time.

Keywords: Genetic diversity; solanum species; EST-SSR (expressed sequence tag-simple sequence repeat); mineral assessment and AAS (atomic absorption spectrophotometry).

1. INTRODUCTION

Solanaceae is a plant family comprising about 2300 species, nearly one-half of which belong to the genus *Solanum* [1]. The genomics in the genus *Solanum* are evolving at a moderate pace compared to other plant species [2]. *Leptostemonum*, subgenus of *Solanum*, comprise almost one third of the genus. They are distributed worldwide and include a number of economically important species of cultivated egg-plants which consist of two sub-groups namely *melongena* and *oliganthes* [1]. The former includes *S. melongena* L. (Brinjal egg-plant, aubergine) and *S. macrocarpon* (Gboma egg-plant), while the latter comprise *S. aethiopicum* L. [3,4]. *S. aethiopicum* and *S. macrocarpon* are known to be domesticated in Africa from their wild relatives namely *S. anguivi* and *S. dasyphyllum* respectively [5]. These eggplants are considered as very important vegetables in Ghana and West Africa [6,7,8,]. Another important wild relative is the *Solanum torvum* (Turkey berry) which can also be found in Mexico, Peru, Venezuela, and in the West Indies including Bermuda [9]. *Solanum* wild species or landraces are rich source of novel genes which can be introgressed into cultivated species [10]. Interestingly, *S. torvum* has been identified to carry resistance genes to serious diseases of eggplant, particularly bacterial and fungal wilts, and infections caused by nematodes [1,11,12,13]. *S. torvum* is used as rootstocks for controlling wilt disease in eggplant [14,15].

Fruits of *Solanum anguivi* can be utilized either in their fresh or dried state. It is believed to have medicinal properties particularly against high blood pressure and diabetes as well as being an appetizer. Nutritional analysis of *S. torvum* reveals that it has special properties of stimulating the production of blood cells and could be helpful in treating anaemia due to its high iron content. It contains about 30 mg of iron, 454mg of calcium and sizeable amounts of protein with other nutrients per 100 g edible portion [9]. Nutritionally the high iron content in *S. torvum* can significantly reduce the incidence of iron (Fe) deficiency [16,17,18] in most parts of the world especially when Fe rich genotypes are identified. Identification of mineral rich genotypes may require mineral characterisation and their diversity.

Photometry and Atomic Absorption Spectroscopy (AAS) has over the years been the most popular techniques used for qualitative and quantitative assessment of trace elements in foods, nanomaterials, biomaterials, forensics and industrial waste [19,20]. However, in recent times, emphasis is given to inductively coupled plasma-atomic emission spectroscopy (ICP-AES) due its high reproducibility and quantitative linear range compared to conventional AAS, and reduces molecular interferences due to a higher temperature (7000–8000 K) in the excitation source (plasma) [20]. Nevertheless, ICP-AES is more expensive than conventional AAS, and in complex samples, emission patterns can be difficult to interpret [21,22]. Moreover, ASS is the method of choice for most routine trace mineral element analyses and very effective for the determination of elements such as magnesium, zinc, and copper [19].

Morphological diversity study has often been employed for genetic diversity studies in crop plants [23]. However variations in environmental conditions such as soil types and fertility levels, light, temperature and moisture regimes affect morphological characterization, particularly when experiments are repeated in time and/or space [24,25,26]. In addition, the genetic make-up of seed, field management practices has been reported to influence the morphology of a crop [27].

The weaknesses of morphological diversity or characterisation paved way for Molecular characterisation which offers the best estimate of genetic diversity since they are independent of the confounding effects of the environmental [28,29]. Among molecular markers, Simple Sequence Repeat (SSR) has proven to be effective due to its high polymorphic and co-dominance nature, reproducibility, affordability, low technical complexity [29] to mention but a few. It has also been widely employed for many *Solanum* sp diversity studies [30,31,32].

SSRs derived from ESTs or cytoplasmic DNA (cDNAs), are mostly genic SSRs and have the potential for application in gene function characterisation [33], association analysis for gene tagging [34,35,36], and quantitative trait loci (QTL) analysis [37,38,39]. EST-SSR designed for *Solanum* sp is currently employed for diversity study [30,40].

In Ghana, despite the role of eggplants on the livelihood of people [7], information on their mineral content and diversity has not been well documented especially, the *S. anguivi*, *S. aethiopicum* and *S. torvum* accessions. This study was therefore designed to establish genetic diversity among three populations of *Solanum* species: *Solanum torvum, Solanum aethiopicum* and *Solanum anguivi* employing EST-SSR primers. It also seeks to assess their mineral contents using the Atomic Absorption Spectrophotometry (AAS) and identify accessions possessing high amount of iron and zinc for enhanced child and maternal nutrition in Ghana and beyond [16,17,18]. The study may assist to identify polymorphic primers useful in *Solanum* species diversity studies or screening in Ghana.

2. MATERIALS AND METHODS

2.1 Plant Materials and Extraction of Genomic DNA

Thirty (30) accessions of *Solanum* sp. were collected from 5 geographical locations in Ghana. The 30 assembled germplasm were grouped into three populations according to their species: *Solanum torvum, Solanum anguivi and Solanum aethiopicum* (Table 1). Two hundred milligram 200mg of each accession young tender leaves were weighed into 2ml eppendorf tubes and grounded with liquid nitrogen to fine powder. The genomic DNA was extracted using the Dellaporta protocol [41] with slight modifications. Thus the protocol was modified through the introduction of 7.5 M Sodium acetate DNA and the addition of the RNAse to digest the RNA present in the genomic DNA.

2.2 DNA Quantification and Gel Electrophoresis

DNA quality was checked on 0.8% agarose in 1x TAE (Tris-acetic EDTA) buffer by gel electrophoresis. Electrophoresis of the DNA was carried at 100V for 40mins and then visualized with a UV transilluminator. The concentration of the DNA was projected by the intensity and comparison to 1kb lambda DNA mass ladder (1kb, Invitrogen). The quantification of the DNA was evaluated by reading absorbance at 260nm and 280nm with the spectrophotometer. The DNA was then diluted to 10 ng/µl for PCR

2.3 PCR Amplification

PCR amplification of DNA was carried out using a 96 well plate AB Applied Biosystem Thermocycler. The PCR conditions were optimized for cycling number, concentrations of the primer, $MgCl_2$ and DNA template. The reaction mixture (10 ul) contained 6.075 µl of Nuclease free sterile water (DNA grade water),1 µl of 10x PCR Buffer,0.9 µl of $MgCl_2$ (25 mM), 0.4 µl dNTPs (10 mM), primer 0.25 µl (50 µg/ml) of each forward and reverse, 0.125 µl of Supertherm Taq Polymerase (1unit) and 1 µl of 10 ng DNA template. The PCR cycling conditions followed an initial denaturing step of 94°C for 5mins, a touchdown PCR protocol of 11 cycles of 94°C for 30 secs, 60°C for 30 secs decreasing by 0.5 per cycle and 72°C for 60 secs followed by 30 cycles of 94°C for 30 secs, 55°C for 30 secs and a final elongation step of 72°C for 60 secs. In each PCR, a negative control was included to detect contaminations. Twelve set of

EST-SSR primers (Table 2) were used in the experiment. PCR products were electropholysed on 6% polyacrylamide gel (Water, 10x TBE, 4% acrylamide (19:1), 10% APS and TEMED). A 100bp ladder (Gene Ruler ^{TM}, Fermentas) was used as a size marker. Amplified bands were visualized and scored.

2.4 Mineral Analysis

Determination of content of mineral elements using atomic absorption spectrophotometry (A.A.S.).

Sample dried seeds of each accession were dehulled (manually) and ground into fine powder. Five hundred milligram (0.5 g) of each sample (accession) was weighed into a Teflon beaker. Amounts of 5ml of HNO_3 (70% HNO_3) and 1ml of H_2O_2 (30% H_2O_2) were added. The Teflon beakers were covered and placed in a rotor for digestion. Each sample was digested in three replicates (Zuink and Planck, 1990). After the digestion, the inside walls of the Teflon beakers were washed with distilled water and made up to a final volume of 20 ml. The samples were then transferred from the Teflon beakers into test tubes. The digested samples were assayed for the presence of iron (Fe) and zinc (Zn) using Atomic absorption spectrophotometry in an acetylene-air flame employing the manufacturers recommended working conditions. Reference standards for the elements of interest, blanks and repeats of the samples were digested the same way as the actual samples. These served as internal positive controls.

2.5 Calculation of Concentrations of Specified Mineral Elements

For each of the two elements; Iron (Fe) and zinc (Zn), concentration in the samples was calculated using the formula:

$$\text{Final Concentration (mg / kg)} = \frac{\text{Conc. (df) x Nominal volume}}{\text{Sample weight in grammes}}$$

Where,

Conc. (df) = AAS Conc. Reading, and
Nominal volume = Final volume after digestion

2.6 Data Collection

The amplified EST-SSR DNA bands representing different alleles were scored. Bands were recorded as homozygous genotypes (AA, BB, CC. . .) or heterozygous genotypes (AB, AC, BC. . .). Genetic parameters such as average observed allele number (*Na*), the percentage of polymorphic loci (*P*), observed heterozygosity (*Ho*), expected heterozygosity (*He*), Nei heterozygosity (*Nei He*), Shannon's diversity index (*I*) , and gene flow (*Nm*) were calculated to estimate the level of genetic diversity [42] . The F-statistics (F_{IS}, F_{IT} and F_{ST}) [42,43] were also computed for polymorphic loci to test for the departure from Hardy–Weinberg equilibrium and to estimate genetic differentiation among the *Solanum species* populations under study. The out-crossing rate [$t = (1-F_{IT})/(1 + F_{IT})$] was calculated based on the F_{IT} values to estimate indirectly the mating pattern of *the* populations [43].

2.7 Statistical Analysis

All the calculations were performed using POPGENE 32 version 1.32 [42,44]. A hierarchical analysis of molecular variance (AMOVA) of the *Solanum* species populations was calculated to partition genetic diversity within and among populations. For AMOVA, the genotype banding patterns were converted into a '1' (present) and '0' (absence) matrix and subjected to analysis, based on Jaccard coefficient using single linked similarity matrix method. Hierarchal AMOVA was analyzed using Genstat software version 9. Correlation between Zinc and iron were generated to determine the degree of association between Zn and Fe of the accessions using Microsoft Excel software (Windows 2007).

3. RESULTS AND DISCUSSION

3.1 Results

3.1.1 Diversity studies

Sixty four alleles were generated for all loci with a mean of 5.333 per locus (Table 2). It was observed that all the EST-SSR markers showed allelic polymorphism (Fig. 1). Primer EEMS 37 recorded the highest number of alleles and showed the highest level of genetic variability (Expected heterozygosity (*He*) = 0.843, Shannon's Information index (*I*) = 1.978). It also revealed Nei's heterozygosity of 0.829 (Table not shown). On the other hand, four primers (EEMS 6, EEMS 13, EEMS 15 and EEMS 16) recorded the least number of alleles (3) with locus EEMS 6 showing the least Shannon diversity index (*I* = 0.4740).

Table 1. Assembled accessions and source of collection

Accession	Region of collection	Lab code	Common name	Accession	Region of collection	Lab code	Common name
Hwidiem *S. torvum*	Brong Ahafo	HST 001	Turkey berry	AsanteAkyem- Wenkyi, *S. torvum*	Ashanti	AAWST 017	Turkey berry
Hwidiem *S. aethiopicum*	Brong Ahafo	HSA 002	African Garden eggs	Bonwire *S. torvum*	Ashanti	BWST018	Turkey berry
Hwidiem *S. anguivi*	Brong Ahafo	HSN 003	Uganda peas	Assin Odumase *S. aethiopicum*	Central	AOSA019	African Garden eggs
Kenyase *S. aethiopicum*	Brong Ahafo	KSA 004	African Garden eggs	Assin Odumase *S. torvum*	Central	AOST020	Turkey berry
Kenyase *S. anguivi*	Brong Ahafo	KSN 005	Uganda peas	Assin Odumase *S. anguivi*	Central	AOSN021	Uganda peas
Kenyase *S. torvum*	Brong Ahafo	KST 006	Turkey berry	Assin Dominase *S. aethiopicum*	Central	ADSA022	African Garden eggs
Ntoturoso *S. torvum*	Brong Ahafo	NST 007	Turkey berry	Assin Dominase *S. anguivi*	Central	ADSN023	Uganda peas
Ntoturoso *S. aethiopicum* 1	Brong Ahafo	NSA 008	African Garden eggs	Assin Dominase *S. torvum*	Central	ADST024	Turkey berry
Ntoturoso *S. aethiopicum* 2	Brong Ahafo	NSA 009	African Garden eggs	Sefwi Beposo *S. torvum*	Western	SBST025	Turkey berry
Kwahu Praso *S. torvum*	Eastern	KPST 010	Turkey berry	Sefwi Asempaneye *S. torvum*	Western	SAST026	Turkey berry
Kwahu Praso *S. aethiopicum*	Eastern	KPSA 011	African Garden eggs	Sefwi Babiani *S. aethiopicum*	Western	SBSA027	African Garden eggs
Nkawkaw *S. anguivi*	Eastern	NKSN012	Ugandan peas	Sefwi Babiani *S. torvum*	Western	SBST028	Turkey berry
Nkawkaw *S. aethiopicum*	Eastern	NKSA013	African Garden eggs	Sefwi Awaso *S. aethiopicum*	Western	SWSA029	African Garden eggs
Amanpe *S. anguivi*	Ashanti	AMSN014	Uganda peas	Sefwi Awaso *S. torvum*	Western	SWST030	Turkey berry
Amanpe *S. aethiopicum*	Ashanti	AMSA015	African Garden eggs				
Amanpe *S. torvum*	Ashanti	AMST016	Turkey berry				

Table 2. List of primers and their genetic variability estimates

Primer	Repeat motif	na *	Forwad primer (5'-3')	Reserve primer (5'-3')	ASR	*ne**	*He**	*I**
EEMS 16	(AC)7	3.000	CAATTTTTCGGTTCACTAATCAAG	CTTCAAGGAAAAAGGAGGCC	135-141	1.575	0.371	0.603
EEMS 7	(T)13	3.000	CCATGCCAGAATGGAAACTT	AACGAAAACACGATCAACCC	250-260	2.135	0.540	0.911
EEMS 10	(A)20	4.000	TCAAGCAGAACGAAGATGGA	GTAGGGGACGTGGATTCAGA	266-290	1.527	0.351	0.708
EEMS 15	(C)12	3.000	GGGACAAATCTGACCTTTGG	CTGGTGGCAAATTCTTCGAT	270-294	2.273	0.570	0.950
EEMS 28	(TAA)21	5.000	GACGATGACGACGACGATAA	TGGACTCACAACTCAGCCAG	180-236	3.416	0.719	1.335
EEMS 46	(ACCAGC)6	4.000	ACCAAACGTGCATGAAACAA	GGAAATGTTGGTGGAATTGG	245-265	1.227	0.188	0.412
EEMS 50	(TA)9(GA)8	5.000	AAATCCCGGCCATTCTGTGTA	ACATCGTTCCGCCTCTATTG	218-226	2.135	0.540	1.012
EEMS 37	(TCC)5	10.000	CCCTTCCTACCCACACTTCA	GTTTTGCACCTTTCCATCGT	114-123	5.844	0.842	1.978
EEMS 6	(T)14	3.000	TCATGCGAAGATTAATTAAATGTGA	GAGTGGATGATCAAGAATGGC	268-274	1.315	0.244	0.485
EEMS 20	(AT)8	9.000	AACATCAGCCAGGGTGTTTC	TGCTGAAAATTACAAGCAAA	221-227	4.615	0.797	1.846
EEMS 48	(TAA)20(CGA)8	8.000	CAATGCAAACAATTATCATTTCG	TCGATGTTGTCGTCGTT	223-241	1.676	0.410	0.980
EEMS 49	(TA)12(GA)7	7.000	TGAAATTGATCAATACCTATAAATTTAG	GAAAGCCAGGATAGCATTCG	145-153	1.539	0.356	0.817
Total		64				29.277	5.928	12.036
Mean		5.333				2.440	0.494	1.003
SD		2.535				1.452	0.211	0.494

The study reveals genetic variability among individuals in each population [*I* = 1.0032, *He* = 0.4942, *Nei's He* 0.4859 and *Ho* = 0.2417 (Table 3)]. Nevertheless, genetic variability was not uniformly distributed across the population. The highest genetic differentiation existed among the *S. aethiopicum* population with the least among the *S. torvum* population. The overall value of F_{ST} reveals that approximately 10% of the total genetic diversity exists among the population. F_{IS} and F_{IT} of less than 0.5 in the *Solanum sp* populations indicates some degree of heterozygosity within the populations studied and a significant departure of allelic frequencies from the Hardy–Weinberg equilibrium. The overall outcrossing rate was extremely high (81.54%) with a gene flow of 2.2071. A dendogram (Fig. 2) based on genetic Unweighted Pair Group Method with Arithmetic Mean (UPGMA) [45] showed that *S. anguivi* and *S. aethiopicum* populations had the greatest similarity and were separated from *S. torvum* which form different clusters at a genetic distance of 5.77594 (Table 4).

3.2 Cluster Analysis

UPGMA cluster analysis based on the 12 EST-SSR loci employing Jaccard Coefficient single link similarity matrix method (Fig. 3) demonstrated genetic differentiation by revealing six distinct clusters from a genetic similarity distance of 37.3% up to 85%. However at a genetic distance of 37.3%, all accessions of the various populations could be grouped into two main clusters, i.e. Ntoturoso *S. aethiopicum1* and the other 29 accessions which further separated into five distinct sub clusters. The accessions with the closest resemblance were revealed at the sixth cluster at a genetic similarity of 85% (Hwidiem *S. torvum,* A/A Wenkyi *S. torvum*, A/Dominase *S. anguivi*, A/ Odumasi *S. torvum* and ntoturoso *S. torvum*). The widest separation occurred between Hwidiem *S. torvum* and Ntoturoso *S. aethiopicum* 1. The clustering did not exactly coincide with geographical origin as well as population or species groupings of the accessions. This shows a clear genetic variation and differentiation pattern among the individuals of the various species.

3.3 Variation Based on Minerals

The result on mineral analysis for the concentrations of Fe and Zn showed that no accession recorded the highest concentrations of both minerals (Table 5). However, one accession (Assin Odumase *S. anguivi*) recorded the least concentration for both Fe (7.26 mg/Kg) and Zn (9.16 mg/Kg). Averagely the *S. torvum* and *S. aethiopicum* population had the highest concentration of Fe and Zn respectively. The highest concentration of Fe was recorded by Assin Odumase *S. torvum* (261.48 mg/kg) and the highest zinc concentration was recorded by Sefwi Awaso *S. aethiopicum* (29.16 mg/kg). Statistically insignificant degree of association was observed between Fe and Zn for each species (Table 5).

4. DISCUSSION

4.1 Diversity Studies

The 100% allelic polymorphism among the individuals of the species generated by the use of the EEMS primers indicates that all the loci are very informative especially EEMS 37, EEMS 20 and EEMS 48 which produced ten, nine and eight alleles respectively with the former producing the highest diversity (*I* = 1.978). The informative nature of the EST-SSR markers is also consistent with similar finding by Aniko et.al. [26] who employed the same primers utilised in this study for genetic mapping and phylogenetic studies of eggplant with EEMS 48 producing the highest number of alleles. This suggests that in selecting primers for biodiversity evaluation for eggplant or species closely related, the three primers mentioned above can be given priority. The present results based on the polymorphic pattern of 12 selected EST-SSR loci confirmed that *Solanum* sp or populations under studied possessed relatively high genetic diversity (*He* = 0.494, *Nei's He* = 0.486, *I* = 1.003). These agrees with results by other workers [26,27,35,46] who employed either SSR, AFLP and/or, RAPD markers as well as morphological descriptors for characterization of eggplant. Except a few loci (EEMS 6, EEMS 16 and EEMS 46), the genetic variation (*I*) observed among the individual accessions of the various species (Table 1) where either comparably close or higher than genetic variation (*I*) obtained for the three populations. According to Muller *et al* (2001) [47], this can be attributed to a high genetic overlap as a result of probable common ancestry. Furthermore, the low values of F_{IS} (0.4188) and F_{IT} (0.4779) as well as high values of *t* (0.8154) and *Nm* (2.2071) from the EST-SSR analysis suggest that the *Solanum* sp under study were more cross pollinated with wide pollen-mediated gene flow although they can undergo self-pollination due to their hermaphroditic nature.

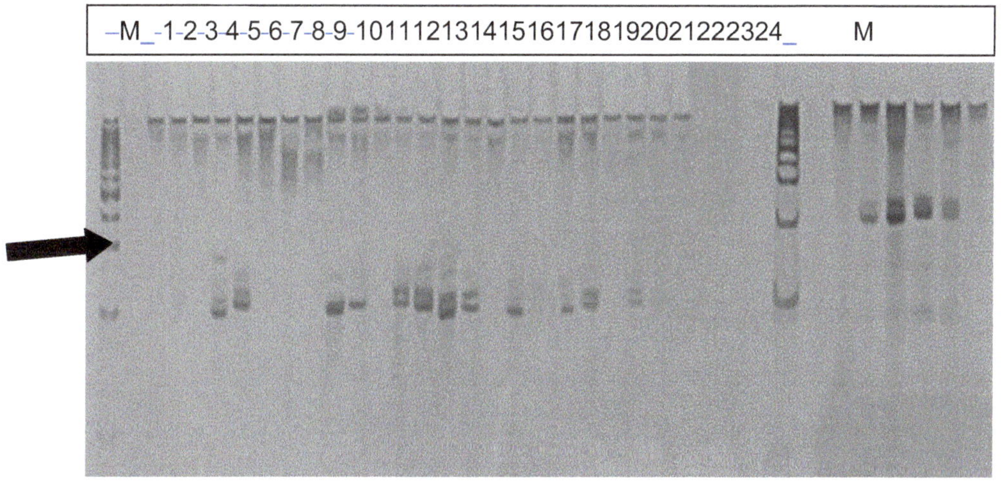

Fig. 1. The banding pattern of primer EEMS 20 on accessions numbered 1-30: Hwidiem *S. torvum*= 1; Hwidiem *S. aethiopicum* = 2, Hwidiem *S. anguivi* = 3; Kenyase *S. aethiopicum* = 4; Kenyase *S. anguivi*= 5; Kenyase *S. torvum* = 6; Ntoturoso *S. torvum* = 7; Ntoturoso *S. aethiopicum* 1 = 8; Ntoturoso *S. aethiopicum* 2= 9; K. Praso *S. torvum* =10; *K. Praso S. aethiopicum* = 11; Nkawkaw *S. anguivi*= 12; Nkawkaw *S. aethiopicum*= 13;Amanpe *S. anguivi* = 14; Amanpe *S. aethiopicum* = 15; Amanpe *S. torvum* = 16; A/AWenkyi *S. torvum* = 17; Bonwire *S. torvum* = 18; A/ Odumase *S. aethiopicum*= 19; A/Odumase *S. torvum*= 20; A/Odumase *S. anguivi* = 21; A/Dominase *S. aethiopicum* =22; A/Dominase *S. anguivi* = 23; A/Dominase *S. torvum* = 24; S/Beposo *S. torvum* = 25; S/ Asempaneye *S. torvum* = 26; S/Babiani *S. aethiopicum*= 27; S/Babiani *S. torvum* = 28; S/Awaso *S. aethiopicum*= 29;S/Awaso *S. torvum*=30, M= Marker, 200 bp

Fig. 2. Dendrogram based on Nei's (1978) genetic distance: method-UPGMA modified from NEIGHBOR procedure of PHYLIP Version 3.5 using 12 EST-SSR markers

Torvum population=*Hwidiem S. torvum, Kenyase S. torvum, Ntoturoso S. torvum, K. Praso S. torvum, Amanpe S. torvum, A/A Wenkyi S. torvum, Bonwire S. torvum, A/Odumasi S. torvum, A/Dominase S. torvum, S/Beposo S. torvum, S/Asempaneye S. torvum, S/Babiani S. torvum, S/Awaso S. torvum*

Anguivi population=*Hwidiem S. anguivi, Kenyase S/anguivi, Nkawkaw S. anguivi, Amanpe S.anguivi, A/Odumase S.anguivi , A/Dominase S.anguivi*

Aethiopicum population=*Hwidiem S. aethiopicum, Kenyase S .aethiopicum, Ntoturoso S. aethiopicum, K. Praso S. aethiopicum, Nkawkaw S. aethiopicum, Amanpe S. aethiopicum, A/Dominase S. aethiopicum, S/Babiani S. aethiopicum, S/Awaso S. aethiopicum*

Table 3. Genetic variability estimates of the genotypes

Population region	N_a	N_e	PL	P %	H_o	H_e	Nei's H_e	I	F_{is}	F_{it}	F_{st}	t	Nm
Solanum torvum	3.5000	1.7284	10	83.33	0.1154	0.6692	0.3180	0.6369					
Solanum anguivi	3.2500	2.3840	9	75.00	0.3194	0.4949	0.4537	0.8444					
Solanum aethiopicum	4.3333	2.5263	12	100.00	0.3485	0.5841	0.5575	1.0582					
Overall mean	5.3333	2.4398	12	100.00	0.2417	0.4942	0.4859	1.0032	0.4188	0.4779	0.1017	0.8154	2.2071

Na= average number of alleles; Ne= average number of effective alleles (Kimura and Crow 1964); PL= number of polymorphic Loci; P= percentage of polymorphic loci; Ho, observed heterozygosity (Levene, 1949); He= expected heterozygosity(Levene, 1949); Nei's He = Nei's expected heterozygosity (Nei, 1973); I, Shannon diversity index (Lewontin 1972); Fis, Fit and Fst, estimates of F-statistics of regional populations (Hartl and Clark 1989); t, out crossing rate = (1- Fit)/ (1 + Fit) (Cao et al., 2006); Nm = Gene flow estimated from Fst = 0.25(1 - Fst)/Fs

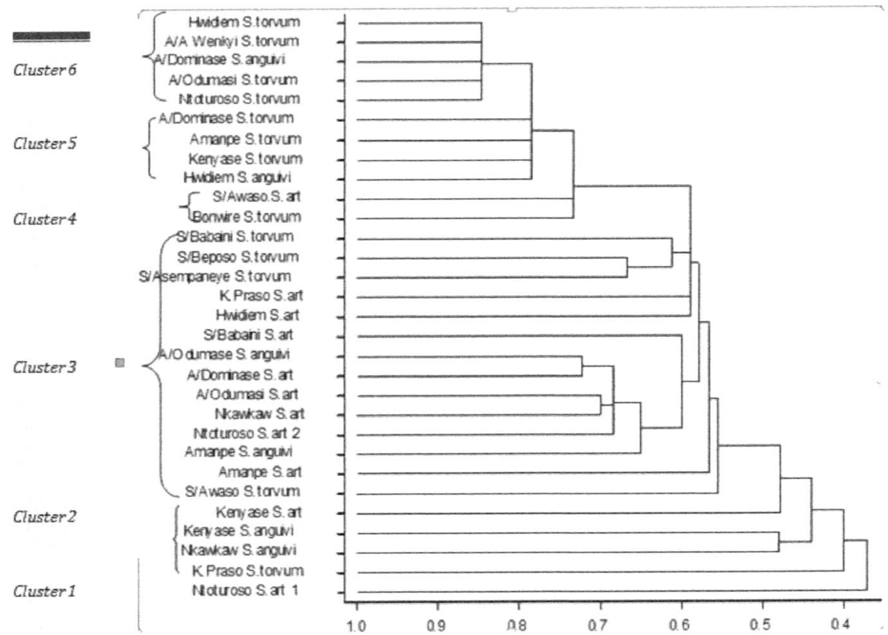

Fig. 3. Dendrogram showing the relationships among the various accessions

Table 4. Genetic distance between populations

Between	And	Length
2	*S. torvum*	7.27970
2	1	5.77594
1	*S. anguivi*	1.50376
1	*S. aethiopicum*	1.50376

Table 5. Mineral content of accessions and simple linear correlation coefficients between the iron and zinc variables studied considering the values expressed in dry matter basis for the 30 accessions of *Solanum* sp. (n=10 for each accessions; *: P<0.05)

Designation	Mineral		Correlation
	Fe (mg/Kg)	Zn (mg/Kg)	
Hwidiem *S. torvum*	68.12	18.76	
Kenyase *S. torvum*	44.24	21.56	
Ntoturoso *S. torvum*	256.36	13	
Kwahu Praso *S. torvum*	126.8	13.28	
Amanpe *S. torvum*	173.88	14.08	
Bonwire *S. torvum*	8.64	12.64	
Asante Akyem Wenchi *S. torvum*	106.92	12.88	
Assin Odumase *S. torvum*	261.48	14.72	
Assin Dominase *S. torvum*	89.16	17.88	-0.4455
Sefwi Beposo *S. torvum*	215.44	10.72	
Sefwi Awaso *S. torvum*	91.8	14.56	
Sefwi Asempaneye *S. torvum*	59.68	14	
Sefwi Babiani *S. torvum*	130.96	14.48	
Mean of *torvum*	125.65	14.78	
Hwidiem *S. anguivi*	30.36	27.12	
Kenyase *S. anguivi*	17.84	18.76	
Nkwakaw *S. anguivi*	15.8	14.32	0.041778
Amanpe *S. anguivi*	127.56	15.08	
Assin Odumase *S. anguivi*	7.26	9.16	
Assin Dominase *S. anguivi*	45.36	28.52	
Mean of *anguivi*	40.70	18.83	
Hwidiem *S. aethiopicum*	155.96	18.52	
Kenyase *S. aethiopicum*	114.24	17.16	
Ntoturoso *S. aethiopicum* 1	114.28	23.6	
Ntoturoso *S. aethiopicum* 2	68.32	21.64	
Nkawkaw *S. aethiopicum*	14.24	20	
Amanpe *S. aethiopicum*	40.28	18.36	-0.67893
Assin Odumase *S. aethiopicum*	23.38	22.8	
Assin Dominase *S. aethiopicum*	165.16	14.2	
Sefwi Awaso *S. aethiopicum*	29.84	29.16	
Sefwi Babiani *S. aethiopicum*	223	14.56	
Mean of *aethiopicum*	100.69	19.79	
Overall Correlation			-0.41491

The population dendogram estimated from Nei's unbiased measures of genetic identity and genetic distance also confirm the genetic variation among the populations with the population of *S. torvum* and *S. aethiopicum* being widely separated from each other. This is more evident in Hwidiem *S. torvum* and Ntoturoso *S. aethiopicum* 1 which were two accessions with widest separation (Fig 3). The close relatedness of *S. aethiopicum* and *S. anguivi* from the result clearly demonstrate that the two populations have a common ancestry as *S. anguivi* is the wild type of egg plant from which *S. aethiopicum* was domesticated [5]. This is consistent with report by Lester [5] that *S. aethiopicum* and *S. macrocarpon* were domesticated in Africa from their wild relatives *S. anguivi* and *S. dasyphyllum* respectively. Furthermore, UPGMA dendrogram by Sakata and Lester [48] also indicated *S. anguivi* and *S. aethiopicum* as duplicates proving their close relatedness. This further explains the reason for wide diversity between *S. aethiopicum* and *S. torvum*. In principle, the wide variation between *S. torvum* and the two other populations offer a broad genetic base with the potential for wider adaptation in a range of agroecosystems. This also indicates the genetic potential of the populations for breeding purposes [42].

4.2 Variation via Cluster Analysis

The dendogram reveals that cluster five and six constitutes mainly *S. torvum* accessions with the exception of A/ Dominase *S. anguivi* and Hwidiem *S. anguivi* in cluster five and six respectively. Cluster five and six were captured at a genetic distance of 0.85 (85%) and 0.77 (77%) respectively. The inclusion of the *S. anguivi* accessions among the *S. torvum* may be attributed to the high outcrossing rate and gene flow (t = 81.54%; Nm = 2.2071). It can be inferred from the results of the study that about 85% of the *S. torvum* genetic characteristics has been introgressed into such *S. anguivi* accessions making them lose their unique genetic characteristics (Fig. 3). This also implies that the *S. torvum* accessions in cluster five and six should be assembled and conserved for breeding purpose to avoid contamination. Nevertheless, the remaining clusters constitute a mixture of the three populations which grouped not based on geographical origin or species as indicated by the low values of F_{IS} and F_{IT} resulting in probable heterozygosity of those accessions [42].

4.3 Variation among Accessions via Mineral Content

The distinct concentration of the Fe and Zn analyzed among the accessions also display variation among the accessions. Diversity in Fe and Zn concentrations in the accessions of the present study agrees with similar findings by several early workers [49,50]. This suggests that micronutrient enrichment traits also exist in the egg plants as evident in other crops such as rice and common beans (*Phaseolus vulgaris L.)* [51,52]

Concentrations obtained for iron is higher than recorded by Chinese red long-grain varieties (64 mg/kg) [53,16] but lower than values obtained by Doku [19] for Awerema brown rice (386.6±0.46 mg/kg), an *Oryza glaberrima* accession in Ghana. However, since eggplants do not undergo polishing before usage compared to brown rice, the eggplant accession with the highest concentration still holds a great prospect in human nutrition. With the exception of accessions with the least iron concentration for the three species of the present study, (Bonwire S. *torvum*, Assin Odumase *S. anguivi* and Nkawkaw S. *aethiopicum*) concentration of iron recorded for the rest of accessions can supply adequate iron to the human body when ingested taking into account the recommended dietary allowance of iron for males and females being 10 g and 15 g respectively [52,54]. These however suggest the usefulness of Assin Odumase *S. torvum*, the accession rich in iron, for crop improvement.

The recommended dietary allowance of zinc is 15 mg and 10 mg for male and females respectively. With the exception of the accession with the least value (9.16 mg / kg) of Zinc, frequent ingestion of the accessions under study with other zinc rich food would help improve the Zn needs of the human body especially Sefwi Awaso *S. aethiopicum*. However, range of Zn concentrations recorded for accessions of the present studies are lower compared to the Zn range content (14-59 mg/kg) recorded by several authors: Frei and Becker, 2005; Kennedy and Burlingame, 2003 as well as Gregorio, 2002 [16,55,56] for different rice varieties.

Nevertheless, the range of Zn concentrations recorded for the present study are higher compared to range of concentrations recorded by other eggplant and closely related vegetables mineral study. Arivalagan et al. [49] for instance recorded 0.73 mg/kg – 2.33 mg/kg for 32 eggplants as the Zn concentrations range and Kamga et al. [50] also recorded 7.63±0.005 µg/g – 102.57±2.589 µg/g as the Zn concentration range for 5 indigenous African vegetables.

The study did not reveal any accession among the three egg plant species possessing high concentration of both iron and zinc neither any accessions with high correlation for Fe and Zn.

Therefore the outstanding concentrations recorded by Assin Odumase *S. torvum* and Sefwi Awaso *S. aethiopicum* for iron and zinc respectively suggest their usefulness in a breeding program whose objective is to improve iron and zinc concentration. Thus the two species can be utilized in a hybridisation program to develop an eggplant rich in iron and zinc.

5. CONCLUSION

In conclusion, the EST-SSR markers and mineral analysis revealed genetic diversity among the individual accessions of the populations studied without duplications. However the population studies suggest the gradual homogeneity of these accessions due the high gene flow and outcrossing rate over time. It is therefore recommended that the accessions should be collected and conserved to avoid further contamination especially with the almost distinct *S. torvum* species. Molecular diversity of the species under study or closely related ones can easily be carried out with EST-SSR primer EEMS 37. Assin Odumase *S. torvum* and Sefwi Awaso *S. aethiopicum* can be recommended as a good source for improved child and maternal health in Ghana because of their high content of iron and zinc respectively.

ACKNOWLEDGEMENTS

I wish to acknowledge the Molecular Department of the CSR-Crops Research Institute for their assistance in the molecular work. I also wish to acknowledge the Chemistry unit of the CSIR-Soil Research Institute for their kind assistance in the mineral analysis. Efforts of the co-authors, Dr. Marian Quain, Mr. Allen Oppong, Henry Akrofi-Doku, Mrs. Adelaide Agyeman and Mr. Offei-Bonsu need to be commended particularly, for acquisition of primers, data analysis and proof reading of the manuscript. My sincere gratitude also goes to African Women in Agricultural and Research Development (AWARD) for providing the initial funds for this study.

COMPETING INTERESTS

Authors have declared that no competing interests exist.

REFERENCES

1. Sękara A, Cebula S, Kunicki E. Cultivated eggplants – origin, breeding objectives and genetic resources, a review. Folia Horticulturae Ann. 2007;19(1):97-114.
2. Doganlar S, Frary A, Daunay MC, Lester RN, Tanksley SD. A comparative genetic linkage map of eggplant (*Solanum melongena*) and its implications for genome evolution in the *Solanaceae*. Genetics. 2002;161:1697-1711.
3. Levin RA, Watson K, Bohs L. A four-gene study of evolutionary relationships in *Solanum* section Acanthophora. Am. J. Bot. 2005;92(4):603-612.
4. Levin RA, Myers NR, Bohs L. Phylogenetic relationship among the "spiny solanums" (*Solanum* subgenus *Leptostemonum*, *Solanaceae*). Am. J. Bot. 2006;93:157-169.
5. Lester RN. Genetic resources of capsicum and eggplants. Xth EUCARPIA meeting on genetic and breeding of capsicum and eggplant. Avignon, France. 1998;25-30.
6. Norman JC. Tropical vegetable crops. Devon: Arthur Stockwell Ltd. L.) with aqueous neem seed extracts. J. Ghana Sci. Assoc. 1992;3:70-84.
7. Owusu-Ansah FK, Afreh-Nuamah D, Obeng-Ofori, Ofosu-Budu KG. 'Managing infestation levels of major insect pests of garden eggs (*Solanum integrifolium*). J. Ghana Sci. Assoc. 2001;3:70-84.
8. Bukenya-Ziraba R. *Solanun anguvi* Lam. In: GJH Grubben, Denton OA (Eds.) Plant resources of tropical africa vegetables. PROTA Foundation, Wageningen, Netherlands. 2004;2:480-482.
9. Leelavathi M, Anitha L, Maheswarlu M, Narayan PN. Standardization of the technology for the development of *Solanum torvum* - Herbal based therapeutic lehyam. Asian Journal of Experimental Biology Sci. 2010;1(4):758-764.
10. Lucca MF Hopp E, Romero-Zaliz R. Comparative transcription profiles of *Solanum* wild species under drought conditions: Preliminary results. Intelligent Systems and Applications (ISDA). 2011;1218-1223.
11. Dauney MC, Lester RN, Laterrot H. The use of wild species for the genetic improvement of brinjal eggplant (*Solanum melongena*) and tomato (*Lycopersicon esculentum*), In: Hawkes JH, Lester RN, Nee M, Estrada N (Eds.) Solanaceae III:

Taxonomy, Chemistry, Evolution, Royal Botanic Gardens Kew and Linnean Society, London. 1991;27:389-413.

12. Stravato VM, Capelli C. Behaviour of *Solanum* spp. on inoculation with different isolates of *Fusarium oxysporum* f. sp. melongenae, Bull. Organisation Europe´enne et Me´diterrane´enne Pour la Protection des Plantes. 2000;30:247-249.

13. Collonnier C, Fock I, Kashyap V, Rotino GL, Daunay MC, Lian Y, Mariska IK, Rajam MV, Servaes A, Ducreux G, Sihachakr D. Applications of biotechnology in eggplant, Plant Cell Tissue Org. 2001;65:91-107.

14. Peregrine WTH, Ahmad K. Grafting a simple technique for overcoming bacterial wilt in tomato. Trop. Pest Manage. 1982;28:71-76.

15. Singh PK, Gopalakrishnan TR. Grafting for wilt resistance and productivity in brinjal (*Solanum melongena* L.). Hort. J. 1997;10:57–64.

16. Frei M, Becker K; 2005. Accessed on 28th October, 2013.
Available:http://www.greenpeace.org/india/PageFiles/128046/Rice_Biodiversity___Nutrients.pdf

17. Kennedy G, Burlingame B, Nguyen, VN. Nutritional contribution of rice and impact of biotechnology and biodiversity in rice-cinsuming countries. Proceeding of the 20th session of the international rice commission; 2006.

18. Welch RM. Micronutrients, Agriculture and nutrition. Linkages for improved health and well being; 2008. Accessed on 30 June, 2013. Available:http://www.css.cornell.edu/Food systems/ Micros %26 AgriManl ref.html on

19. Milne D B. Laboratory assessment of trace element and mineral status. In: Bogden JD, Klevay LM (ed.) Clinical nutrition of the essential trace elements and minerals nutrition and health, Humana Press (Springer Science+Business Media), New York, U.S.A. 2000;69-90.

20. González-Tenorio R, Fernández-Diez A, Caro I, Mateo J. Comparative assessment of the mineral content of a Latin American raw sausage made by traditional or non-traditional processes, Atomic Absorption Spectroscopy, (Farrukh, M. A. Ed.); 2012. Accessed on 22th October, 2013. Available: http://www.Intechopen.com/books/atomic-absorptionspectroscopy/comparativeassessment-of-the-mineral-content-of-a-latin-american-raw-sausage-made-by-traditional-or

21. Ibáñez E, Cifuentes A. New analytical techniques in food science. Critical Reviews in Food Science and Nutrition. 2001;41(6):413–450.

22. Luykx DMAM, van Ruth SM. An overview of analytical methods for determining the geographical origin of food products. Food Chemistry. 2008;107(2):897-911.

23. Doku HA. Characterisation of some accessions of indigenous rice (*Oryza glaberrima* Steud.) in Ghana. M.Phil Thesis, University of Ghana, Legon, Accra. 2011;121.

24. Steel WM. Cowpeas in Nigeria. Ph. D thesis, University of Reading. 1972;241.

25. Summerfield RJ, Huxley PA. Day length and night temperature sensitivity screening of selected cowpea and soybean cultivars. Reading University/IITA International Communication Reading, England. 1973;5.

26. Morakinyo JA, Ajibade SR. Characterization of the segregants of an improved cowpea line IT84K-124-6. Nig. J. Sci. 1998;32:27-32.

27. Singh SR, Rachie KO. Cowpea Research, production and utilization. John Wiley and Sons, New York. 1985;460.

28. Nawaz M, Hussain SA, Ullah I, Younus M, Iqbal MZ, Rana SM. Estimation of genetic diversity in wheat using dna markers. American-Eurasian Journal of Sustainable Agriculture. 2009;3(3):507-511.

29. Park Y, Lee J, Kim N. Simple Sequence Repeat Polymorphisms (SSRPs) for evaluation of Molecular Diversity and Germplasm Classification of Minor Crops. Molecules. 2009;14:4546-4569.

30. Aniko S, Ezio P, Laura T, Guiseppe LR, Sergio L. BMC Genomics. 2008;9:357.

31. Demir K, Bakır M, Sarıkamış G, Acunalp S. Genet. Mol. Res. 2010;9(3):1568-1576.

32. Ali Z, Xu ZL, Zhang DY, He XL, Bahadur S, Yi JX. Molecular diversity analysis of eggplant (*Solanum melongena*) genetic resources. Genet. Mol. Res. 2011;10(2):1141-1155.

33. Ronning CM, Stegalkina SS, Ascenzi RA, Bougri O, Hart AL, Utterbach TR, Vanaken SE, Riedmuller SB, White JA, Cho J. Comparative analyses of potato expressed sequence tag libraries. Plant Physiol. 2003;131:419–429.
34. Szalma SJ, Buckler ES, Snook ME, McMullen MD. Association analysis of candidate genes for maysin and chlorogenic acid accumulation in maize silks. Theor. Appl. Genet. 2005;110:1324–1333.
35. Shin JH, Kwon SJ, Lee JK, Min HK, Kim, NS. Genetic diversity of maize kernel starch synthesis genes with SNAPs. Genome. 2006;49:1287–1296.
36. Crossa J, Burgueno J, Dreisigacker S, Vargas M, Herrera-Foessel SA, Lillemo M, Singh RP, Trethowan R, Warburton M, Franco J. Association analysis of historical bread wheat germplasm using additive genetic covariance of relatives and population structure. Genetics. 2007;177:1889–1913.
37. Buerstmayr H, Lemmens M, Hartl L, Doldi L, Steiner B, Stierschneider M, Ruckenbauer P. Molecular mapping of QTLs for Fusarium head blight resistance in spring wheat. I. Resistance to fungal spread (Type II resistance). Theor. Appl. Genet. 2002;104:84–91.
38. Breseghello F, Sorrells ME. Association analysis as a strategy for improvement of quantitative traits in plants. Crop Sci. 2006;46:1323–1330.
39. Zeng Y, Yang S, Cui H, Yang X, Xu L, Du J, Pu X, Li Z, Cheng Z, Huang X. QTLs of cold-related traits at the booting stage for NIL-RILs in rice revealed by SSR. Genes Genom. 2009;31:143–145.
40. Tümbilen Y, Frary A, Mutlu S, Doğanlar S. International Research Journal of Biotechnology. 2011;2(1):016-025.
41. Egnin M, Mora A, Prakash CS. Factors enhancing agrobacterium tumefaciens-Mediated Gene Transfer in Peanut (*Arachis hypogaea* L.). In vitro Cell. Dev. Biol.-Plant. 1998;34:310-318.
42. Cao Q, lu B, Xia H, Rong J, Sala F, Spada A, Grassi F. Genetic diversity and origin of weedy rice (*Oryza sativa* f. spontanea) populations found in North-eastern China revealed by Simple Sequence Repeat (SSR) markers. Annals of Botany. 2006;98:1241–1252.
43. Wright S. Variability within and among natural populations. In: Evolution and the genetics of populations, Chicago, University of Chicago Press. 1978;590.
44. Yeh FC, Yang RC, Boyle T. Microsoft window-based freeware for population genetic analysis (POPGENE), version 1_31; 1999.
45. Nei M. Estimation of average heterozygosity and genetic distance from a small number of individuals. Genetics. 1978;89:583–590.
46. Saraye B, Ranghoo-Sanmukhiya VM. Morphological and molecular characterisation of eggplant varieties and their related wild species in Mauritius, In: Harnessing the potential of biotechnology for food security and socioeconomic development in Africa Nzuma JM, M'mboyi F (eds.). Proceedings of the 1st all Africa congress on biotechnology 22nd – 26th september. 2008;60-68.
47. Muller MH, Prosperi M, Santoni S, Ronfort J. How mitochondrial DNA diversity can help to understand the dynamics of wild-cultivated complexes; the case study of *Medicago sativa* in Spain. Mol. Ecol. 2001,10:2753 2763.
48. Sakata Y, Lester RN. Chloroplast DNA diversity in brinjal eggplant (*Solanum melongena* L.) and related species. Euphytica. 1997;97:295-301.
49. Arivalagan M, Bhardwaj R, Gangopadhyay KK, Prasad TV, Sarkar SK. Mineral composition and their genetic variability analysis in eggplant (*Solanum melongena* L.) germplasm. Journal of Applied Botany and Food Quality. 2013;86:99–103.
50. Kamga RT, Kouamé C, Atangana AR, Chagomoka T, Ndango R. Nutritional evaluation of five African indigenous vegetables. Journal of Horticultural Research. 2013;21(1):99-106.
51. IRRI. Accessed on June, 5; 2011. Available: http://irri.org/our-science/better-varieties/high-nutrient433rice
52. Welch RM, Graham RD. Breeding for micronutients in staple food crops from a human nutrition perspective. Journal of Experimental Botany. 2004;55:353–364.
53. Datta S, Bouis HE. Application of biotechnology to improving the nutritional

quality of rice. Food and Nutrition Bulletin. 2000;21:451-456.

54. FAO/WHO. Preliminary report on recommended nutrient intakes. Joint FAO/WHO expert consultation on human vitamin and mineral requirements, FAO, Bangkok, Thailand, Food and Agricultural Organization of the United Nations Rome, Italy and World Health Organization, Geneva, Switzerland. Revised July 13, 2000; 1998.

55. Kennedy G, Burlingame B. Analysis of food composition data on rice from a plant genetic resources perspective. Food Chemistry. 2003;80:589-596.

56. Gregorio GB. Progress in breeding for trace minerals in staple crops. Journal of Nutrition. 2002;132:500-502.

20

A Fast Experimental Electrical Conductivity Survey for Agricultural Applications, Case of Small Crop Area of Otoes (*Xanthosoma* sp.) in Central Region of Panama

Alexis Mojica[1*], J. Solís[1,2], N. Bernal[2], F. Vergara[3], B. Duarte[3], C. A. Ho[1], J. Ortega[4] and L. Pastor[5]

[1]*Laboratorio de Investigación en Ingenieríay Ciencias Aplicadas, Centro Experimental de Ingeniería, Universidad Tecnológica de Panamá, 0819-07289 El Dorado, Panamá.*
[2]*Facultad de Ingeniería Civil, Universidad Tecnológica de Panamá, 0819-07289 Zona 6, Panamá.*
[3]*Laboratorio de Análisis Industrialy Ciencias Ambientales, Centro Experimental de Ingeniería, Universidad Tecnológica de Panamá. 0819-07289 El Dorado, Panamá.*
[4]*Instituto Nacional de Agricultura, Ministerio de Desarrollo Agropecuario, Divisa, Panamá.*
[5]*Sorbonne Universités, UPMC Université Paris 06, UMR 7619 METIS, F-75005, Paris, France.*

Authors' contributions

This work was carried out in collaboration between all authors. Author AM designed the study, wrote the protocol and wrote the first draft of the manuscript. Authors AM, JS, NB and JO conducted measurements and reviewed the experimental design. Authors FV and BD managed the geochemical analyses of the study. Author CAH worked the protocol for data acquisition and reviewed all drafts of the manuscript. Author LP performed the statistical analysis. All authors read and approved the final manuscript.

Editor(s):
(1) Özge Çelik, Department of Molecular Biology and Genetics, Istanbul Kultur University, Turkey.
Reviewers:
(1) Saad Bakkali, Earth sciences department, University Abdelmalek Essaadi, Tangier , Morocco.
(2) Klára Kosová, Division of Crop Genetics and Breeding, Crop Research Institute, Prague, Czech Republic.

ABSTRACT

A quasi-continuous automatic recording system of apparent electrical conductivity data of soil (EC_a) was assembled to obtain at a reliable and fast rate, spatial distribution maps of this physical parameter. The device was moved manually throughout several profiles in a small zone of 3753 m^2

**Corresponding author: E-mail: alexis.mojica@utp.ac.pa;*

used for the farming of otoes (*Xanthosoma* sp.) in Panama, Central America, obtaining 401 voltage values in approximately 20 minutes. These voltages were turned into EC_a values and mapped. This result was compared with total dissolved solids (TDS) and pH maps of 29 soil samples at 20 cm of depth; salinity was also measured. The EC_a map presented a zone of high electrical conductivity (130 - 500 µS/cm) on the Eastern part which is consistent with the zones in the TDS and pH maps whose value ranges fall within the optimal threshold for *Xanthosoma* sp. (45-95 mg/l and 5.5-5.8 respectively). Salinity levels were weak (<0.10%) and are consistent with the zone of high EC_a typical of non-saline soils. All these results show a significant potential in the use of this mobile system to achieve a fast assessment of agricultural exploitation zones. Finally, a highly moderate correlation is showed among the values of EC_a and TDS, and pH data.

Keywords: Apparent electrical conductivity (EC_a); square quadrupole array; pH; total dissolved solids (TDS); Xanthosoma sp.

1. INTRODUCTION

The study of soils constitutes a topic that has been highly addressed by the scientific community throughout the world, because the understanding of their physical properties has a significant impact in the positive exploitation of natural resources. Therefore, geophysical methods play an important role in the fast evaluation of a determined zone of interest. One of these techniques is the electrical method which has had a key role in solve various problems linked to the detection of archeological features, groundwater and zones of high agricultural productivity (just to mention a few examples). This electrical technique has its beginnings in the second decade of the last century and it was not until the last decades of the 20th century that electronic and computerized systems (with AC sources) capable to automate the recording of field data, and in addition to the election of special electrodic configurations with the feature of determining the magnitude and the course of the electrical anisotropy. In this sense, the square quadrupole array is adapted to a great number of demands that includes the issues caused by such effect until the coverage a large extension of land in a relatively short period of time. The use of such configuration started in the 60's of the last century, where it was used to solve geological problems [1,2], as well as archaeological problems [3-7] and in explorations of groundwater [6,8,9]. Currently, this technology is being used successfully in zones of agricultural cultivation [10].

The aim of this work is focused on the study of the effectiveness of the assembling of a mobile system that is made up of metallic wheels traction that supports a platform with an AC source, a data logger and a GPS. A case study related to a small crop area of otoes (*Xanthosoma* sp.) located in the central sector of the Isthmus of Panama, is presented. Such results include a mapping of apparent electrical conductivity (EC_a) and some geochemical parameters (Total Dissolved Solids - TDS and pH) obtained by the analysis of a set of samples collected in the site of interest. A Pearson correlation among these data is also presented in the study.

2. SOME ADVANTAGES OF THE SQUARE QUADRUPOLE ARRAY IN ELECTRICAL PROSPECTION AND TECHNICAL DESCRIPTION OF THE DEVICE

The knowledge of physical properties of the soils is of fundamental importance since according to soil mechanics it is possible to predict the future behavior of a soil under loads when it contains different moisture contents. On the other hand, in archaeology and agriculture, the accurate interpretation of these properties will involve the optimization of material and human resources. Today farms require a complete knowledge of the growing areas, which would lead to increase the data density and time reduction of laboratory sample analysis; this implies an additional problem related to the changes experienced by soil properties if the process of collection of samples is developed for several days. Because of this, the use of a mobile system for making faster and more detailed EC_a data acquisition can play an important role.

The electrical method is based on the measurement of the apparent electrical resistivity of the soil or its inverse value, the EC_a. The methodology is characterized by the use of an alternating or direct current passing through the ground between two electrodes (A and B); this electrical current generates a voltage which can be recorded by another two electrodes (M and N) inserted few centimeters into the soil; then, the knowledge of electrical current (*i*), the voltage (ΔV) and the geometry of this quadrupole (which is defined through a geometrical coefficient), allow the calculation of EC_a. Through a simple form and according to the physical laws, the equation obtained is:

$$EC_a = bR^{-1}$$

In this equation, $R = \Delta V / i$ and the proportionality constant between EC_a and R varies according to the quadrupole array, which depend on the research objectives. Of all the possible electrode arrays, the square quadrupole is the most symmetrical array since the electrodes are located in the vertices of a square and the effects of the apparent anisotropy are reduced significantly [5]. For such array, the previous equation is represented by:

$$EC_a = \left(\sqrt{2} - 1\right)\left(\pi a \sqrt{2}\right)^{-1} R^{-1}$$

Where *a* represents the dimension of sides of the square and is part of the geometrical coefficient.

The device was assembled following three important concepts: (i) the mobile part, (ii) the quasi-continuous injection system of electrical current and data-logger, and (iii) GPS. For the first part, four metallic wheels of 0.15 m of diameter were built; 15 equi-spaced pins of metallic stainless steel of 5 cm long and 0.5 cm of diameter were inserted in each wheel, which also contribute to the traction in the soil for a continuous movement along a profile. Each set of wheels (front and rear) was linked to an aluminum platform of (1.20 x 0.60) m of surface through an insulating materials; such platform serves as support for the devices that will be described later. In the second part, a prototype of alternating current source was used; this device transforms a continuous signal of 12 V supply in a signal of 40 Hz and an output of 50 V. In the output, a square electrical pulse of 5 mA is generated; the outputs of the injection system are connected to the base of the front wheels represented by A and B of depicted in Fig. 1(a).

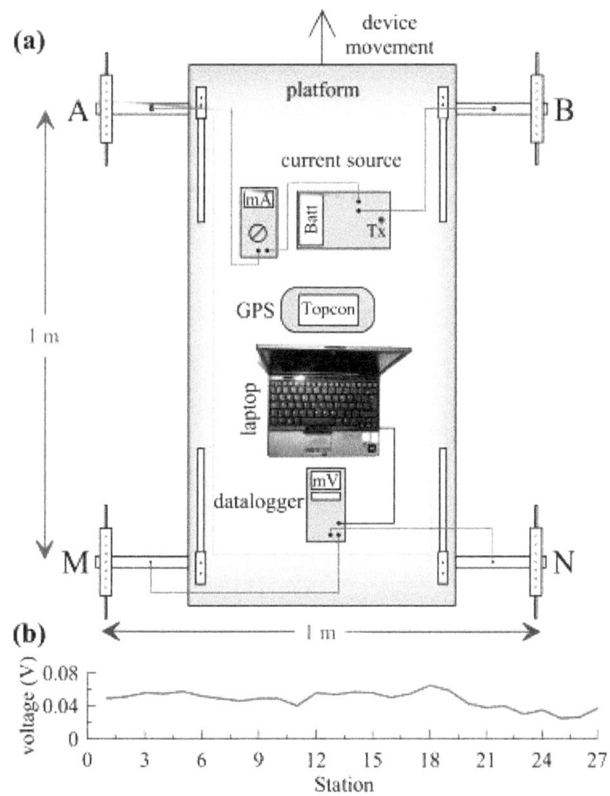

Fig. 1. (a) General scheme of mobile device (square quadrupole array) for automatic recording of voltage and (b) graph of voltage data recorded long a profile

On the other hand, the rear wheels, represented by M and N of the Fig. 1(a), are connected to a data logger through connection cables, which record in each time interval, the voltage generated in the soil when the electrical current is injected through the front wheels. The data logger is connected to a laptop for the automatic storage of voltages. The acquisition program can be configured in a range of a temporary period of time that is between 1 and 500 seconds according to the study. Eventually, GPS is incorporated to the mobile system for recording coordinates of the measurement points along a given profile. These coordinates, after undergoing a post-process using GPS data (from a fixed base) can be associated with voltage data recorded, which are then translated into EC_a values. Fig. 1(b) shows an example of a graph of voltage recorded along a given profile.

One way to check the effect of anisotropy is to measure the value of EC_a for different angles of the square quadrupole array and then represent this data in a polar graph. In this work, the tests were carried out for a value of a = 1.0 m which corresponds to the AB, MN, AM and BN distances (see Fig. 1(a)). In this result, the variations in EC_a for different angles with respect to North, are low (106.5 and 118.1 μS/cm); this reflects the low sensitivity of the system to the apparent anisotropy (see Fig. 2).

3. EC_a MEASUREMENTS IN QUATERNARY SEDIMENTS OF PANAMA CENTRAL ZONE - AGRICULTURAL APPLICATIONS

In general, the electrical methods have played an important role in agricultural issues [11-15]. The mobile device was used in a small test area located in the central sector of the Isthmus of Panama (Central America) which is characterized by the presence of quaternary alluvial sediments. Fig. 3(a) shows the geographical location of the site and Fig. 3(b) and (c) detailed satellite images of the interest area. This test zone has 3753 m^2 and is used today for growing otoes (*Xanthosoma* sp.).

A total of 21 profiles between crops lines separated a distance of 1.5 m were surveyed in a period of about 20 minutes. Then the data logger recorded a total of 401 voltage values while the system is in a continuous movement; this data was translated into EC_a values and associated with the corrected global positioning data. Fig. 4 shows the mobile system with accessories used in this test.

This information was finally processed geostatistically with Surfer 12 of Golden Software to generate a map of spatial variations of EC_a, see Fig. 5(a). Additionally, TDS, pH and salinity

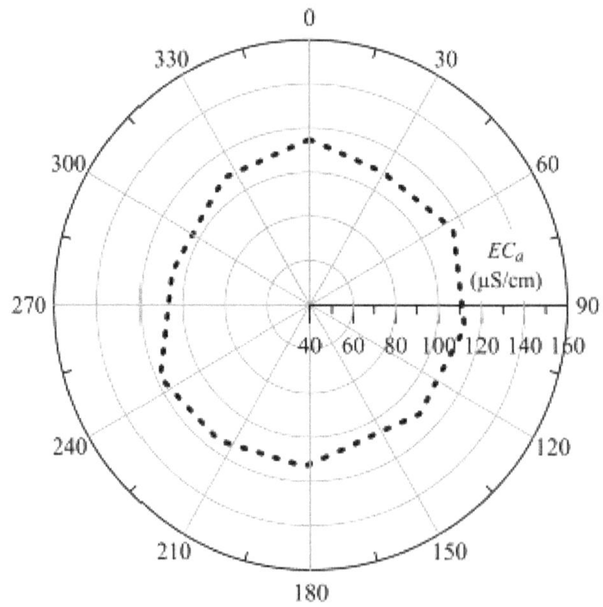

Fig. 2. Polar graph showing the EC_a values by the dotted line using the square quadrupole array

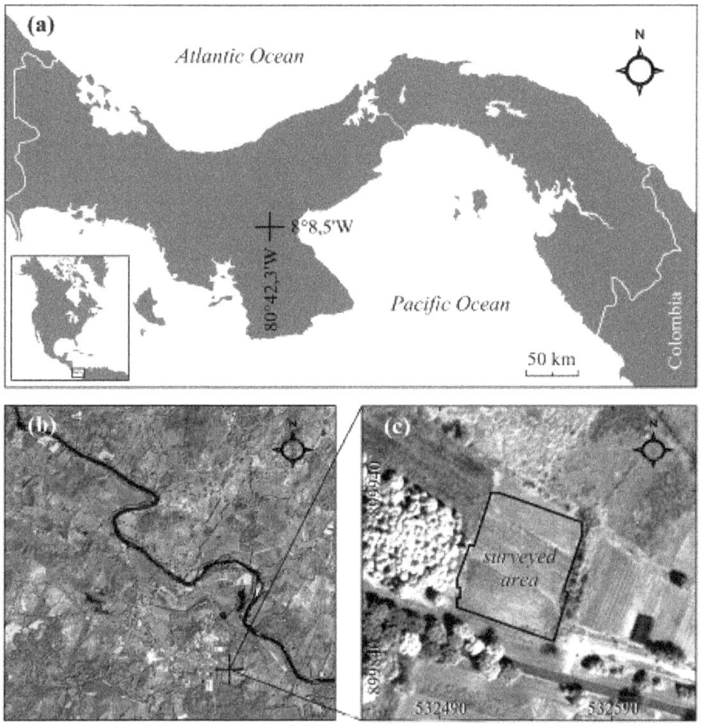

Fig. 3(a). Location of the test site, central sector of the Isthmus of Panama and (b), (c) satellite images of the prospected area

Fig. 4. Mobile quadrupole square system and accessories used for automatic recording of voltages in the 21 profiles established in the test area

percentages of 29 soil samples to a depth of 20 cm were measured. pH measurements were performed according to the AASHTO Designation T 289-91 (1996) using a Corning 320 pH electrode. TDS measurements and the percentage of salinity were conducted through the use of an Orion 145 + Thermo-Scientific device; the results were based on the method developed by [16].

Only TDS and pH were geostatistically processed because the salinity percentages presented a low variation in their values (< 0.10%). A Pearson correlation analysis between the geophysical and geochemical data was carried out by Matlab program. The results of the mapping process of TDS and pH values are presented in Fig. 5(b) and (c), respectively; the difference between the map of Fig. 5(a) and Fig. (b) and (c) is the density in the EC_a data (401) versus TDS and pH data (29).

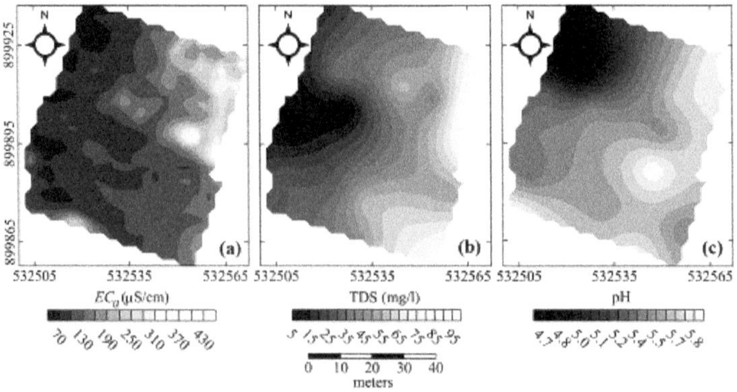

Fig. 5. Map of spatial variation of (a) EC_a, (b) TDS and (c) pH at the *Xanthosoma* sp. growing area

Table 1. Statistical and correlation coefficients for 29 measurements of EC_a, TDS and pH

						Correlation	
Property	Min	Max	Mean	Median	Stan. Dev.	TDS	pH
EC_a (µS/cm)	58.7	355.1	130.1	111.6	64.6	0.5857	0.5446
TDS (mg/l)	8	92	43	41	23		
pH	4.62	6.70	5.49	5.52	0.42		

The result of the electrical prospecting using the mobile system shows a high electrical conductivity zone in the Eastern sector of the map, with values between 130 and 500 µS/cm; this range is not more than 2000 µS/cm which represents the upper limit for non-saline soils. This fact is confirmed by the low salinity percentages of geochemical analysis obtained. The area represented by a light gray hue of Fig. 5(a) agrees well with shaded areas clear hue of Fig. 5(b) and 5(c), the characteristic ranges of these areas for the TDS and pH ranging between 45-95 mg/l and 5.5-5.8, respectively. The ranges of these parameters fall within boundaries which are characterized by high agricultural yield areas for *Xanthosoma* sp. (otoes), which is where most of the nutrients are concentrated.

About the statistical analysis and Pearson correlations, the mean value of EC_a data obtained by mobile system in the 29 positions (where the samples were obtained), was 130.1 µS/cm (see Table 1 above) with a positively skewed in the sample population distribution. Furthermore, the values of TDS have a mean of 43 mg/l, with a distribution positively skewed. The pH values have a mean of 5.49 and the distribution was also positively skewed. These results show a moderately high correlation between EC_a and TDS data (0.5857), and between EC_a and pH data (0.5446), see Table 1.

4. CONCLUSION

In order to generate spatial variation maps of EC_a in tropical areas like Panama, it has been common to measure this parameter by using different electrode arrangements manually displaced along several profiles (spatial maps). The main disadvantage of this methodology is the measuring time when we want a good density of data (high resolution maps) over relatively large areas. Thus, the use of a square quadrupole array adapted to a mobile system comprising: (i) an electrical current source, (ii) a quasi-continuous automatic voltage recording system, and (iii) a GPS, is an advantage to generate EC_a maps in a very short period of time (187.6 m^2/min for lines spaced 1.5 m), also with a low cost and low sensitivity to the effect of apparent anisotropy. The use of this system in a test area for growing otoes (*Xanthosoma* sp.) located in the central sector of the Isthmus of Panama could be accomplished in a relatively short time (about 20 minutes). The area of high EC_a values matches whose geochemical properties (TDS and pH) that are characteristic of soils that present high growing potential for *Xanthosoma* sp. This high degree of reliability offered by the geophysical survey has also been highlighted by the correlation results of geophysical and geochemical data obtained in the 29 positions of the test site.

AKNOWLEDGEMENTS

The authors would like to thank the SENACYT for funding this project.

COMPETING INTERESTS

Authors have declared that no competing interests exist.

REFERENCES

1. Habberjam GM, Watkins GE. The use of a square configuration in resistivity prospecting. Geophysical Prospecting. 1967;15:445-467.
2. Habberjam GM. Apparent resistivity observations and the use of square array techniques. Geoexploration Monographs, Series. 1979;1(9).
3. Clark AJ. A square array for resistivity surveying. Prospezioni Archeologiche. 1968;3:111-114.
4. Bolton F. Archaeological resistivity surveys using the square array: A case history. 54th Annual International Meeting and Exploration, Society of Exploration Geophysics (Atlanta), extended abstract. 1984;204-206.
5. Hesse A, Jolivet A, Tabbagh A. New prospects in shallow depth electrical surveying for archaeological and pedological applications. Geophysics 1986;51:585-594.
6. Merlanti F, Pavan M. Some considerations on the use of the geoelectric square array. Annali di Geofisica. 1996;39:141-157.
7. Panissod C, et al. Archaeological prospecting using electrical and electrostatic mobile arrays. Archaeological Prospection. 1998;2:239-251.
8. Degnan JR, Moore RB, Mack TJ. Geophysical investigation of well fields to characterize fractured-bedrock aquifers in Southern New Hampshire. Water-Resources Investigations Report 01-4183. U.S. Geological Survey; 2001.
9. Ravindran A. Azimuthal square array resistivity method and groundwater exploration in Sanganoor, Coimbatoore District, Tamildanu, India. Research Journal of Recent Science. 2012;1:41-45.
10. Dabas M, Jubeau T, Rouiller D, Larcher J, Charriere S, Constant T. Using high-resolution electrical resistivity maps in a watershed vulnerability study. First Break. 2012;30:51-56.
11. Corwin DL, Lesch SM. Apparent soil electrical conductivity measurements in agriculture. Computers and Electronics in Agriculture. 2005;46:11-43.
12. Corwin DL. Past, present, and future trends of soil electrical conductivity measurement using geophysical methods. In: Allred BJ, Daniels JJ, Ehsani MR, editors. Handbook of Agricultural Geophysics. United States of America: Taylor & Francis Group; 2008.
13. Allred BJ, Groom M, Ehsani MR, Daniels JJ. Resistivity Methods. In: Allred BJ, Daniels JJ, Ehsani MR, editors. Handbook of Agricultural Geophysics. United States of America: Taylor & Francis Group; 2008.
14. Toushmalani R. Application of Geophysical Methods in Agriculture. Australian Journal of Basic and Applied Sciences. 2010; 4:6133-6439
15. Motavalli PP, Udawatta RP, Bardhan S. Apparent soil electrical conductivity used to determine soil phosphorus variability in poultry litter-amended pastures. American Journal of Experimental Agriculture. 2013; 3:124-141.
16. Jackson ML. Soil Chemical Analysis. Prentice-Hall Inc., United States of America; 1958.

21

An Outline for Economical and Technical Analysis of Solar Panels for Agricultural uses: A Case Study on Texas Weather Conditions

David Lacey[1], Jersson Placido[1*] and Sergio Capareda[1]

[1]Department of Biological and Agricultural Engineering, Texas A&M University, College Station, Texas, 303D Scoates Hall, 2117 TAMU, College Station, TX 77843, United States.

Authors' contributions

DL, constructed the solar panels stands, configured the data recorded, and performed experimental work.JP performed experimental work, data analysis and wrote the first draft of the manuscript and SC designed the study, managed the analyses of the study and supervised the manuscript. All authors read and approved the final manuscript.

Editor(s):
(1) Aleksander Lisowski, Department Agricultural and Forestry Engineering, Warsaw University of Life Sciences, Poland.
Reviewers:
(1) Huan-Liang Tsai, Department of Electrical Engineering, Da-Yeh University, Taiwan.
(2) LOGERAIS Pierre-Olivier, CERTES , Université Paris-Est Créteil, IUT de Sénart, France.
(3) Anonymous, Egypt.
(4) Anonymous, South Africa.
(5) Anonymous, India.

ABSTRACT

Aims: To outline the necessary steps in the evaluation of solar panel systems in agriculture.
Study Design: Using the outline proposed, GaAs and Si solar panels were evaluated under central Texas weather conditions.
Place and Duration of Study: Bioenergy Testing Laboratory (BETALab) between January and August 2013 in College Station, Texas.
Materials and Methods: The study includes an outline on how to do a technical and simple economic comparison between solar panels. The outline includes the solar panel efficiency, required area, installations and power generation costs and simple payback period. To exhibit the application of the outline, two different photovoltaic systems were compared (silicon (Si) PV panels and the gallium arsenide (GaAs) PV panels). The solar panels were compared simultaneously, taking measurements of voltage and current automatically in College Station, Texas.

**Corresponding author: E-mail: jerssonplacido@gmail.com;*

Results: The GaAs solar panel showed conversion efficiency (18.36%) higher than Si panels (14.02% Si); however, it also has a capital cost ($3/Wp) higher than Si panels ($2/Wp). The study suggested that both panels are viable alternatives for energy independence in a small farm or ranch application. The final selection will depend on the economic alternatives and power necessities of the farmer or rancher. In this study, the GaAs have provided less payback period (6.8 years) compared with 7.5 years for the Si units using various assumptions. The larger efficiency of the GaAs (4%) unit translated into a better payback period despite its higher initial installed cost.
Conclusion: The technical and economical outline proposed in this paper was useful to decide between GaAs and Si panels using basic economic assumptions. However, this paper did not recommend a specific solar panel over the other in every situation. Each situation needs to be analyzed individually taking in consideration geographical situation, government subsidies, rebates and tax credits.

Keywords: Economic analysis; technical analysis; gallium arsenide photovoltaic panels; silicon photovoltaic panels; efficiency.

1. INTRODUCTION

The energy potential in Texas is one of the largest in the United States, with abundant wind, solar, and biomass resources in different regions across the state. In 2010, the renewable energies in Texas accounted 3.9% of the total energy consumed in the state. From the renewables energies used in the U.S., solar energy was responsible for 2% of the renewable energy total in 2011 [1]. Besides this low percentage, the solar energy industry is growing rapidly, example of that is the growing in 2010 (29%) and 2011 (25%) [1]. The state of Texas is ranked nationally as No. 1 in solar energy potential and No. 13 in installed solar photovoltaic capacity [2]. The large potential that the state of Texas possesses for solar energy opens many opportunities to develop new types of technologies and improve conventional ones.

In Texas, agriculture is one of the most important industries being the No. 1 state for total livestock and livestock products receipts, and the No. 2 state for agricultural receipts [3]. This high agricultural production is related with economical expenses which are principally connected with water, chemicals, and energy consumption. The energy consumption in a cattle farm in one year can be between 200 kWh/cow and 600 kWh/cow, this calculation is made by adding the total number of kWh units used in a year and divided it by the number of cows in the ranch [4]. The energy expenses can be reduced in the agricultural practices by the usage of solar energy as principal energy resource or as a supplementary source. The implementation of solar energy as power source in an agricultural system is a possibility for Texan farmers because the high solar potential and the federal and state benefits allow the implementation of this technology [1].

Solar panels can be applied in agriculture for water pumping, lighting, irrigation, aeration, electric fencing, ventilation, greenhouse production and building needs [5-7]. Additionally, farmers have great opportunities to be favored for the solar energy in terms of governmental benefits and environmental and economic revenues. However, they need to select the best type of solar technology that best fits their economic necessities. In that way, this paper outlines the necessary steps to perform a technical an economic evaluation of photovoltaic solar panels. This outline can be used as a tool by agricultural and biological engineers, Agricultural Systems Management or another agriculture related profession around the world without need of extensive knowledge in the operation of photovoltaic solar panels.

As example of the implementation of this outline; in this paper, two types of photovoltaic panels were evaluated. PV panels are glass covered semiconductor panels which can convert sunlight directly into electric energy. The PV panels can be produced from different materials, generating different characteristics, advantages, and prices [8]. The traditional PV panels have been formed using crystalline silicon, a less expensive material with reasonable performance [9]. Today, new PV panels have been developed trying to achieve higher efficiencies than silicon PV panels. The Gallium Arsenide (GaAs) solar panel is recognized as a very efficient panel [9]; however, the material prices make GaAs more expensive than traditional silicon. The differences in performance and price of these two panels are clear. However, before make a selection between these solar panels; it is necessary to perform economic and technical studies that take under consideration the weather conditions, the location and the power necessities.

The principal aim of this research was to outline the steps in the technical and economical evaluation of solar PV systems, comparing GaAs and Si solar panels under Texas weather conditions.

2. MATERIAL AND METHODS

2.1 Solar Panels Configuration

The experiment was performed using three commercial Silicon (Si) (0.9××0.28 m) and two Gallium Arsenide (GaAs) panels (0.59×0.59 m) donated by USDA-ARS, in Lubbock, Texas. The panels' configuration (Fig 1.) was developed to create identical testing situations for both panels under temperatures between 29 and 33 °C. The panels were mounted on separate rolling platforms, both at the same distance from the ground and tilted at an angle of 35 degrees from the horizontal. The angle used in this research corresponds to the original angle of the commercial Si solar panels. To obtain the closest conditions between both solar panels, the surface area of one of the Si panels was covered to match the surface area of the GaAs panel. After the adjustments the final exposed area for each solar panel was 0.6975 m^2.A research-grade pyranometer (CMP 22, Kipp and Zonen) was used to measure the input solar irradiance (W/m^2) that the panels were exposed to. The pyranometer was placed near the panels and tilted at the same 35 degree angle as the solar panels. During experimentation the panels were placed in the same location inside the Texas A&M campus. All data were recorded on an Omega 320 data logger (Omega Engineering Inc., Stamford, Connecticut). The logger was programmed to record and calculate the current, solar irradiance, power output and efficiency of both panels simultaneously taking points each minute. The data logger and the pyranometer used in this research were previously acquired by the laboratory for other researches.

2.2 Solar Panels Efficiency

To calculate the efficiency of the panels, it is necessary to know the irradiance, the power output and the surface area. The irradiance was measured by a pyranometer in W/m^2. To calculate the power output of the panels, a resistor was connected into the panel's circuit to induce a current. The resistor had a known constant resistance and the voltage was recorded across the resistor. According to Ohms law $V=I\times R$, the current was calculated from the measured voltage and known resistance (15 ohms). Once current was calculated, Watt's law (Power = $I\times V$) calculation yields the power output of the panel with the given resistance.

The power output from the panel is divided by the product of the irradiance and solar cell area to calculate the efficiency of the unit. Note that for the solar PV to operate under maximum power output at all times, the product of current (or amperage) and voltage must be at its maximum. Thus, the first sets of tests were made to determine the open circuit voltage (V_{oc}) of both the GaAs PV cell and the Si cell as solar irradiance varies throughout the day. This will establish the maximum output voltage as a function of solar irradiance. Unfortunately, during actual operation, current could not be measured if there is no load for the system hence, the set-up where a fixed or varying load is necessary. This may be done by varying the resistance across the load.

The short circuit current (I_{sc}) was established for each system and the maximum current drawn from each system is determined under varying solar irradiance. Thus, with the short circuit current known as well as the open circuit voltage, the maximum power output may be estimated or measured since a solar cell may operate over wide ranges of voltages and current. Under ideal scenario, maximum power output could always be achieved, with proper combination of current and voltage, or in this case, by adjusting the load for the solar panel. In actual ideal applications, the solar PV panels are exposed under the sunlight and are used to charge a battery using a MPPT which generates the combination of current and voltage that would generate the maximum power, while the load is simply connected to the battery during usage. This configuration will achieve optimum utilization of energy and power from the sun. In this research, a fixed load was used to show how to perform the technical and economic analysis of solar panels using basic conditions.

The calculation of efficiencies for this study was based on the maximum combination of current and voltage to achieve maximum actual power from each PV unit as shown in Equation 1.

$$\eta = \frac{P_m}{A_c \times S} \times 100\% \quad (1)$$

Where, η=solar PV panel efficiency, %, P_m=maximum power point, Watts, A_c= area of solar collector, m^2 (0.6975 m^2 in this study), S=incident solar irradiance on the collector, W/m^2. P_m is the maximum power output that could be generated at a given incident solar irradiance. This value will change according to the open circuit voltage and the current. In that way, at a given incident solar irradiance, the output voltage maybe projected and the proper load may be adjusted to generate the maximum power. Electronically, this is simply done by varying the resistance or the resistive load.

2.3 Solar Panels Economical Evaluation

To develop an economic comparison between the two solar panels, a simulation of the yearly performance of each solar panel was done. In this case, the environmental conditions used corresponded to College Station, Texas; Latitude (ϕ): 30.6°, Longitude: 96.3°, Solar Constant (G_{sc}): 1367 W/m^2, and Panels slope (β): 35°. The consumption assumed was a small cattle farm with 10 to 40 animals. Using 40 cows and supposing that each cow consumes 300 kWh/year [4], the farm will consume 1000 kWh/month. The efficiency of the systems was calculated using the experimental data and the life of systems supposed was 20 years. The economic assumptions used for the economical simulation are shown in Table 1. Balance of system includes all additional cost to have the unit in full operation such as the batteries, converters, controllers, wiring and other switches.

To make the comparison between the two solar panels it is necessary to calculate the daily revenue (D_R) using equation 2:

$$D_R = P_{day} \times grid \quad (2)$$

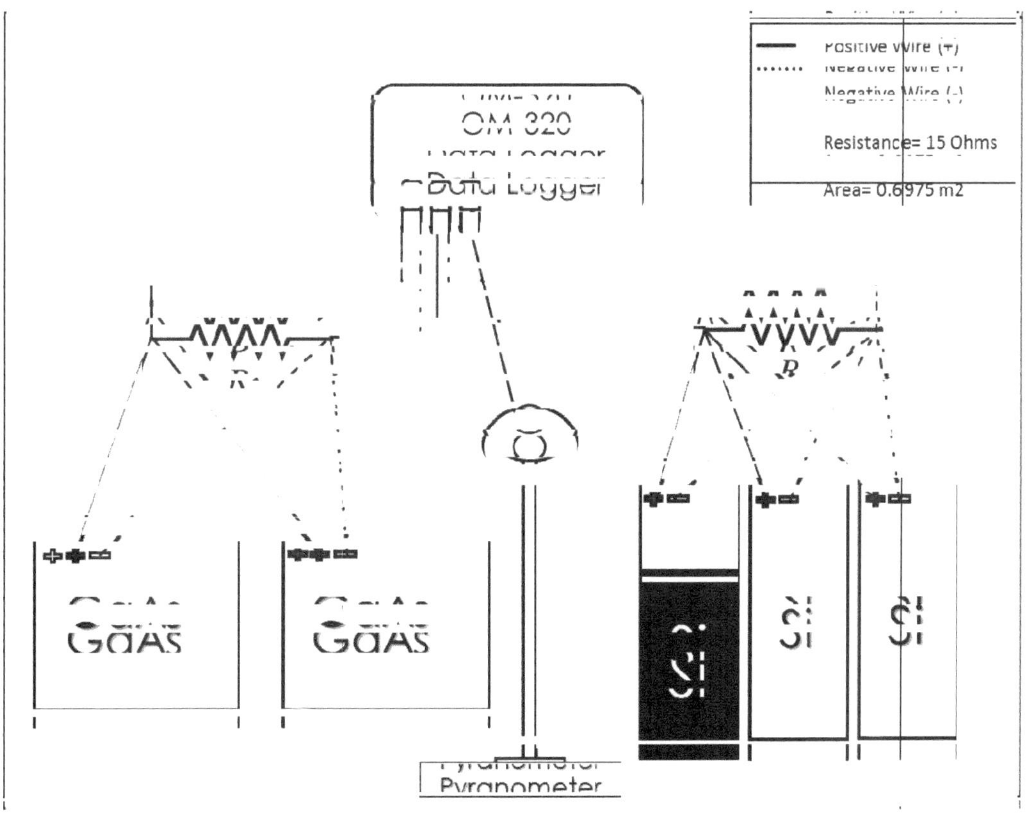

Fig. 1. Experimental circuit used in the solar panels data acquisition. The black colour in the Si panel represents the covered area

Table 1. Economic assumptions

Variable	Costs (Dollars)
Gallium Arsenide panels costs	$3/Wp
Silicon panels costs	$2/Wp
Gallium Arsenide system cost (Balance of system included)	$7/Wp
Silicon system cost (Balance of system included)	$6/Wp
Grid	$0.12/kWh

Where *grid*= Cost of the energy in dollars/kWh, P_{day}= power generated by day kWh. P_{day} is calculated using the next equation:

$$P_{day} = \frac{A_c \times S \times \eta \times N_h}{1000} \quad (3)$$

Where A_c=solar panel area, m^2, S=incident solar irradiance, W/m^2, η=solar panel efficiency, N_h= number of daylight hours, h. N_h was calculated using equation 4 and S was calculated using the equation 6:

$$N_h = \frac{2}{15}\cos^{-1}(-\tan\phi\tan\delta) \quad (4)$$

Where ϕ is the Latitude, and δ is the declination angle. The declination angle is calculated using the equation 5:

$$\delta = 23.45 \times \sin\left(360 \times \frac{284 + n_d}{365}\right) \quad (5)$$

In equation 5 n_d correspond to the day of the year which is simulated from 0 to 365. The incident solar irradiance was calculated using the next equation:

$$S = H_0 / N_h \quad (6)$$

Where H_0= daily solar irradiance number, MJ/m^2 and N_h= number of daylight hours, h. To obtain the solar irradiance units (W/m^2) is necessary to use conversion factors in the formula. The daily solar irradiance number is calculated using equation 7.

$$H_o = \frac{24\times3600 G_{sc}}{\pi}\left(1 + 0.033\cos\frac{360nd}{365}\right) \times \left(\cos\emptyset\cos\omega_s + \pi\omega s 180 \sin\emptyset\sin\delta \quad (7)$$

Where G_{sc}= Solar Constant (1367 W/m^2), n_d = day of the year, ϕ = Latitude, δ= declination angle, and ω_s= hour angle. The hour angle can be calculated by using equation 8.

$$\omega_s = \cos^{-1}\left(-\frac{\sin\phi\sin\delta}{\cos\phi\cos\delta}\right) \quad (8)$$

Where ϕ is the latitude and δ is the declination angle. Demonstrations and other equations used in this research can be found in solar energy references [10].

3. RESULTS AND DISCUSSION

3.1Solar Panels Efficiency

The open circuit measurements were performed using the circuit of the solar panels without the resistor. The GaAs solar panel open circuit voltage measurements are shown in Fig. 2a. The maximum open circuit voltage was found to be around 33.82 Volts for the maximum solar irradiance (1,227 W/m^2) measured during the experiment. Likewise, the Si panel's open circuit voltage relationship is shown in Fig. 2b. In this type of solar panel, the maximum voltage was found to be around 22.5 Volts for the highest solar irradiance in the experiment (1,227 W/m^2). The sun irradiance varied between 350-1227 W/m^2. The open circuit voltage relationship for the Si solar panel was not quite distinct and was assumed to be of second order with an R^2 of around 0.7167. For the GaAs unit, the R squared was found to be 0.9779 for a fourth-order relationship using a polynomial distribution.

The short circuit current (I_{sc}) measurements for the GaAs and Si solar panels are shown in Figs. 3a and 3b. The solar panels I_{sc} was obtained graphically from Figs. 3a and 3b, from the point with greatest current. The GaAs' I_{sc} was found to be 2.01 Amps, and the Si panel's I_{sc} was 2.44 Amps. These values will be the basis for the calculation of maximum power point (P_m).

Fig. 4 shows the calculated conversion efficiencies for the GaAs and Si solar panels on a typical sunny day in College Station, Texas. The average daily efficiency for the GaAs unit was found to be 18.36% compared with 14.02% for the Si solar panels or a difference of over 4%. Equation 1 was used for the calculation of maximum power for both systems. The plot of efficiencies as a function of solar irradiance variations is shown in Fig. 5. Note the decrease in the conversion efficiency as solar irradiance is increased. This can be a consequence of the increment in the surface temperature which increase with the increment in the solar irradiance, and the load effects [11]. The efficiency reduction related with an increment in the temperature has been reported by different authors showing reduction between 60-70% in Si solar panels [12]. The temperature effect in GaAs panels is lower than Si [13], which is an explanation for the best performance observed by GaAs panels. Data in Fig. 5 will be utilized to make a year-long inventory of the solar PV output using daily solar irradiance data for the experiment location. The trend line for each graph is also shown with very high degree of correlation. Thus, if solar irradiance is known each day for a given location, the yearly output of the solar PV cell may be easily estimated. Further, if the initial capital costs as well as operational costs are also known, the overall economic benefit of the unit may be evaluated appropriately. This is done by simply multiplying the solar irradiance received per unit area on a given day throughout the year with the solar collector area. If the unit is metered-in at the local utilities and the utilities cost is known ($/kWh), then the yearly revenue may be is estimated. These analyses will be presented in the next section.

3.2 Economic Evaluation of PV Systems Studied

To compare the economic return of the solar PV systems, it was necessary to simulate the solar panels' yearly performance from actual solar irradiance data available on a given location. Numerous cities worldwide have reported their solar energy potential and perhaps the basic data is the one similar to that shown in Fig. 6. This graphs shows the average solar energy received each month in units of MJ/m^2. This type of data was used to make a yearlong initial estimate of the performance of a solar PV panel when used in this location. The actual values of solar irradiance received are shown in Table 2. Included in Table 2 are the theoretical solar energy received each month (H_o) and the clearness index (K_t) in College Station, Texas. The clearness index is a factor that was used to calculate the actual solar irradiance received compared with the theoretical maximum possible solar irradiance received, if the sky were very clear and absolutely free of obstructions. K_t data was obtained from our laboratory data files in solar energy using the theoretical and actual radiation (actual solar irradiance received (H) = H_o x K_t). The solar panel units were facing directly south. The theoretical number of hours of sunshine is also needed in the calculations to convert a power output (kW) into energy (kWh).

Further assumptions include the use of about 15 units of the solar panel with an initial capital cost of $8,400 (Kyocera 2150 W system, multi crystalline Si) or an assumed 21 m^2 of a solar panel area. The unit will be net-metered in a utility that has a utility cost of approximately $0.12/kWh (or 12 cents per kWh). This will be compared with a similar GaAs solar PV panel with an assumed initial capital cost of $10,000. The bases for these costs are the following: $3/Wp for GaAs unit versus $2.04/Wp for the crystalline Si unit [14]. The total installed cost was assumed to be around triple the value for the crystalline Si or an assumed $4/Wp additional balance-of-system (BOS) cost for both systems [15]. BOS includes all additional cost to have the unit in full operation such as the batteries, converters, controllers, wiring and other switches. Thus, the overall cost used for the analysis was $7/Wp for GaAs and $6/Wp for the Si unit. On the average, a small farm with 40 cows will have a consumption of approximately 1,000 kWh/month of power and the assumptions were made to have this requirement supplied by the solar panels. This translates to a solar module with a rating of around 1,400 Wp which for the GaAs unit will have an initial installed cost of around $9,800 (assumed $10,000 above) and $8,400 for the crystalline Si.

The procedure for calculation is a simulation over a year using the product of K_t and H_o and the expected power output for a given day. This daily output is then added to provide a yearly estimate of the revenue over an assumed 20 year life of the units with no operating and maintenance cost. The revenue is simply spread out over the capital cost to determine the simple payback period (PBP). In this study, the yearly revenue for the Si system is around $1,120 versus $1,470 for the GaAs unit. One should do a much more complicated economic analysis depending upon the state and other fixed and variable cost data including use of tax rebate for the purpose. The calculations of this study showed that it would take about 7.5 years for the Si PV panels to recover the initial investment cost as against the net-metered revenue while 6.8 years for the GaAs unit, without any subsidy. Thus, there is a slight advantage for the GaAs PV units due to its higher conversion efficiency even though the initial capital cost is slightly higher.

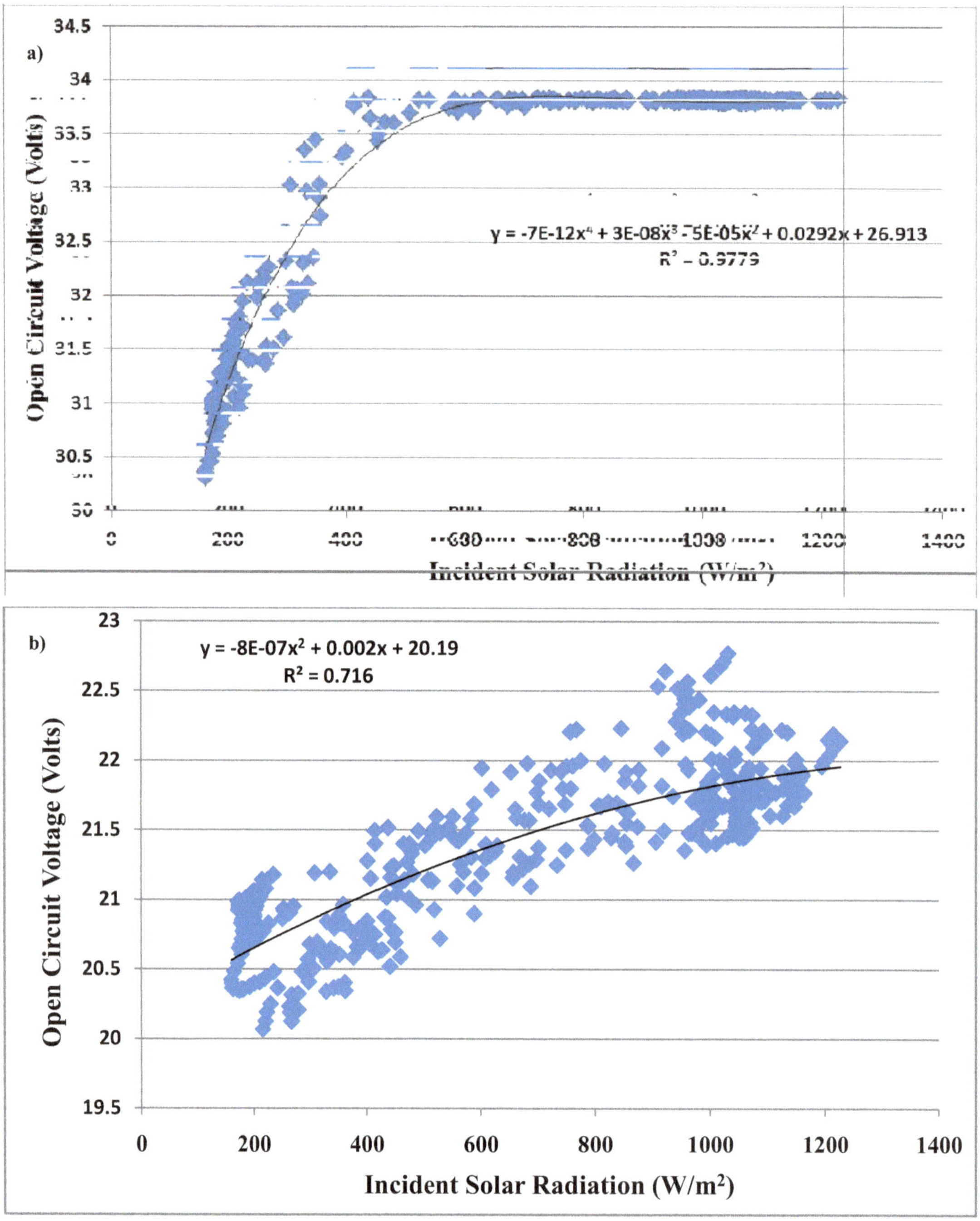

Fig. 2. Relationship between open circuit voltage and solar irradiance a) GaAs Solar Panel, b) Si Solar Panel

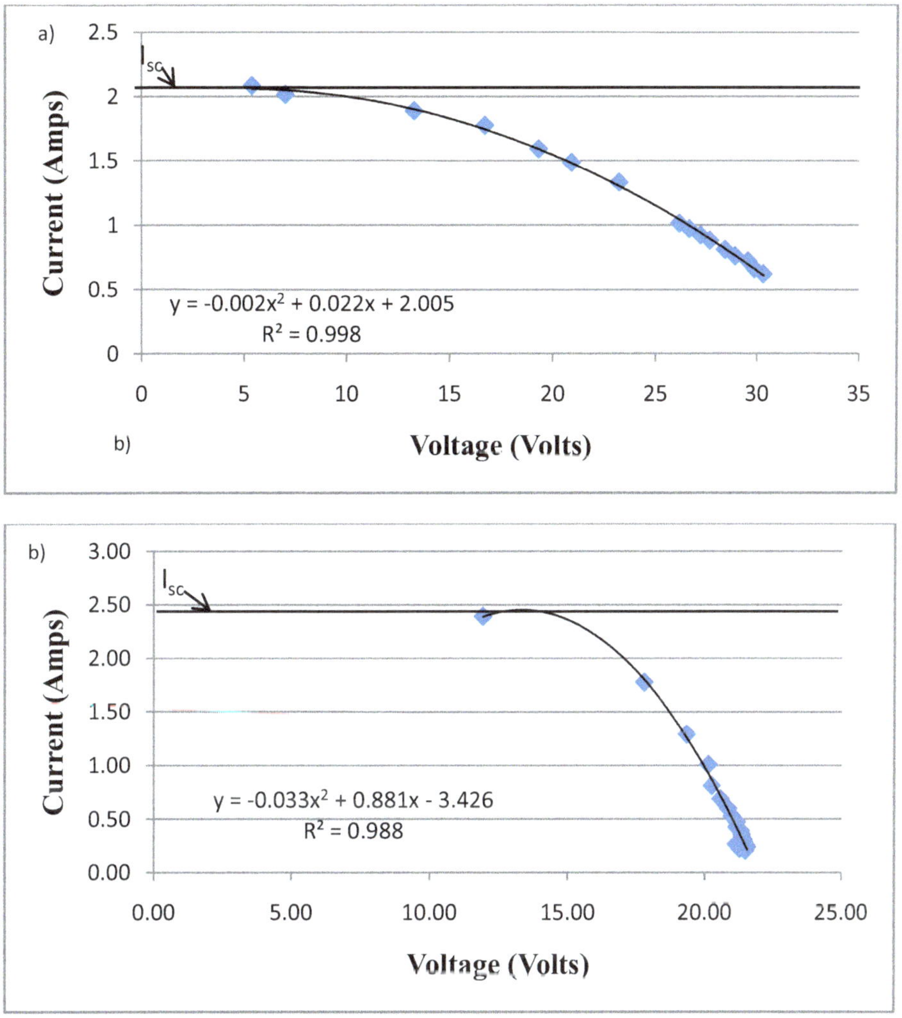

Fig. 3. Relationship between voltage and current, a) GaAs Solar Panel, b)Si Solar Panel

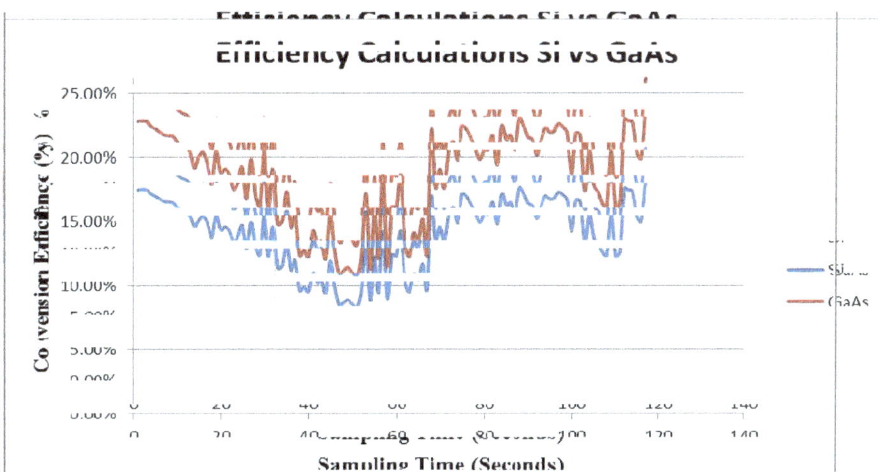

Fig. 4. Conversion Efficiencies for GaAs and Si Solar Panels

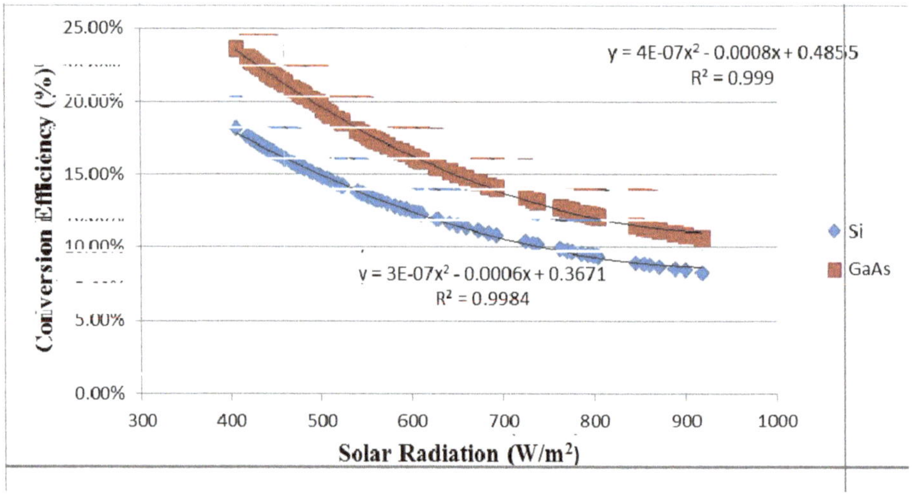

Fig. 5. Conversion Efficiencies for GaAs and Si Solar Panels as a function of solar irradiance

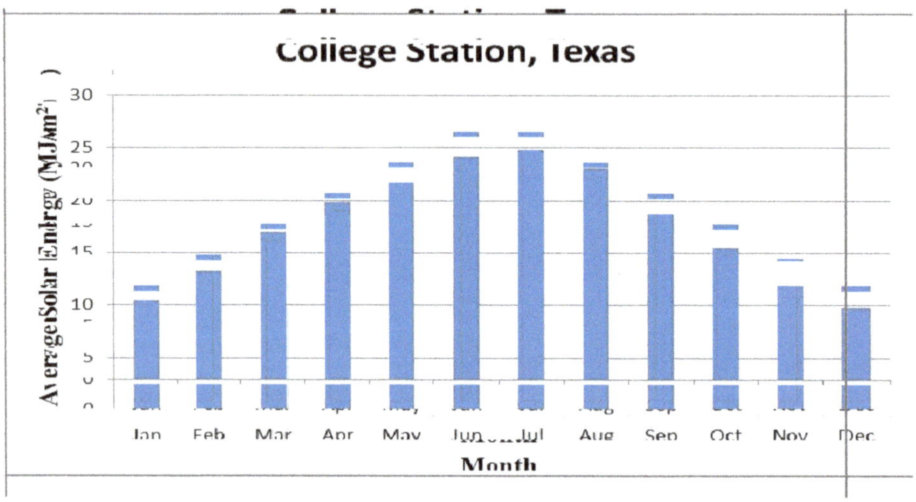

Fig. 6. Average Solar irradiance Received in College Station, Texas for a Year

Table 2. Solar irradiance Data in College Station, Texas

Month	H (MJ/m^2)	H_o (MJ/m^2)	K_t	Monthly revenue	
				Si	GaAs
January	10.4	20.91	0.498	63.68	83.39
February	13.3	25.86	0.519	72.60	95.08
March	16.9	31.56	0.536	95.86	125.53
April	19.8	36.74	0.537	108.44	142.01
May	21.6	39.97	0.539	121.90	159.63
June	24.1	41.18	0.585	121.23	158.76
July	24.8	40.56	0.613	122.97	161.03
August	23.0	38.10	0.609	114.56	150.02
September	18.7	33.77	0.566	96.77	126.73
October	15.5	28.07	0.573	81.82	107.15
November	11.5	22.53	0.523	64.05	83.87
December	9.7	19.67	0.497	59.29	77.65
Yearly Mean	17.4	31.58	0.550	93.60	122.57

The economics analysis will change if additional data are considered. What is shown in this study is a procedure for comparing solar PV units based on their actual field performance and the step-by-step procedure in comparing actual commercial PV units. This paper in no way recommends one type of solar PV system to another since the prices and individual state tax incentives will vary. For example, if there is a 50% subsidy on the total initial capital cost, the PBP will simply be reduced by half, making the system very attractive to farmers and ranchers, especially if the PBP can be reduced to less than 5 years (i.e. 3.4 years for the GaAs and 3.75 years for the Si system). The operating and maintenance costs for the solar PV system are very minimal and will not provide significant changes into the economic analysis over the life of the system. Perhaps, the most important maintenance cost for this system is the replacement of batteries. Deep cycle batteries are quite reliable and may not be replaced for at least 5 years. They are just quite expensive compared with automotive batteries. Thus, one would expect the purchase of new deep cycle batteries around every 5 years.

If the GaAs unit is to be operated on a concentrated mode (i.e. highly intense sunlight), the advantage may be more significant than with the multi-crystalline Si panels. The rebates and tax incentives will change the overall economics if they are included in the calculations. If for example the prices of the units are cut in half (say due to capital cost subsidy), then the payback period will be cut in half as well. The final application of the solar panels in a cattle farm will depend of the farmer necessity, this research showed the calculation for all the power requirements in a small farm, however the same calculation can be done to utilize solar energy independently in different application inside the farm such as water pumping, ventilation, electric fencing and building needs.

4. CONCLUSION

The technical and economical outline proposed in this paper was useful to decide between GaAs and Si panels using basic economic assumptions. In that way, it was calculated the conversion efficiency of the GaAs (18.36%) and Si (14.02%) panels under East Central Texas conditions. The economic study exhibited that Si panels has a recovery time (7.5 years) of the initial capital larger than the GaAs solar panels (6.8years). This paper did not recommend a specific solar panel over the other in every situation. In other locations will be necessary to correct the prices and individual state tax incentives. Future economic calculations should take into consideration government subsidies, rebates and tax credits to make the economic and technical analyses a more accurate tool to select solar panels.

ACKNOWLEDGEMENTS

The TAMU high impact activities grant from the Biological and Agricultural Engineering Department. The Texas AgriLife Research of Texas A&M University. The conservation and production research laboratory of the USDA - ARS, Bushland, Texas; which kindly donate the GaAs Solar PV Panels. The Colombian government and the Fulbright organization are acknowledged for the financial support via the granting of FULBRIGHT-COLCIENCIAS for the Ph.D. studies of Jersson Plácido in the USA.

COMPETING INTERESTS

Authors have declared that no competing interests exist.

REFERENCES

1. The state of texas. Texas renewable energy industry report; 2012. Accessed 5 September 2014. Available:http://governor.state.tx.us/files/ecodev/renewable_energy.pdf
2. House Research Organization. Solar energy in Texas. Focus; 2010. Report 4/4/2013. Accessed 5 September 2014. Available:http://www.hro.house.state.tx.us/pdf/focus/Solar81-13.pdf
3. USDA. Texas annual statistics bulletin. 3/24; 2009. Accessed 5 September 2014. Available:http://www.nass.usda.gov/Statistics_by_State/Texas/Publications/Annual_Statistical_Bulletin/
4. Granted consultancy. Energy in farming. Energy in farming 3/27:04/1/2013; 2012. Accessed 5 September 2014. Available:http://www.morrisons-corporate.com/Global/0_FarmingPage/Morrisons_Renewables_Lo_Res.pdf
5. Taki M, Ajabshirchi Y, Mobtaker HG, Abdi R. Energy consumption, input–output relationship and cost analysis for greenhouse productions in Esfahan Province of Iran. American Journal of

Experimental Agriculture. 2012;3(2):485-501.
6. USDA. Solar Energy use in U.S. Agriculture – Overview and Policy. 2011;83.
7. Ohajianya D, Mgbada J, Onu P, Enyia C, Henri-Ukoha A, Ben-Chendo N, et al. Technical and economic efficiencies in poultry production in Imo State, Nigeria. American Journal of Experimental Agriculture. 2013;4(3):927-38.
8. Razykov TM, Ferekides CS, Morel D, Stefanakos E, Ullal HS, Upadhyaya HM. Solar photovoltaic electricity: Current status and future prospects. Solar Energy. 2011;8(85):1580-608. Available:http://dx.doi.org/10.1016/j.solener.2010.12.002
9. Andreev VM, Grilikhes VA, Khvostikov VP, Khvostikova OA, Rumyantsev VD, Sadchikov NA, et al. Concentrator PV modules and solar cells for TPV systems. Solar Energy Mater Solar Cells. 2004;1–4(84):3-17. DOI:10.1016/j.solmat.2004.02.037.
10. Duffie J, Beckman W. Solar engineering of thermal processes. 4th ed. United states: John Wiley & Sons, Inc; 2013.
11. Radziemska E. Thermal performance of Si and GaAs based solar cells and modules: A review. Progress in Energy and Combustion Science. 2003;5(29):407-24. Available:http://dx.doi.org/10.1016/S0360-1285(03)00032-7
12. Burger B, Rüther R. Inverter sizing of grid-connected photovoltaic systems in the light of local solar resource distribution characteristics and temperature. Solar Energy. 2006;1(80):32-45. Available:http://dx.doi.org/10.1016/j.solener.2005.08.012
13. Kalogirou SA. Solar thermal collectors and applications. Progress in Energy and Combustion Science. 2004;3(30):231-95. Available:http://dx.doi.org/10.1016/j.pecs.2004.02.001
14. Giannouli M, Yianoulis P. Study on the incorporation of photovoltaic systems as an auxiliary power source for hybrid and electric vehicles. Solar Energy. 2012;1(86):441-51. Available:http://dx.doi.org/10.1016/j.solener.2011.10.019
15. Aanesen K, Heck S, Pinner D. Solar power: Darkest before dawn; 2012. Accessed 5 September 2014. Available:http://www.mckinsey.com/client_service/sustainability/latest_thinking/solar_powers_next_shining

22

The Impact of Weather Conditions on Response of Sorghum Genotypes to Anthracnose (*Colletotrichum sublineola*) Infection

Louis K. Prom[1*], Ramasamy Perumal[2], Thomas Isakeit[3], Ghada Radwan[3], William L. Rooney[4] and Clint Magill[3]

[1]USDA-ARS, Southern Plains Agriculture Research Center, College Station, TX 77845. United States.
[2]Western Kansas Agricultural Research Center, Kansas State University, Hays, KS 67601, United States.
[3]Department of Plant Pathology and Microbiology, Texas A&M University, College Station, TX 77843, United States.
[4]Department of Soil and Crop Sciences, Texas A&M University, College Station, TX 77843, United States.

Authors' contributions

This work was carried out in collaboration between all authors. Author LKP conceived and designed the study, performed the statistical analysis, and wrote the first draft of the manuscript. Authors RP and GR assisted in inoculating the plants and data collection. Authors TI, CM and WLR assisted in field operations, including planting, inoculating, and disease assessments, data collection, and reviewed the experimental design and all drafts of the manuscript. All authors read and approved the final manuscript.

Editor(s):
(1) Juan Yan, Sichuan Agricultural University, China.
(2) Masayuki Fujita, Departmen. of Plant Sciences, Kagawa University, Japan.
Reviewers:
(1) Shelley Gupta, Pune University, India.
(2) Klára Kosová, Department of Plant genetics, Breeding and Product Quality, Crop Research Institute, Czech Republic.
(3) L. Gitonga, School of Animal, Plant and Environmental Sciences, University of the Witwatersrand, South Africa.

ABSTRACT

Rainfall is a major climatic factor influencing anthracnose development. In this study, 68 sorghum accessions were evaluated for anthracnose resistance under dry and wet growing conditions at the Texas A&M Agricultural Experiment Station, near College Station, Texas. Accessions, planted in a randomized complete block design with three replications, were inoculated with a mixture of

*Corresponding author: E-mail: louis.prom@ars.usda.gov;

Colletotrichum sublineola isolates 30 days after planting. Under dry growing conditions, three accessions showed a susceptible response across replications, whereas 41 accessions exhibited susceptibility under wet growing conditions. Also, 15 accessions that showed variation in susceptibility across replications under dry conditions were rated as susceptible across replications under wet growing conditions. Nineteen accessions consistently showed a resistant response under both dry and wet growing conditions. There was no significant correlation between weather variables and anthracnose development during the dry growing season, suggesting that climatic conditions were unfavorable for disease development. In contrast, there was a significant positive correlation between total rainfall and anthracnose infection and moderately significant relationships between number of days with rain and minimum relative humidity, with anthracnose infection during wet growing conditions. These results indicate that frequency and cumulative rainfall, as well as relative humidity are critical factors for disease development.

Keywords: Sorghum anthracnose; sorghum bicolor; foliar disease; Colletotrichum sublineola.

1. INTRODUCTION

Anthracnose, caused by *Colletotrichum sublineola* P. Henn, in Kabat and Bubák (syn. *C. graminicola* (Ces.) G. W. Wilson), is one of the most damaging diseases of sorghum and it is presently found in most growing areas [1-3]. The leaves, stalks, and panicles can be infected by the fungus although foliar infection is most common [4]. Foliar infection can occur at any stage of plant development, but symptoms are generally observed 40 days after seedling emergence. Symptom development on the leaves will depend on the cultivar and environmental conditions [4]. Symptoms may appear as small circular to elliptical spots or elongated lesions, and as the fungus sporulates, fruiting bodies (acervuli) appear as black spots in the center of the lesions [4]. Estimating grain yield losses due to foliar anthracnose infection can often be difficult [5], but losses as high as 50% have been reported in susceptible cultivars [6-8]. The occurrence of different pathotypes and levels of pathogenicity within the pathogen population require the identification of different sources of resistance [2,9-11]. Weather conditions, in particular rainfall, have been shown to play a critical role in the incidence and severity of sorghum anthracnose [4,12-16]. In this communication, we report the influence of weather conditions on the severity of anthracnose on sorghum accessions inoculated with a mixture of *C. sublineola* isolates.

2. MATERIALS AND METHODS

2.1 Experimental Materials and Design

Sixty-eight accessions were randomly selected from the Ethiopia, Mali, Sudan, and Uganda sorghum collections maintained by the US National Germplasm System. Seed samples for the anthracnose evaluation were provided by the USDA-ARS, Plant Genetic Resources Conservation Unit, Griffin, Georgia. Sorghum genotype BTx623 was included in the evaluation as the susceptile control while SC748-5 was included as a resistant control. Accessions were planted in a randomized complete block design with three replications in 2009 at the Texas A&M Agricultural Experiment Station, in Burleson county near College Station, Texas. During the 2009 growing season, summer (April-August) growing conditions were very dry followed by wet growing conditions in the fall (September-November), which provided a unique opportunity to evaluate the effect of weather variables on anthracnose disease development. Weather data during the two growing periods are presented in Table 1. Seed was planted in 6 m rows at 0.31 m spacing between rows. Field preparation included fall plowing and incorporation of NPK according to local recommendation. To control weeds and seedling insects, a pre-emergent insecticide 'Counter 20 CR' (BASF Group, Southfield, MI) and herbicide 'Atrazine' (Syngenta Crop Protection Inc. Greenboro, NC) were applied before planting. The trial was planted in April and disease assessment conducted in June and July and in late August, plants within each plot were cut about 15.2 cm above the ground, the debris removed, and plants allowed to tiller for the second disease assessment.

2.2 Inoculation Technique

The inoculation technique and disease assessment method followed the procedures described by Prom et al. [17]. Briefly, sorghum plants were inoculated 30 days after planting by placing 10 *C. sublineola*-colonized grains onto the leaf whorls. A mixture of anthracnose isolates was used. For the evaluation conducted under wet growing conditions, tillers were inoculated 30 days after the plants were cut. Disease assessments were conducted 30 days post-inoculation and thereafter, on a weekly basis for four weeks. Ratings were based on a scale of 1 to 5 [14,17], where 1 = no symptoms or chlorotic flecks on leaves; 2 = hypersensitive reaction (reddening or red spots) on inoculated leaves but no acervuli formation and no lesion development on other leaves; 3 = lesions on inoculated and bottom leaves with acervuli in the center; 4 = necrotic lesions with acervuli on inoculated leaves and infection spreading to bottom and middle leaves; and 5 = most leaves dead with abundant acervuli on the flag leaf. Accessions were considered resistant if plants in the row were rated as 1 or 2 and susceptible if rated as 3, 4, or 5.

2.3 Statistical Analysis

Pearson's correlation coefficients were calculated between anthracnose rating and weather parameters such as maximum average temperature, minimum average temperature, maximum average relative humidity, minimum average relative humidity, total rainfall, and number of days with rain for the two seasons. All these weather parameters were noted for the dry and wet growing conditions starting a week before inoculation and ending 8 weeks post-inoculation. Using PROC REG (SAS Institute, SAS version 9.2, Cary, NC) with stepwise as the selection option, the best model for the dry and wet periods anthracnose score were determined.

3. RESULTS AND DISCUSSION

In this study, anthracnose response was significantly affected by accession ($P<0.001$) under both dry and wet growing conditions. For the dry growing conditions, three accessions showed a susceptible response with 15 accessions showing a variation in susceptibility (Table 2).

While under the wet growing conditions, 41 accessions exhibited susceptibility across replications and eight showed a variation in susceptibly across replications. Nineteen accessions that were rated as resistant and three accessions that were rated as susceptible showed a similar disease response across the two evaluations. Erpelding and Prom [14] evaluated 270 accessions from the Mali working collection in Isabela, Puerto Rico and noted higher number of susceptible accessions during the wet season than the dry season. Studies have shown that the environmental conditions during evaluation of sorghum germplasm have profound influence on anthracnose infection response [4,12,13,15]. Néya and Le Normand [12] reported a higher number of susceptible genotypes and higher anthracnose severities in locations with favorable weather conditions, such as high relative humidity, warmer temperature, and higher total cumulative rainfall. Higher anthracnose severities on sorghum germplasm were also noted in wetter regions of Nigeria [13]. In this study, 23 accessions with a resistant response to anthracnose under dry growing conditions exhibited susceptibility when evaluated under wet conditions (Table 2). Five accessions (PI563301, PI568373, PI570743, PI217674 and PI570873) from Sudan, 7 accessions (PI609009, PI609044, PI608974, PI608986, PI609991, PI609947, and PI609008) from Mali, and 7 accessions (PI297093, PI154758, PI330977, PI154966, PI584214, PI584284, and PI330985) from Uganda exhibited resistant responses in both the dry and wet periods (Table 3). Further, 15 accessions with a variable response across replication under dry conditions were found to be susceptible under wet growing conditions.

Table 1. Weekly averaged weather data for the dry (summer) and wet (fall) growing conditions in 2009[a]

Weather variable	Dry	Wet
21 days after planting (DAP)		
Total rainfall (mm)	0	3.81
Number of days with rain	0	4
Maximum temperature (°C)	32.6	28.3
Minimum temperature (°C)	18.4	19.4
Maximum relative humidity (%)	89.2	91.8
Minimum relative humidity (%)	29.2	56.9

Weather variable	Dry	Wet
37 DAP (First week post-inoculation)		
Total rainfall (mm)	0	19.8
Number of days with rain	0	4
Maximum temperature (°C)	33.2	24.7
Minimum temperature (°C)	18.8	17.3
Maximum relative humidity (%)	89.7	94.4
Minimum relative humidity (%)	33.8	71.8
44 DAP (Second week post-inoculation)		
Total rainfall (mm)	0	0
Number of days with rain	0	0
Maximum temperature (°C)	36.1	26.7
Minimum temperature (°C)	24.0	14.4
Maximum relative humidity (%)	87.5	92.9
Minimum relative humidity (%)	36.0	46.7
51 DAP (Third week post-inoculation)		
Total rainfall (mm)	0	13.7
Number of days with rain	0	4
Maximum temperature (°C)	36.9	22.6
Minimum temperature (°C)	23.2	10.8
Maximum relative humidity (%)	90.0	95.3
Minimum relative humidity (%)	29.9	53.3
58 DAP (Fourth week post-inoculation)		
Total rainfall (mm)	0	0.3
Number of days with rain	0	3
Maximum temperature (°C)	39.2	24.6
Minimum temperature (°C)	24.3	8.4
Maximum relative humidity (%)	82.1	94.3
Minimum relative humidity (%)	24.8	38.5
1st Disease assessment week [65 DAP (5th week post-inoculation)]		
Total rainfall (mm)	0.23	1.27
Number of days with rain	1	4
Maximum temperature (°C)	36.4	25.3
Minimum temperature (°C)	24.3	11.6
Maximum relative humidity (%)	86.4	94.4
Minimum relative humidity (%)	34.0	47.2
2nd Disease assessment week [72 DAP (6th week post-inoculation)]		
Total rainfall (mm)	0.10	3.6
Number of days with rain	2	5
Maximum temperature (°C)	38.9	23.6
Minimum temperature (°C)	24.7	9.2
Maximum relative humidity (%)	86.6	95.8
Minimum relative humidity (%)	24.8	46.9
3rd Disease assessment week [79 DAP (7th week post-inoculation)]		
Total rainfall (mm)	0.76	4.1
Number of days with rain	2	5
Maximum temperature (°C)	37.9	19.4
Minimum temperature (°C)	23.5	10.6
Maximum relative humidity (%)	88.5	91.4
Minimum relative humidity (%)	28.1	58.0
Final disease assessment week [86 DAP (8th week post-inoculation)]		
Total rainfall (mm)	0.13	0.13
Number of days with rain	1	1
Maximum temperature (°C)	36.1	17.8
Minimum temperature (°C)	24.3	2.3
Maximum relative humidity (%)	88.2	94.7
Minimum relative humidity (%)	31.5	25.7

[1a] *Data obtained from: http://researchfarm.tamu.edu/test.txt*

A significant positive correlation between total rainfall and anthracnose infection was noted during the wet growing conditions, indicating that rainfall plays a critical factor in the disease development (Table 4). The number of days with rain and minimum relative humidity also appear to be vital in anthracnose development as indicated by the moderately significant relationships. In the study, 1.22 mm of total rain on 6 rainy days were noted during the dry season, while 46.7 mm of rain on 30 rainy days were noted during the wet season (Table 1). Modeling disease score for the dry season showed that none of the measured variables met the significant level for entry into the model. While in the wet season, minimum average temperature with $R^2 = 0.94$ and precipitation with $R^2 = 0.99$ were the two variables with significant levels for entry into the model. This would suggest that minimum average temperature and rainfall are critical elements in anthracnose development within sorghum fields. Several studies have shown the importance of rainfall in creating high levels of anthracnose infection in sorghum [4,12,13,15,16,18,19]. Rain plays an integral part in the dispersal and subsequent germination of *C. sublineola* conidia [4]. Thus, anthracnose is more severe in areas with higher annual rainfall [4,12,13,15]. Hess et al. [18] observed higher anthracnose severity on susceptible sorghum genotypes in regions of higher annual rainfall in Mali. Ngugi et al. [19] noted that higher rainfall and humidity are important factors in creating epiphytotics in sorghum. Rainfall has also been shown to be a critical factor in other sorghum diseases. Tarekegn et al. [20] noted significant relationship between total rainfall and frequency of rainfall with the incidence of grain mold fungi in sorghum seeds. Under dryland conditions, Hennessy et al. [21] observed little or no leaf blight infection on sorghum. Similarly, no significant correlations between the weather variables and anthracnose development were observed under the dry growing conditions in this study.

Table 2. The number of anthracnose resistant and susceptible sorghum accessions observed during the dry (summer) and wet (fall) growing conditions in 2009[a]

Reaction type	Number of accessions	Number of accessions
	Dry	Wet
Resistant	50	19
Susceptible	3	41
Variable response*	15	8

*[a]Sorghum accessions were planted at the Texas A&M University Experiment Station near College Station, Texas. Ratings were based on a scale of 1 to 5 (Erpelding and Prom, 2004; Prom et al. 2009), where 1 = no symptoms or chlorotic flecks on leaves; 2 = hypersensitive reaction (reddening or red spots) on inoculated leaves but no acervuli formation and no spreading to other leaves; 3 = lesions on inoculated and bottom leaves with acervuli in the center; 4 = necrotic lesions with acervuli on the bottom and middle leaves; and 5 = most leaves dead with abundant acervuli on the flag leaf. Accessions were considered resistant if plants in the row were rated as 1 or 2 and susceptible if rated as 3, 4, or 5. *At least one replication was resistant*

Table 3. Disease reaction of 68 sorghum accessions and two controls inoculated with *Colletotrichum sublineolum* and evaluated under dry and wet growing conditions in 2009 at the brazos bottom near college station, texas[a]

Accession	Country	Dry	Wet
PI569076	Sudan	4[b]	5[b]
PI297192	Uganda	3	4
PI305056	Ethiopia	3	3
PI330794	Ethiopia	3/3/2*	4
PI305022	Ethiopia	3/3/2	4
PI568288	Sudan	2/3/3	5
PI297218	Uganda	3/2/3	4
PI568388	Sudan	2/3/3	4
PI568485	Sudan	2/3/3	5
PI568403	Sudan	3/2/3	4
PI276797	Ethiopia	2/3/2	4
PI297212	Uganda	3/2/2	3
PI152687	Sudan	2/3/2	4
PI154804	Uganda	2/2/3	3
PI568477	Sudan	3/2/2	4
PI563321	Sudan	2/3/2	4
PI350287	Uganda	2/2/3	5
PI305044	Ethiopia	2/2/3	4
PI609009	Mali	2	2

Accession	Country	Dry	Wet
PI154973	Uganda	2	3
PI297093	Uganda	2	2
PI154901	Uganda	2	3
PI217891	Sudan	2	3
PI305034	Ethiopia	2	4
PI196054	Ethiopia	2	5
PI297204	Uganda	2	4
PI152634	Sudan	2	5
PI297215	Uganda	2	2/3/3
PI154758	Uganda	2	2
PI330977	Uganda	2	2
PI154802	Uganda	2	3
PI297196	Uganda	2	3
PI454096	Ethiopia	2	3/3/2
PI454164	Ethiopia	2	4
PI563145	Sudan	2	4
PI563301	Sudan	2	2
PI330819	Ethiopia	2	4
PI563328	Sudan	2	5
PI568284	Sudan	2	4
PI305035	Ethiopia	2	4
PI568300	Sudan	2	4
PI568373	Sudan	2	2
PI297128	Uganda	2	4
PI297139	Uganda	2	2/2/3
PI297144	Uganda	2	3/2/2
PI568406	Sudan	2	3
PI570743	Sudan	2	2
PI568660	Sudan	2	4
PI154966	Uganda	2	2
PI609044	Mali	2	2
PI217674	Sudan	2	2
PI570211	Sudan	2	2/2/3
PI570873	Sudan	2	2
PI571386	Sudan	2	3
PI584214	Uganda	2	2
PI584284	Uganda	2	2
PI608974	Mali	2	2
PI608986	Mali	2	2
PI608992	Mali	2	2/3/2
PI609991	Mali	2	2
PI267624	Ethiopia	2	4
PI330983	Uganda	2	2
PI609008	Mali	2	2
PI569049	Sudan	2	4
PI609947	Mali	2	2
PI569033	Sudan	2	2/3/3
PI608990	Mali	2	2/3/2
PI569066	Sudan	2	4
BTx623	USA	3	5
SC748-5	USA	2	2

[a]Sorghum accessions were planted at the Texas A&M University Research Farm, Brazos Bottom near College Station, Texas. Ratings were based on a scale of 1 to 5 (Prom et al. 2009; Erpelding and Prom, 2004), where 1 = no symptoms or chlorotic flecks on leaves; 2 = hypersensitive reaction (reddening or red spots) on inoculated leaves but no acervuli formation and no spreading to other leaves; 3 = lesions on inoculated and bottom leaves with acervuli in the center; 4 = necrotic lesions with acervuli on the bottom and middle leaves; and 5 = most leaves dead due to infection with infection on the flag leaf containing abundant acervuli. A rating of 1 or 2 is considered a resistant response, whereas a rating of 3, 4, or 5 is considered a susceptible response. [b]Rows with a single value indicate no variation for disease response within an experiment. An accession with more than one disease response value indicates variation within an experiment (data presented for the replications) or within-accession variation (*)

Table 4. Pearson's correlation coefficients between anthracnose rating and weather parameters for the dry (summer) and wet (fall) growing conditions: maximum average temperature (MaxT), minimum average temperature (MinT), maximum average relative humidity (MaxRH), minimum average relative humidity (MinRH), total rainfall (TRF) and number of days with rain (NDR)[a]

Season	MaxT		MinT		MaxRH		MinRH		TRF		NDR		WetS	
	r	*P*	*r*	*P*	*r*	*P*	*r*	*P*	*r*	*P*	*r*	*P*	*r*	*P*
DryS[b]	0.08352	0.9165	-0.67201	0.3280	0.75531	0.2447	-0.19313	0.8069	0.85584	0.1442	0.79143	0.2086	0.93091*	0.0691
WetS[c]	-0.83929	0.1607	0.39525	0.6048	-0.50945	0.4905	0.94365*	0.0564	0.96889**	0.0311	0.94497*	0.0550		

[a]*Anthracnose score for the two growing seasons (dry and wet) is based on the sum of the scores across the accessions from each disease assessment date. The data for the weather parameters were based on the weekly average starting at the 4th week post-inoculation and ending at the 7th week post-inoculation.* [b]*DryS=anthracnose score for the dry season.* [c]*WetS=anthracnose score for the wet season.**or * = denotes significant at the 5% or moderately significant at the 10% probability level, respectively*

4. CONCLUSION

From these results, cumulative rainfall, frequency of rainfall and relative humidity are critical factors for disease development. Thus, the lack of rainfall during critical periods of disease development under the dry growing conditions would be responsible for the lack of susceptibility observed for some of the accessions in the study. Nineteen accessions from the Mali, Sudan, and Uganda collections were found to be resistant to anthracnose when evaluated during the dry and wet periods. These accessions could be used as parental lines in anthracnose breeding programs.

DISCLAIMER

Mention of trade names or commercial products in this manuscript is solely for the purpose of providing specific information and does not imply recommendations or endorsement by the U.S. Department of Agriculture.

COMPETING INTERESTS

Authors have declared that no competing interests exist.

REFERENCES

1. Pastor-Corrales MA. Variation in pathogenicity of *Colletotrichum graminicola* (Cesati) Wilson and in symptom expression of anthracnose of *Sorghum bicolor* (L.) Moench. Ph.D. dissertation, Texas A&M University, College Station, TX; 1980.
2. Ali MEK, Warren HL. Physiological races of *Colletotrichum graminicola* on sorghum. Plant Dis. 1987;71:402-404.
3. Sherriff C, Whelan MJ, Arnold GM, Bailey JA. rDNA sequence analysis confirms the distinction between *Colletotrichum graminicola* and *C. sublineolum*. Mycol. Res. 1995;99:475-478.
4. Thakur RP, Mathur K. Anthracnose. Pages 10-12 in: Compendium of sorghum diseases. Frederiksen RA, Odvody GN. eds. The American Phytopathological Society. St. Paul, MN, USA; 2000.
5. Ngugi HK, Julian AM, King SB, Peacocke BJ. Epidemiology of sorghum anthracnose (*Colletotrichum sublineolum*) and leaf blight (*Exserohilum turcicum*) in Kenya. Plant Pathol. 2000;49:129-140.
6. Harris HB, Johnson BJ, Dobson Jr JW, Luttrell ES. Evaluation of anthracnose on grain sorghum. Crop Sci. 1964;4:460-462.
7. Powell P, Ellis M, Alaeda M, Sotomayer AM. Effect of natural anthracnose epiphytotic on yield, grain quality, seed health and seed borne fungi in *Sorghum Bicolor*. Sorghum Newsl. 1977;20:77-78.
8. Thomas MD, Sissoko I, Sacco M. Development of leaf anthracnose and its effect on yield and grain weight of sorghum in West Africa. Plant Dis. 1995;80:151-153.
9. Cardwell KF, Hepperly PR, Frederiksen RA. Pathotypes of *Colletotrichum graminicola* and seed transmission of sorghum anthracnose. Plant Dis. 1989;73:255-257.
10. Pande S, Mughogho LK, Bandyopadhyay R, Karunakar RI. Variation in pathogenicity and cultural characteristics of sorghum isolates of *Colletotrichum graminicola* in India. Plant Dis. 1991;75:778-783.
11. Casela CR, Ferreira AS, Schaffert RE. Physiological races of *Colletotrichum graminicola*in Brazil. In *Sorghum* and Millets diseases: A second world review. Milliano de WAJ, Frederiksen RA, Bengston GD. eds. Patancheru, A. P. 502 324, India: International Crops Research Institute for the Semi-Arid Tropics. 1992;209-212.
12. Néya A, Le Normand M. Response of *sorghum* genotypes to leaf anthracnose (*Colletotrichum graminicola*) under field conditions in Burkina Faso. Crop Prot. 1998;17:47-53.
13. Marley PS, Ajayi O. Assessment of anthracnose resistance (*Colletotrichum graminicola*) in *sorghum* (*Sorghum bicolor*) germplasm under field conditions in Nigeria. J. Agric. Sci. 2002;138:201-208.
14. Erpelding JE, Prom LK. Evaluation of Malian sorghum germplasm for resistance against anthracnose. Plant Pathol. J. 2004;3:65-71.
15. Gwary DM, Bwatanglang N, Bdliya BS. Integrated management of *Sorghum* anthracnose through the use of fungicides, crop variety and manipulation of sowing dates in Sudan Savanna of Nigeria. Int. J. Agric. & Biol. 2008;10:661-664.
16. Chala A, Alemu T, Prom LK, Tronsmo AM. Effect of host genotypes and weather variables on the severity and temporal dynamics of sorghum anthracnose in Ethiopia. Plant Pathol. J. 2010;9:39-46.
17. Prom LK, Perumal R, Erpelding J, Isakeit T, Montes-Garcia N, Magill C. A pictorial

technique for mass screening of *Sorghum* germplasm for anthracnose (*Colletotrichum sublineolum*) resistance. The Open Agric. J. 2009;3:20-25.

18. Hess DE, Bandyopadhyay R, Sissoko I. Pattern analysis of sorghum genotype x environment interaction for leaf, panicle, and grain anthracnose in Mali. Plant Dis. 2002;86:1374-1382.
19. Ngugi HK, King SB, Abaya GO, Reddy YVR. Prevalence, incidence, and severity of *Sorghum* diseases in western Kenya. Plant Dis. 2002;86:65-70.
20. Tarekegn G, McLaren NW, Swart WJ. Effects of weather variables on grain mould of *Sorghum* in South Africa. Plant Pathol. 2006;55:238-245.
21. Hennessy GG, de Milliano WAJ, McLaren CG. Influence of primary weather variables on *Sorghum* leaf blight severity in Southern Africa. Phytopathology. 1990;80:943-945.

23

Mycoflora Analysis of Hybrid Sorghum Grain Collected from Different Locations in South Texas

Louis K. Prom[1*], Ramasamy Perumal[2], Zheyu Jin[3], Ghada Radwan[4], Thomas Isakeit[4] and Clint Magill[4]

[1]Crop Germplasm Research Unit, USDA-ARS, Southern Plains Agricultural Research Center, College Station, TX 77845, USA.
[2]Western Kansas Agricultural Research Center, Kansas State University, Hays, KS 67601, United States.
[3]Baicheng City Academy of Agricultural Science, Jilin Province, P.R. of China, 137000, China.
[4]Department of Plant Pathology and Microbiology, Texas A&M University, College Station, Texas 77843, United States.

Authors' contributions

This work was carried out in collaboration between all authors. Author LKP prom conceived and designed the study, performed the statistical analysis and wrote the first draft of the manuscript. Authors RP, ZJ and GR performed the mycoflora analysis. Authors TI and CM assisted in the collection of samples and reviewed the experimental design and all drafts of the manuscript. All authors read and approved the final manuscript.

Editor(s):
(1) Lixiang Cao, Department of Biotechnology, Sun Yat-sen University, Guangzhou, P. R. China.
(2).
Reviewers:
(1) Idowu Oluyemi Olawumi, Obafemi Awolowo University, Ile-Ife, Osun State. Nigeria.
(2) Klára Kosová, Department of Plant genetics, breeding and Product Quality, Crop Research Institute, Prague, Czech Republic.
(3) Anonymous, Côte d'Ivoire.
(4) Anonymous, Egypt.

ABSTRACT

Mycoflora characterization of sorghum grain collected from different locations in Texas during the 2008 and 2009 growing seasons revealed *Alternaria* spp. as the dominant genus. At the beginning of grain development to soft dough stage, *Alternaria* spp. accounted for 66% of the recovered fungal species in 2008 and 55.8% in 2009. At grain maturity in 2009, *Alternaria* spp. represented 66.8% of the recovered fungal species across locations, followed by *Bipolaris* spp. (12.8%), *C.*

**Corresponding author: E-mail: louis.prom@ars.usda.gov;*

lunata (6.8%) and other *Fusarium* spp. (6.8%). Other fungal taxa recovered from sorghum grain included *F. semitectum*, *F. thapsinum*, *Aspergillus* spp. *Phoma* spp. *Colletotrichum* spp. and *Rhizopus* spp. In conclusion, *Alternaria* spp. was the most dominant fungal genus recovered from sorghum grain collected from different locations in southern Texas. However, there is little or no information on the reaction of sorghum hybrids in the region against *Alternaria* spp.

Keywords: Sorghum grain mold; Sorghum bicolor; fungal pathogens; mycoflora analysis.

1. INTRODUCTION

One of the major obstacles to sorghum [*Sorghum bicolor* (L.) Moench.] productivity and seed quality is grain mold, a complex fungal disease. The disease is most prevalent and severe in areas with moist conditions later in the growing season [1,2]. Numerous fungal taxa, including *Fusarium thapsinum* Klittick, Leslie, Nelson et al. Manasas; *Fusarium semitectum* Berk. & Ravenel; *Curvularia lunata* (Wakk.) Boedijn; *Colletotrichum graminicola* (Ces.) G. W. Wilson; *Alternaria alternata* (Fr.:Fr.) Keissl and *Phoma sorghina* (Sacc.) Boerema, Dorenbosch & Van Kesteren have been shown to be involved with the disease complex [1,3,4]. Generally, *F. thapsinum*, *F. nygamy* and *C. lunata* are considered the most common grain molding fungi [1,3,5]. Pathogen infection can occur during flowering to grain harvest [3,6]. Grain mold symptoms include small or deteriorated seed, with discoloration ranging from pink to reddish-pink or black to grayish-black [3,6]. The establishment of excellent plant stand in the field requires high quality seed that produce vigorous seedlings. Seeds infected with grain mold fungi exhibit poor germination and seedling vigor [6-10]. Diseased grain may also not be suitable for food or feed since some of the infecting fungi have the capacity to produce mycotoxins [11,12].

In this study, we evaluated the frequency and diversity of fungal genera and species infecting or contaminating sorghum grain collected from several counties in southern Texas.

2. MATERIALS AND METHODS

Hybrid sorghum seeds were collected from a number of commercial fields in south Texas during the 2008 and 2009 growing seasons. In 2009, samples were collected when plants were at the beginning of grain development to soft dough stage (June 4) and at maturity (July 10). The sampled fields were located in Bee, Goliad, Hidalgo, Kleberg, Nueces, Refugio and Willacy Counties. In each County, four fields were sampled using a W-shaped pattern. Five to 10 panicles were collected arbitrarily from each location. The panicles from each location were threshed and grains bulked. The mycoflora analysis used in this study was previously described by Prom [10]. One hundred and fifty randomly selected seeds from each location were placed in vials and soaked in tap water and then immersed in 10% Clorox bleach (NaOCl) for 1 minute, rinsed three times in distilled water and dried under a laminar flow hood. The seeds were plated on half-strength potato dextrose agar and incubated at 25°C with a 12 hours photoperiod for 7 days. Identification of fungal genera and species was based on the conidia, conidiophores, colony morphology and color and, in the case of *Fusarium* spp. the descriptions provided by Booth [13] and Nelson et al. [14].

PROC GLIMMIX (SAS version 9.3, SAS Institute, Cary, NC) was used to determine the main effect of fungal species using the beta distribution as the conditional distribution and a logit link function. To include extreme values (0, 1) into the analysis, the constant 0.001 was added to the lowest value and subtracted from the highest before performing the analysis. Least-squares means for the fungal species were compared using the Tukey-Kramer *P*-value (0.05) adjustment for multiple comparisons.

3. RESULTS AND DISCUSSION

In both years, the main effect of fungal species was highly significant ($P<0.0001$). In 2008, *Alternaria* spp. (66%) were the most frequently isolated fungal genus across locations followed by *Bipolaris* spp. (17%) and other *Fusarium* spp. (11.3%) (Table 1). *Curvularia lunata* accounted for 2.4%, *F. semitectum* 1.2% and *F. thapsinum* 0.5% of the recovered fungal species. *Aspergillus* spp. and *Colletotrichum* spp. were also isolated but with much lower frequency. Castor and Frederiksen [15] also reported that *Alternaria* spp. were the most frequently isolated fungal species in naturally infected sorghum grains collected in Texas followed by

F. semitectum, *C. lunata* and *C. protuberate*. Similarly, Prom [10] showed that *Alternaria* spp. and *F. semitectum* were the most frequently recovered fungal species from naturally-infected seeds of sorghum grown in Burleson County in central Texas. In Isabela, Puerto Rico, *F. semitectum* was the most frequently isolated fungal species from seeds naturally infected with grain mold, followed by other *Fusarium* spp. *C. lunata* and *Bipolaris* spp. *Alternaria* spp. and *Colletotrchum graminicola* were also isolated, but at lower frequencies [16]. Sorghum grain collected from Santa Fe Province in Argentina had *Fusarium* spp. (*F. moniliforme*, *F. subglutinans*, *F. chlamydosporum* and *F. semitectum*), *A. alternata*, *C. lunata* and *Bipolaris cynodontis*as the most frequently recovered fungal species [11]. Mycoflora analysis of sorghum grain collected from different locations in West and Central Africa revealed *Phoma* spp. and *Fusarium* spp. as the dominant fungal taxa followed by *Curvularia* spp. [17].

In 2009, sorghum grains were collected from the same fields at the beginning of grain development to soft dough stage and also at maturity (Tables 2 and 3). Mycoflora analysis of sorghum grain conducted at beginning of grain development to soft dough stage across locations revealed *Alternaria* spp. with a mean frequency of 55.8% as the dominant genus, followed by non-colonized or clean seed, 22.5%, and *Bipolaris* spp. 14.3%. *C. lunata* and other *Fusarium* spp. accounted for 3.3% and 2.3%, respectively (Table 2). Other fungal genera and species, *F. thapsinum*, *F. semitectum*, *Aspergillus* spp. and *Phoma* spp. were isolated at lower frequencies.

At harvest, *Alternaria* spp. accounted for 66.8% of the recovered fungal genera and species across locations, followed by *Bipolaris* spp. 12.8%, *C. lunata* 6.8% and other *Fusarium* spp. 6.8% (Table 3). Other fungal genera isolated in low frequencies included *Rhizopus* spp. and *Phoma* spp.

Table 1. Frequency of recovery (%) of various fungal genera and species isolated from sorghum grain collected from fifteen locations in southern Texas, 2008[1]
fungal species

County	Alt[2]	Bipo	Coll	Cur	FT	FS	Fus	Unkn	Asper
Bee	60	12	2	10	0	0	8	6	2
Bee	74	14	0	0	0	0	10	0	0
Refugio	64	20	0	8	0	0	8	0	0
Refugio	78	16	0	0	0	0	6	0	0
San Patricio	60	28	0	0	0	2	6	4	0
San Patricio	42	24	0	0	0	0	34	0	0
Nueces	64	18	0	0	0	4	8	6	0
Nueces	84	4	0	0	4	0	6	2	0
Kleberg	82	4	0	2	0	0	12	0	0
Kleberg	38	48	0	0	0	4	8	2	0
Willacy	80	20	0	0	0	0	4	0	0
Willacy	74	4	0	0	0	0	14	6	0
Goliad	90	2	0	6	0	0	20	2	0
Hidalgo	46	24	0	0	0	0	22	0	0
Hidalgo	60	16	0	10	4	8	4	0	0
Overall[3] mean	66a	17b	0.1c	2.4c	0.5c	1.2c	11.3b	1.7c	0.1c

[1]*Sorghum grains were collected at harvest. one hundred and fifty seeds from each location were analyzed.*
[2]*Alt=Alternaria species, bipo=Bipolaris species, coll=Colletotrichum species, cur=C. lunata, FT=Fusarium thapsinum, FS=Fusarium semitectum, Fus=other Fusarium species, unkn=unknown, asper=Aspergillus species, Ph=Phoma species, Cln=clean seed (not colonized).*
[3]*Means within a row followed by the same letter(s) are not significantly different at the 5% probably level based on tukey-kramertest for multiple comparisons*

Table 2. Frequency of recovery (%) of fungal genera and species isolated from sorghum grain collected from several locations in southern texas, 2009[1]fungal species

County	Alt[2]	Bipo	Coll	Cur	FT	FS	Fus	Unkn	Asper	Ph	Cln
Bee	24	2	0	16	0	0	2	4	0	10	42
Refugio	68	20	0	4	0	0	8	0	0	0	0
Hidalgo	88	10	0	0	0	0	0	0	0	0	0
Hidalgo	74	0	2	2	2	2	4	4	4	4	20
San patricio	60	12	0	0	0	0	2	0	0	0	26
Nueces	6	2	0	0	0	0	0	0	0	0	92
Willacy	46	50	0	4	0	0	0	0	0	0	0
Kleberg	80	18	0	0	0	0	2	0	0	0	0
overall[3]mean	55.8a	14.3b	0.3b	3.3b	0.3b	0.3b	2.3b	1.0b	0.5b	1.8b	22.5b

[1]*Sorghum grains were collected at the beginning of kernel development to soft dough stage, June 4. One hundred and fifty seeds from each location were analyzed.* [2]*Alt=Alternaria species, bipo=Bipolaris species, coll=Colletotrichum species, cur=C. Lunata, FT=Fusarium thapsinum, FS=Fusarium semitectum, Fus=other Fusarium species, Unkn=unknown, Asper=Aspergillus species, Ph=Phoma species, Cln=clean seed (not contaminated).* [3]*Means within a row followed by the same letter(s) are not significantly different at the 5% probably level based on tukey-kramertest for multiple comparisons*

Table 3. Frequency of recovery (%) of various fungal genera and species isolated from sorghum grain collected from several locations in texas, 2009[1] Fungal species

County	Alt	Bipo	Cur	FS	Fus	Unkn	Ph	Cln	Rhz
Bee	56	6	2	0	4	4	4	12	12
Refugio	58	18	20	2	2	0	0	0	0
Hidalgo	72	8	6	8	6	0	0	0	0
Hidalgo	62	22	6	2	8	0	0	0	0
San patricio	62	16	6	4	14	0	0	0	0
Nueces	78	10	0	6	6	0	0	0	0
Willacy	66	16	6	4	8	0	0	0	0
Kleberg	80	6	8	0	6	0	0	0	0
Overall[3]mean	66.8a	12.8b	6.8b	3.3b	6.8b	0.5b	0.5b	1.5b	1.5b

[1]*Sorghum grains were collected at maturity, July 10. one hundred and fifty seeds from each location were analyzed.*[2]*Alt=Alternaria species, bipo=Bipolaris species, Cur=C. Lunata, FS=Fusarium semitectum, Fus=other Fusarium species, Unkn=unknown, Ph=Phoma species, Cln=clean seed (not contaminated), Rhz=Rhizopus species.*[3]*Means within a row followed by the same letter(s) are not significantly different at the 5% probably level based on tukey-kramer test for multiple comparisons*

The recovery of *Alternaria* spp. in Bee, Nueces and Willacy counties, *C. lunata* in six of the seven counties, *F. semitectum* in Hidalgo and Nueces counties and other *Fusarium* spp. in five of the seven counties increased at harvest, whereas the percentage of non-colonized or clean seeds and, to a lesser degree, *Bipolaris* spp. and *Phoma* spp. decreased. Generally, this indicates that some fungal species that are present early on the developing grain tended to be the dominant species in the grain mycoflora later in the season. The frequency of isolation of different fungal species from sorghum grain collected in a number of locations in South Korea showed *A. alternata*, *A. flavus*, *C. lunata*, *F. moniliforme* and *Phoma* sp. as the dominant species [18], while, *F. thapsinum*, *F. nygamy* and *C. lunata* are the most commonly-isolated grain mold fungi at other locations [1,3,5]. However, this study and others have shown, *Alternaria* spp. *F. semitectum* and other fungal species are usually the dominant species associated with naturally- infected grain [10,11,15,16,18]. Table 4 shows the weather data for the 2008 and 2009 growing seasons. In 2008, mean temperature for the months of June and July ranged from 28°C to 38°C and mean precipitation ranged from 1.6 mm to 6.4 mm across locations, while in 2009, mean temperature for the same months ranged from 32°C to 37°C and mean precipitation ranged from 0.03 mm to 0.94 mm The lack of rain and high temperatures during the 2008 and 2009 growing seasons were not conducive for grain mold development in the regions where the sorghum grain was collected.

Table 4. Weather data for the 2008 and 2009[1]

County	2008		2009	
	MaxT (°C)[2]	Prec (mm)[3]	MaxT (°C)	Prec (mm)
Bee	34.4	2.03	37.0	0.64
Goliad	34.8	2.7	36.5	0.64
Hidalgo	34.1	6.4	36.4	0.38
Kleberg	28.0	1.8	32.0	0.03
Nueces	33.1	3.8	35.6	0.05
Refugio	32.6	1.6	33.7	0.04
San Patricio	33.4	2.8	36.0	0.94
Willacy	34.1	5.2	36.6	0.25

[1]*Weather data for juneand July (plant development stages from flowering to maturity).*
[2]*MaxT (°C) =mean maximum temperature in celsius.*
[3]*Prec (mm)=mean precipitation in millimeters*

4. CONCLUSION

In conclusion, *Alternaria* spp. was the most dominant fungal genus recovered from sorghum grain collected from different locations in southern Texas in 2008 and 2009. However, there is little or no information on the reaction of sorghum germplasm against this fungal species.

DISCLAIMER

Mention of trade names or commercial products in this publication is solely for the purpose of providing specific information and does not imply recommendations or endorsement by the U.S. Department of Agriculture

COMPETING INTERESTS

Authors have declared that no competing interests exist.

REFERENCES

1. Bandyopadhyay R, Chandrashekar A. Biology and management of sorghum grain mold in: Proceedings of Consultative Group Meeting on Technical and Institutional Options for Sorghum Grain Mold Management. 18-19 May. Chandrashekar A, Bandyopadhyay R, Hall AJ, eds. International Crops Research Institute for the Semi-Arid Tropics, Patancheru, India; 2000.
2. Garud TB, Sawant LV, Ismail S. Development of grain mold in sorghum-pigeonpea intercropping systems. Int. Sorghum Millets Newsl. 2000b;41:55.
3. Singh SD, Bandyopadhyay R. Grain mold. Pages 38-40 in: Compendium of Sorghum Diseases. Frederiksen RA, Odvody GN, eds. The American Phytopathological Society. St. Paul, MN, USA; 2000.
4. Esele JP, Frederiksen RA, Miller FR. Importance of plant colour and modifier genes in grain mould resistance in sorghum. J. E. Afr. Agric. For. 1995;61:31-37.
5. Forbes GA, Bandyopadhyay R, Garcia G. A review of sorghum grain mold. Pages 253-264 in: Sorghum and Millets Diseases: A Second World Review. De Milliano WAJ, Frederiksen RA, Bengston GD, eds. International Crops Research Institute for the Semi-Arid Tropics, Patancheru, India; 1992.
6. Castor LL. Grain mold histopathology, damage assessment and resistance screening within *Sorghum bicolor* (L.) Moench lines. Ph. D. Dissertation Texas A. & M. Univ. College Station, TX; 1981.
7. Hepperly PR, Feliciano C, Sotomayor A. Chemical control of seedborne fungi of sorghum and their association with seed quality and germination in Puerto Rico. Plant Dis. 1982;66:902-904.
8. Garud TB, Ismail S, Shinde BM. Effect of two mold-causing fungi on germination of sorghum seed. Int. Sorghum Millets Newsl. 2000a;41:54.
9. Girish AG, Deepti S, Rao VP, Thakur RP. Detection of seedborne grain mold fungi in sorghum and their control with fungicidal seed treatment. Int. Sorghum Millets Newsl. 2004;45:31-33.
10. Prom LK. The effects of *Fusarium thapsinum* and *Curvularia lunata* and their combination on sorghum germination and seed mycoflora. Journal of New Seed. 2004;6:39-49.

11. Saubois A, Piontelli Laforet E, Nepote MC, Wagner ML. Mycological evaluation of a sorghum grain of Argentina, with emphasis on the characterization of *Fusarium* species. Food Microbiol. 1999;16:435-445.
12. Isakeit T, Prom LK, Wheeler M, Puckhaber LS, Liu J. Mycotoxigenic potential of ten Fusarium species grown on sorghum and in vitro. Plant Pathol. J. 2008;7:183-186.
13. Booth C. The Genus *Fusarium*. Commonwealth Mycological Institute. Surrey, England; 1971.
14. Nelson PE, Toussoun TA, Marasas WFO. Fusarium species: An Illustrated Manual for Identification. Pennsylvania State University Press, University Park, PA; 1983.
15. Castor LL, Frederiksen RA. Fusarium head blight occurrence and effects on sorghum yield and grain characteristics in Texas. Plant Dis. 1980;64:1017-1019.
16. Erpelding JE, Prom LK. Seed mycoflora for grain mold from natural infection in sorghum germplasm grown at Isabela, Puerto Rico and their association with kernel weight and germination. Plant Pathol. J. 2006;5:106-112.
17. Ratnadass A, Marley PS, Hamada MA, Ajaya O, Cissé B, Assamoi F, Atokple IDK, Beyo J, Cissé O, Dakouo D, Diakite M, Dossou-Yovo S, Le Diambo B, Vopeyande MB, Sissoko I, Tenkouano A. Sorghum head-bugs and grain molds in West and Central Africa: I. Host plant resistance and bug-mold interactions on sorghum grains. Crop Prot. 2003;22:837-851.
18. Yago JI, Roh JH, Bae S, Yoon YN, Kim HJ, Nam M. The effect of seed-borne mycoflora from sorghum and foxtail millet seeds on germination and disease transmission. Mycobiol. 2011;39:206-218.

24

The Influence of Strong Static Magnetic Field on the Germination and Growth of Garden Cress (*Lepidium sativum* L.)

Filipina J. Krótki[1], Barbara K. Kościelniak[2] and Przemysław J. Tomasik[3*]

[1]Students Research Society, Witkowski High School, Cracow, Poland.
[2]Students Research Society, College of Medicine, Jagiellonian University, Cracow, Poland.
[3]Tutor of Students Research Society, College of Medicine, Jagiellonian University, Wielicka St. 265, Cracow, Poland.

Authors' contributions

This work was carried out in collaboration between all authors. Authors FJK and BKK designed the study, wrote the protocol, performed study and statistical analysis and wrote the first draft of the manuscript. Author PJT supervised all stages of the study and managed the literature searches. All authors read and approved the final manuscript.

Editor(s):
(1) Vladimir K. Mukhomorov, Agrophysical Institute, 14 Grazhdanskiy Ave., St-Petersburg, 195220, Russia.
(2) Daniele De Wrachien, State University of Milan, Italy.
Reviewers:
(1) Anonymous, Bulgaria.
(2) Sadar Uddin Siddiqui, Plant Genetic Resources Program, National Agricultural Research Centre, Pakistan.
(3) Anonymous, Malaysia.
(4) Tarlochan Singh Mahajan, Physics Department, Khalsa College, Patiala (Pb), India.

ABSTRACT

Aims: The aim of the study was to analyze the effect of static magnetic field generated by a neodymium-iron-boron magnets on the growth of garden cress (*Lepidium sativum* L).
Methodology: At the bottom of vessels with garden cress seeds (sample tested) magnets with an established direction of the vector of magnetic field were placed. In the first phase the magnets were put with the south pole facing up, and in the second phase the magnets were put with the north pole facing up. The control group consisted of garden cress cultivated on nonmagnetic cylindrical bars of magnets size. Every day the diameter of the growing area were measured and simultaneously the photographic documentation was collected. On adjusted photos a percentage of green was analyzed.

**Corresponding author: E-mail: p.tomasik@uj.edu.pl;*

Results: In both experiments *Lepidium sativum* seeds began to sprout on the first day. After three days the plants cultivated on the magnets with the south pole facing up had the field of growth significantly larger than the control probes on almost all days whereas the plants cultivated on the magnets with the north pole facing up had similar growth area to control except for day 7.
Conclusion: Magnetic field generated by neodymium magnets orientated similarly to Earth` magnetic field stimulated growth of garden cress.

Keywords: Neodymium-iron-boron magnet; plant growth; germination.

1. INTRODUCTION

In the second half of the 19th century, first attempts on impact of magnetic field on plants was published by J. Reinke and by affirmed by Tolomeidis [1]. They showed the positive effect of magnetic fields on sprout and plant development. Further studies typically analyzed effect of magnetic fields on the seeds of grains and fruit trees. As a result, diminished requirement of oxygen during sprout was observed as well as an increase in the synthesis of proteins, nucleic acids and carbohydrates [2]. Results of these experiments are not consistent, mainly because different magnetic fields used (static vs. pulsed) with different intensities (weak, low, strong) and different schedules as well plants were used [3-5]. The influence of strong magnetic fields on the growth of medicinal herbs has not been analyzed yet. *Lepidium sativum* (garden cress) is one of the fastest growing, and most edible, annual herbs. Garden cress due to a bitter taste and specific scent is used as popular a compound for salads. What is more this herb is valuable natural source of vitamins and minerals and has anti-inflammatory, diuretic, anti-oxidative and hepatoprotective properties [6-9]. The aim of this study was to evaluate the effect of the stable magnetic field generated by neodymium-iron-boron (NIB) magnets on the growth of garden cress.

2. MATERIALS AND METHODS

2.1 Materials

In the experiment the seeds of garden cress (Polan, Poland) and permanent $Nd_2Fe_{14}B$ (N38), disc magnets (33 x 10 mm) (Xinxin, Poland) were used. The magnets were coated with Ni-Cu-Ni layer, their magnetic field vector was axial, with remanence (Br): 1.21 – 1.25 T (12,1-12,5 kG), coercivity (bHc): min. 11,3 kOe, coercivity (jHc): min. 12,0 kOe and maximum energy density (BH) max.: 286-302 kJ/m^3 or 36-38 MgsOe.

2.2 Methods

A serial of identical round vitrioceramic sprouters was prepared as follows. A magnet with an established direction of the vector of magnetic field was placed at the bottom of every vessel. In the first phase (I), the magnets were put with the south pole facing up, and in the second phase (II) the magnets were put with the north pole facing up. Cylindrical bars made of aluminum (paramagnetic element; magnetic susceptibility 2.2) were located in the control vessels identical in dimension. To avoid any influence of compounds from the magnets or aluminium bars all of them were firmly closed in plastic ziploc bags. The bars were covered with strictly weighted (Sartorius MD1; Germany) 5.43 g of lignin and 5 g of garden cress seeds were smoothly sowed in each vessel. At the beginning, 40 ml of tap water was added to all sprouters. Three vessels with magnets and three control vessels were put in the dark room under a fluorescent lamp of 120 cm in length (T8, Mesko, Poland). The lamp was switched on 12 hours per day (8 am – 8 pm), using an electronic controller (Dahl, Germany). All seeds and later plants had the same access to the light. The temperature in the room was controlled using a thermometer (Sunbean, USA). During the light phase in the location of sprouters, the temperature was 25±1 centigrade, and in the dark phase 21±1 centigrade. Every day, at the same time, 15 ml of tap water (collected in the container at the beginning of experiment) was delivered to each sprouter. Also at the same time, once per day every vessel was photographed using a tripod and digital camera (Olympus, Japan), and when the cress sprouted the diameter of visible growth was measured using a digital caliper (PowerfixProfi, UK). Minimal, maximal and visually mean diameters were measured to 0.1 cm precision (Fig. 1). Measurements were stopped when vegetation ceased.

The photos were adjusted with the dimension of the vessels as a benchmark, using Microsoft

Office 2010. The percentage of green in the pictures was obtained using Ink Calculator 5.0.1.35 (freeware; by D. Najgiebauer). The experiment was duplicated.

2.3 Statistics

The statistical analysis was performed in Microsoft Office Excel 2010 (Microsoft, USA). The results were presented using descriptive statistic functions, such as arithmetic mean, median, minimum and maximum values and standard deviation. Differences between groups were tested for statistical significance by Student's t-test. $P \leq .05$ was considered significant.

Fig. 1. Measurements of garden cress` growth

3. RESULTS AND DISCUSSION

The neodymium magnets give the opportunity to explore the influence of strong magnetic fields (over 1 T) on plant development. These magnets despite their relatively small dimensions generated magnetic field thousands more intense than Earth`s magnetic field (in Europe 40-50 μT). In both phases of this study, seeds of *Lepidium sativum* began to sprout on the first day and no differences were observed in this matter between studied and control groups. There were also no significant differences in the green of field (pictures) between magnetized groups and control groups within first 2 days. There are a lot of papers showing that strong magnetic field boost sprout [10-12]. Observations in this study were performed once per day. It would be not sufficient to show a delicate distinct differences in germination. However lack of effect of magnetic field on the germination of *Lepidium sativum* are confirmed by no differences in green on the pictures taken in the first three days of the growth between studied and control probes.

After 3 days, the field of growth was visible, differentiated and measureable. In the case of the plants cultivated on the magnets with the south pole facing up, the area of growth was significantly larger than in control probes on almost all days (except day 6) (Fig. 2 I). Adversely, the growth of the plants cultivated on the magnets with the north pole facing up was similar to the growth in control probes except for day 7 (Fig. 2 II). This day the mean diameter of growth in the studied magnetized group was higher than in the control group. On day eight and later, the field of growth in all groups successively diminished. On the 11th - 13th day plants become weak, and overgrown with mold.

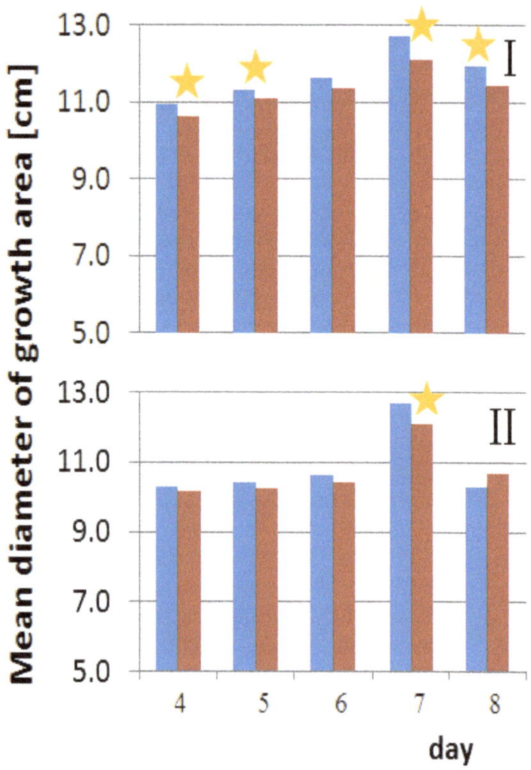

Fig. 2. Mean diameter of growth area in the magnetic field (blue bars) – I. magnets with south pole facing up, and II – magnets with north pole facing up. Red bars – control probes. Asterisks indicated statistically significant differences

Analysis of the mean percentage of green on the pictures of cultivations confirm better growth of plants on south pole of magnets (Fig. 3 I), however observed differences were not statistically significant. The mean green of probes cultivated on north pole of magnets was very similar to the control group (Fig. 3 II).

Magnetotropism of plants has been explored since the sixties. The oriented growth of plants in magnetic fields was observed, as well as the root development toward the regions of the decreasing field strength of magnetic field [13]. In older studies by Sperber et al. [14] magnets enhanced growth despite their orientation, but the recent study of Poinapen et al. [15], similarly to the present study showed better growth of the plants cultivated on the north pole. The differences in growth between cultivation on north and south pole of strong magnets could be related to the orientation of magnets in regard to the Earth`s magnetic field. The South pole on the top oriented magnets had synergistic effect with magnetic field of the planet whereas the North pole on the top oriented magnets had an adverse effect on the magnetic field of the planet (the experiments were done on the north hemisphere). The enhancement of Earth`s magnetic field by magnets could be responsible for a bigger area of growth as the plant roots developed on the sides of magnets toward to diminished magnetic field [13]. The other explanation is based on the action of Lorentz force in magnetic or electromagnetic fields. Lorentz force influences ions as well dipole molecules as water [16]. Lorentz force modifies the location of these molecules and ions between biological compartments, inside subsequent compartments and in biological membranes [17]. All of this, results in changes intransmembrane potentials and Hall effects. These changes would be beneficial, but also could result in the destruction of delicate molecular balance [18]. Another interesting perspective for future research on the effects of magnetic fields on plants is a hypothesis that similarly to some animal species plants also have magnetoreceptors [19].

In summary, the effect of a natural, geomagnetic field, as well as artificial magnetic fields, on plants are not yet fully understood. This study showed that the orientation of permanent neodymium magnets has significance on garden cress growth, but not on sprout.

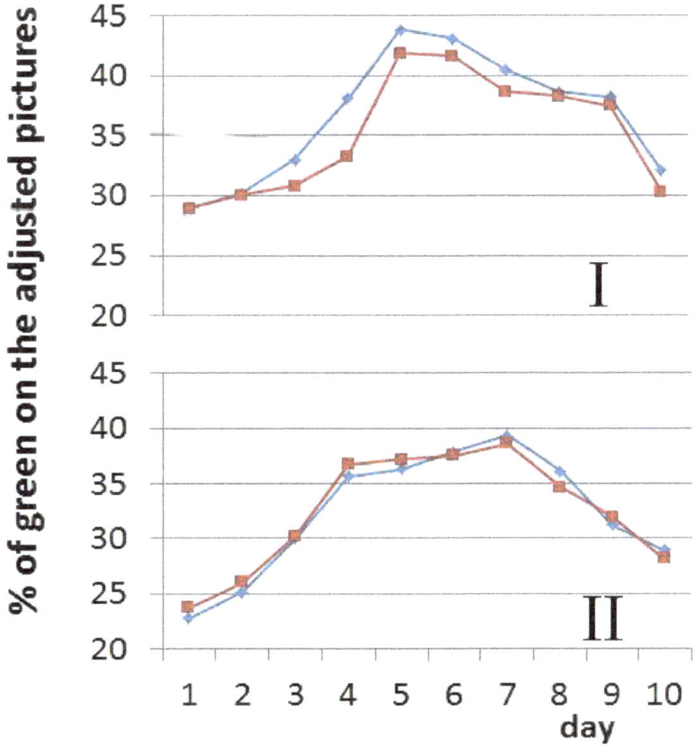

Fig. 3. Mean percentage of green on the unified pictures of the cultivated probes. I – phase I: magnets with south pole facing up, and II – II phase: magnets with north pole facing up. Blueline – magnetized probes, red line – controls

4. CONCLUSION

Magnetic field generated by neodymium magnets orientated similarly to Earth` magnetic field stimulated growth of garden cress.

ACKNOWLEDGEMENTS

Authors would like to thank Mrs Elżbieta Opozda-Zuchmańska (Witkowski High School, Cracow, Poland) for kind scientific support and valuable comments.

COMPETING INTERESTS

Authors have declared that no competing interests exist.

REFERENCES

1. Pietruszewski S. Electromagnetic field, impact on seed germination and plant growth, in Encyclopedia of Agrophysics, Gliński J, Horabik J, Lipiec J. (Eds.), Springer 2011;267.
2. Maffei ME. Magnetic field effects on plant growth, development, and evolution. Frontiers in Plant Sci. 2014;5:1-15.
3. Bilalis DJ, Katsenios N, Efthimiadou A, Karkanis A, Efthimiadis P. Investigation of pulsed electromagnetic field as a novel organic pre-sowing method on germination and initial growth stages of cotton. Electromagn Biol Med. 2012;31:143-50. DOI: 10.3109/15368378.2011.624660.
4. Betti L, Trebbi G, Fregola F, Zurla M, Mesirca P, Brizzi M, Borghini F. Weak static and extremely low frequency magnetic fields affect in vitro pollen germination. Sci W J 2011;19:875-890. DOI:10.1100/tsw.2011.83.
5. Hutova E, Bartusek K, Mikulka J. Study of the influence of magnetic fields on plants tissues. Progress In Electromagnetics Research Symposium Proceedings, Taipei. 2013;57-60.
6. Rehman N, Khan AU, Alkharfy KM, Gilani AH. Pharmacological Basis for the Medicinal Use of *Lepidium sativum* in Airways Disorders. Evid Based Complement Alternat Med; 2012. Article ID 596524, 8 pages.
   DOI:10.1155/2012/596524.
   Available:http://www.hindawi.com/journals/ecam/2012/596524/cta/
7. Kessler CS, Pinders L, Michalsen A, Cramer H. Ayurvedic interventions for osteoarthritis: A systematic review and meta-analysis. Rheumatol Int. 2015;35: 211-32. DOI:10.1007/s00296-014-3095-y.
8. Raval ND, Ravishankar B, Ashok BK. Anti-inflammatoryeffect of Chandrashura (*Lepidium sativum* Linn.) an experimental study. Ayu. 2013; 34:302-304.
   DOI: 10.4103/0974-8520.123132.
9. Al-Mohizea AM, Ahad A, El-Maghraby GM, Al-Jenoobi FI, Alkharfy KM, Al-Suwayeh SA. Effects of Nigella sativa, *Lepidium sativum* and *Trigonella foenum*-graecum on sildenafil disposition in beagle dogs. Eur J Drug Metab Pharmacokinet; 2014. (In press) DOI: 10.1007/s13318-014-0199-4.
10. Radhakrishnan R, Kumari BD. Influence of pulsed magnetic field on soybean (*Glycine max* L.) seed germination, seedling growth and soil microbial population. Indian J Biochemv Biophys. 2013;50:312-317.
11. Vashisth A, Nagarajan S. Effect on germination and early growth characteristics in sunflower (*Helianthus annuus*) seeds exposed to static magnetic field. J Plant Physiol. 2010;167:149–156.
12. Vashisth A, Nagarajan S. Exposure of seeds to static magnetic field enhances germination and early growth characteristics in chickpea (*Cicer arietinum* L.). Bioelectromagnetics. 2008;29:571-578. DOI: 10.1002/bem.20426.
13. Penuelas J, Llusia J, Martinez B, Fontcuberta J. Diamagnetic susceptibility and root growth responses to magnetic fields in Lens culinaris, Glycine soja, and *Triticum aestivum*. Electromagn Biol Med. 2004;23:97–112.
14. Sperber D, Dransfeld K, Maret G, Weisenseel MH. Oriented growth of pollen tubes in strong magnetic fields. Naturwissenschaften. 1981;68:40-41.
15. Poinapen D, Brown DC, Beeharry GK. Seed orientation and magnetic field strength have more influence on tomatoseed performance than relative humidity and duration of exposure to non-uniform static magnetic fields. J Plant Physiol. 2013;170:1251-1258.
16. Dhawi F. Why magnetic fields are used to enhance a plant`s growth and productivity? ARRB. 2014;4:886-896.
17. Kasyanov GI, Syazin IE, Grachev AV,

Davidenko TN, Vazhenin EI. Features of usage of electromagnetic field of extremely low frequency for the storage of agricultural products. J Electromagn Anal Appl, 2013;5:236-241.
DOI:10.4236/jemaa.2013.

18. Kovacs PE, Valentine R, Alvarez PJJ. The effect of static magnetic fields on biological systems: Implications for enhanced biodegradation. Crit Rev Environ Sci Tech 1997;27:319-382.
DOI:10.1080/10643389709388503.

19. Occhipinti A, De Santis A, Maffei ME. Magnetoreception: An unavoidable step for plant evolution? Trends Plant Sci. 2014;19:1-4.

25

Adoption of Improved Maize Farming Technologies by Women Farmers in Zambia

Christopher Sebatta[1*] and Mukata Wamulume[2]

[1]Department of Agribusiness and Natural Resource Economics, Makerere University, P.O.Box 7062, Kampala, Uganda.
[2]Institute of Economic and Social Research, University of Zambia, P.O.Box 30900, Lusaka, Zambia.

Authors' contributions

This work was carried out in collaboration between both authors. Author MW designed the study and oversaw the whole process of data collection and entry. Author CS managed the analysis of the data and wrote the first draft. Both authors read and approved the final manuscript.

Editor(s):
(1) Jamal Alrusheidat, Extension Education Department, National Centre for Agricultural Research and Extension (NCARE), Jordan.
(2) Anonymous.
Reviewers:
(1) Anonymous, Zimbabwe.
(2) Joseph N. Nzomoi, Kilimo Trust, Uganda.
(3) Anonymous, Thailand.
(4) Anonymous, Zimbabwe.

ABSTRACT

The aims of this study were to identify key factors that affect women household heads' awareness and knowledge of existing improved farming technologies and to determine the factors that influence women farmers' adoption of animal draught (ox plough) and agroforestry in Zambia. The study was a survey in which semi structured questionnaires were used to interview women household heads. The focus of this study was mainly on women farmers who are household heads cultivating maize under draught animal (ox) and agroforestry farming practices. The study was conducted in Mkushi and Mazabuka districts in the Central and Southern provinces of Zambia respectively. Data were collected between June 2012 and September 2012.

Mkushi and Mazabuka districts were purposively selected because they lie within the "maize belt" of Southern Africa where maize production intensification technologies have been implemented since the 1990's. The study was conducted at micro level (village level) in the selected agricultural blocks, after which stratification of women farmers was done among adopters and non adopters.

**Corresponding author: E-mail: sebattac@gmail.com;*

Finally, simple random sampling was used in selecting 50 adopters and 114 non-adopters from the two strata. Data were analysed using STATA software. Descriptive statistics and the multi-stage logit regression model formed the basis of analysis.
Access to communication gadgets like phones and membership to farmers' groups was found to increase the probability of awareness of technologies among women farmers by 90% and 35% respectively. The size of land under cultivation was found to increase the likelihood of adoption of ox plough technology by 29% while at the same time reducing the likelihood of adopting agroforestry technologies by 68%.Results revealed that women adopters had higher maize yields at 949 Kg/ha compared to 861.9 Kg/ha for non-adopters. This is mainly attributed to the advantages that come with sustainable agricultural practices like agroforestry.
Communication related factors are key in creating and increasing awareness of existing technologies among women farmers. Similar factors seem to promote adoption of both animal draught (ox plough) and agroforestry technologies among women farmers though expanding scale of production leads to more likelihood of ox plough adoption and later transition to more sophisticated machinery like tractors.

Keywords: Women household heads; technology adoption; multi-stage logit model; Zambia.

1. INTRODUCTION

1.1 Women and Agriculture in Zambia

The agricultural sector in Zambia supports about 80% of the population that is exclusively dependent on agricultural related livelihoods many of whom are poor people in the rural country side [1]. The sector contributed 18% and 20% to GDP in 2011 and 2013respectively and continues to be the largest employer of the Zambian labour force [1,2]. It employs about 52.2% of the working population [3].

Women comprise 50.7% of the total population of Zambia and form 51.6% of the economically active segment of the working population [3]. The majority of the women in Zambia (50.2%) live in the rural areas with 24.1% of rural households headed by females [3]. Many households have less than two hectares of land in the rural area with 71% falling under this category [4]. This is a clear indication that land being a limiting factor in agricultural production, the farmers many of whom are smallholders have to employ intensive farming technologies in order to raise their farm productivity to enable them produce for home consumption but also have a marketable surplus [5]. The main crops grown by the rural people by household numbers in 2011/2012 season in Zambia are maize (86%), groundnuts (39.2%), cassava (37.5%), cotton (20%) and sweet potatoes (12.7%) [6].

In addition, the government has been implementing the food security pack and the fertilizer support program (FISP) through the ministry of agriculture and cooperatives (MACO) since the year 2000 through distributing fertilizer and hybrid maize seed to the farmers in a bid to improve food security. Currently over 85% of all farmers in the study area apply fertilizer (though in inadequate amounts) and use hybrid maize seed [3]. [7] reported that only 29% of female headed households had adopted improved tree fallows after ten years of promoting the technology. Therefore the core problem to sustain production is adoption of other production enhancing technologies like agroforestry for soil fertility and fuel and ox ploughs for affordable smallholder mechanisation.

The main agricultural intensification technologies applied in maize production are use of fertilizer and hybrid seed, use of ox-ploughs for land preparation, zero tillage in addition to agroforestry. Agroforestry is one of the sustainable agricultural systems being promoted in Sub Saharan Africa [8]. However, the main challenge is that the levels of adoption of these technologies are still low among smallholder farmers. Despite the fact that 55% of the farmers use nitrogen fertilizer on their fields, the rate is still at 170 kg/ha below the recommended 200 kg/ha [6]. However, the observed low levels of technology adoption are as a result of a number of issues that are sometimes under the farmer's control and in other cases external to them and out of their control [9]. Therefore to ensure sustainable and increased adoption of yield enhancing technologies, there is need to understand the dynamics and determinants of the area specific characteristics of the farmers and the factors affecting technology uptake by such farmers.

1.2 Factors Affecting Women Technology Adoption

Women household heads face a number of challenges in their bid to sustain their families. In the first case, they have to cater for the household food security and then produce a marketable surplus to earn some income [10]. All this calls for extra efforts in organising labour and other inputs to be able to produce enough. However, a number of factors hamper women's ability to acquire the necessary technologies which include access to extension services, gender disparities with their male peers and ability to associate with a social club like a farmer's group or association for collective action [10]. In other cases, labour constraints and inaccessibility of technologies rather than their availability hinder women from adopting them [11].

Though many rural women lack opportunities for off farm work, studies have always pointed to earning extra income outside the farm as one of the factors promoting technology adoption. This is due to the fact that such incomes are invested in promoting new or better technologies on the farm like buying oxen, ploughs, improved seeds and fertilizers [12].

In many instances, improved technologies tend to be laborious, in that man power is needed to undertake them. This tends to put off many women from taking up such technologies given their age or energy and the fact that women's time is also devoted to other household chores such as cooking, fetching water, gathering fuel wood and child care [13]. [8] found that the number of active household members who can ably provide labour on the farm positively and significantly promoted adoption of agroforestry technologies in Malawi. In addition, the same authors found age of the household head to have a negative effect on adoption.

In many parts of Africa, women have limited rights on possession, ownership and access to land. Their security of tenure on land is usually not guaranteed which incapacitates them from taking on or expanding use of certain agricultural technologies that are tenure sensitive [14,15]. [16] stated that perceived stability of access to land, via stability of personal and social relationships is a more important determinant of technology adoption than mode of access. Therefore, to find out more about such challenges, this study set out to identify the factors that affect women farmers' awareness and knowledge about improved farming technologies, and their adoption for maize cultivation under ox plough and agroforestry farming practices.

2. METHODOLOGY

2.1 Study Area and Sampling Procedure

Mkushi and Mazabuka districts of Zambia in the central and southern provinces respectively formed the study area where women headed households were interviewed in 2012 as part of an ongoing project on pro-poor agricultural growth. The selection of the two districts was purposive so as to ensure sufficient variation in factors assumed crucial for women related agricultural development within the "maize belt". The districts were also selected because of the government restocking program that has introduced more cows some used for ox ploughing and some for draught power in addition to agroforestry programs in the area since the 1990's.

The districts were divided into agricultural blocks (stratification), which in turn are divided into agricultural camps, which is the unit catered for by an extension officer. A camp is further divided into zones corresponding to a number of villages/communities located in the same neighbourhood. For the village diagnostics and household survey, five camps (Musakamba, Musofu, Kalombe, Chitina and Nkumb) in Mkushi and three Camps (Dumba Settlement, Oliver Settlement, Nega Nega) in Mazabuka were selected as "village units"/sampling frames. Thereafter, women headed households were randomly selected where 100 were from Mkushi and 64 from Mazabuka giving a total sample of 164 households.

2.2 Analytical Methods

To outline the factors that influence women farmers' knowledge and adoption of given farming technologies a multi-stage logit model as used by [17,18] with modifications was employed to analyse the data. While in the normal double hurdle the second stage uses a continuous variable (non-binary), in this multi-stage logit, the second stage on adoption uses a dichotomous dependent variable but with only the sub-sample of farmers who are aware of the technology [17]. The other advantage of this model is that the treatment groups can be compared with control groups using predicted probability created through Logit regression [18]. The model is specified as;

$$\Pr_{aware}(KNOW = 1, ADOPT = 1) = 1 / 1 + e^{-z} \ldots\ldots \quad (1)$$

Where, KNOW and ADOPT are dichotomous dependent variables (1 if the farmer has knowledge or adopts a farming technology, 0 otherwise) and e = base of natural logarithms.

The farmer first gets knowledge of the technology as represented in equation (2).

$$Y = \beta_o + \sum_{i=1}^{i=n} \beta_i X_i + \mu_j \ldots\ldots\ldots \ldots\ldots\ldots \quad (2)$$

Where X_i includes a vector of variables in the model that affect a woman farmer's awareness of the existence of available farming technologies, β_i parameters to be estimated and μ_j = error term. Y represents the access to extension services (awareness=1, 0 otherwise), used in this study as a proxy for awareness of existence of a technology on the premise that extension agents act as a link between technology centres and technology users (farmers). Extension access has for long been advocated as instruments that accelerate adoption of technologies by small farmers especially when adverse effects of economies of scale in technology use have been mitigated [5].

Where X_i is defined by the following variables; Age of the woman household head (HH_AGE). Education level of the woman household head specified as the number of years completed in formal school (HH_EDUC). Household size measured as the total number of people living in the household including children (HH_SIZE); engagement in a micro businesses by a woman head of household (MICRO_BIZ). Distance from farm to the nearest village centre (km) (DIST_VILCENTR); main maize selling point (MAIZE_SALPT) which may be the farm gate or outside the farm which is an indicator of the woman farmer's knowledge of what happens outside her farm. Possession of a communication gadget like a phone or radio/TV (COMM_GADGET) taken here as a proxy to indicate that a farmer can access information and adverts from the media and text messages. Membership to a farmer group or association (GRP_MEMBER); ability to make voluntary savings (VOL_SAVINGS) which measures the farmer's ability to invest in new or better farming technologies and total maize harvest in the most recent year (MAIZE_PRODN1).

After the woman farmer gets knowledge of the existence of a technology, she makes a decision to adopt or not to adopt that technology subject to the economic and demographic factors faced. Hence, the second stage of the model estimates the variables responsible for adoption or non-adoption as indicated in equation (3) where T is a binary dependent variable for adoption of a technology (T=1 for adopters, T=0 for non adopters).

$$T_i = \beta_o + \sum_{i=1}^{i=n} \beta_i X_i + \mu_j \ldots\ldots\ldots\ldots\ldots\ldots \quad (3)$$

Intensification of a single component of farm system (with little change to the rest of the farm) - such as home garden intensification with vegetables and/or tree crops has been found to improve food production among households and to generally improve food security and sustainability of the farming business [19].

Where X_i is defined by the variables in Table 1. Age of the woman household head (HH_AGE). Age is also taken as a measure of farming experience in this study, education level of the woman household head specified as the number of years completed in formal school (HH_EDUC). Household size measured as the total number of people living in the household including children (HH_SIZE), Non-farm income (NONFAM_INCOME) earned by the woman head of the household was measured as all income from activities other than farming including micro businesses, remittances and pensions in US dollars. Distance from farm to the nearest village centre (km) (DIST_VILCENTR), highest unit price earned for maize in the most recent season (UPRICE_HIGH1) measured in US dollars, possession of a communication gadget like a phone or radio/TV (COMM_GADGET), membership to a farmer group or association, taken as an indication of collective action participation by a woman farmer but also signifying ability to share knowledge and information with others (GRP_MEMBER) and total land under cultivation in the most recent year, measured in hectares (TOTCULTIV_LAND).

2.3 Empirical Estimation Procedure and Hypothesis Testing

Estimation of the model outlined above in equations followed a series of regression diagnostics. Variables used in both stages of the model were first checked for normality using

Table 1. Variable description

Variable	Description (units)	Expected sign	
		Awareness	**Adoption**
CROP_EXTSVC	**Extension services related to crop production (binary dependent variable)**		
ADOPTER	Woman household head who adopted either an ox-plough or Agro forestry technology		
HH_AGE	Age of the woman head of household (years)	_	_
MICRO_BIZ	Woman engages in running a micro business (dummy)	+	
COMM_GADGET	Access to a communication gadget like a phone or radio (dummy)	+	+
HH_EDUC	Education level of woman head of household (years in formal school)	+	+
HH_SIZE	Size of the household (Number of people living in the household)	+	+
DIST_VILCENTR	Distance to the nearest village centre (Km)	_	_
GRP_MEMBER	Woman being a member of any farmer group or association (dummy)	+	+
MAIZE_SALPT	Maize selling point regularly used by the woman farmer (farm gate=1, otherwise=0)	_	
VOL_SAVINGS	Woman farmer's ability to make voluntary savings (dummy)	+	
OX_PLOUGH	**Woman farmer adopted use of an ox plough (binary dependent variable)**		
AGROFORESTRY	Woman farmer adopted agroforestry as a technology (dummy)		
NONFAM_INCOME	Annual income a woman head of household earns outside the farm in non-farm related work (USD).		+
UPRICE_HIGH1	Highest price earned per Kg of maize in current year (USD)		+
TOTCULTIV_LAND	Total land the woman's household put under cultivation in current year (ha)		+

Exploratory Data Analysis using the coefficient of kurtosis and skewness. Regression diagnostics included tests for multicollinearity, self-selection bias and heteroscedasticity. Multicollinearity was tested using the variance inflation factor (VIF) while heteroscedasticity was checked using Breusch-Pagan/Cook-Weisberg tests [20].

3. RESULTS AND DISCUSSION

3.1 Characteristics of Farmer Women Household Heads

Tables 2 and 3 show the social economic characteristics of the sampled women headed households who adopted either the ox plough or agroforestry technologies in maize production. Results in Table 1 indicated that adopter households had a significantly (P = .05) lower number of household members providing labour with four persons compared to five persons for the non adopters. This is because in some cases like the ox plough technology, farmers hire labour rather than do the ploughing themselves. This is however counter to the results by [8] that those households with more household labour are more likely to adopt a technology than those with less labour force.

Table 1 results further indicated that adopters had smaller land under maize cultivation (2.73 ha) compared to non-adopters (3.5 ha). Adopters also applied less fertilizer annually on average (318.2 Kg) compared to non-adopters (342.61 Kg). However results indicated that adopters produced more maize per unit area at 949.1 Kg/ha compared to the non-adopters at 861.88 Kg/ha. This is an important result that indicates how technologies like ox ploughs and agroforestry can increase yields while reducing costs on fertilizer as well as reducing risks associated with unpredictable crop failure for the resource constrained women farmers. [21] noted that the low fertilizer use in many African maize farming systems are a result of risk aversion, lack of agricultural credit and fertilizer availability and timely delivery.

Results in Table 3 show that there is a significant difference (P=.05) between the percentage of women household heads who engaged in micro businesses who are adopters of the ox-plough or agroforestry technologies and those are non-adopters. It is indicated that 76% of women adopters of these technologies were engaging in micro businesses as well compared to 57.89% of the non-adopters. The reason for this is that incomes from outside the farm are used to invest in farm technologies that enhance agricultural production. This is consistent with the findings of [22] that earning off farm income, in this case income from the micro businesses enhances and positively affects the likelihood of natural resource management practices.

Results also indicated that a significantly (*P* = .05) higher percentage of technology adopters (84%) possessed a communication gadget like a mobile phone, radio or television compared to the women non adopters at 64.91%. This result points to the fact that access to information through a number of media concerning agricultural technologies can augment adoption among smallholder farmers. [23] noted that mobile phones significantly reduce search costs for information concerning input and output prices in addition to complementing existing extension services where they are used by farmers.

Results in Table 3 also highlighted the scenario that a significantly (*P* <.01) higher percentage of non-adopter women household heads (43.86%) had access to credit compared to only 20% of the technology adopter women household heads. This is consistent with the previous result that more adopters are involved in micro business to earn income outside the farm than non-adopters and hence adopters face a higher production risk than their counterparts given the investment they make in taking up the new technologies. Therefore, the non-adopters resort to borrowing as a safety net to smooth their consumption through paying school fees, buying consumables and other services. In the same line, more non-adopters made voluntary savings (85.96%) compared to adopters (84.00%) though the two were not statistically significant at the three conventional significance levels. The result is in support of [24] who found that actually production risk suppresses demand for credit among smallholder farmers who would want to invest in production technologies.

Table 2. Characteristics of farmer women household heads and technology adoption

Variable	Overall mean (Std. dev)	Mean (Std. dev)		t-value	P-value
	(164)	Adopter(50)	Non adopter (114)		
Age of the woman head of household (years)	51.20 (11.67)	52.10 (10.90)	50.77 (11.98)	-0.70	0.48
Education level of woman head of household (years in school)	4.93 (3.59)	4.28 (3.28)	5.21 (3.70)	1.54	0.13
Size of the household (Number of people living in the household)	7.36 (3.42)	7.00 (2.97)	7.51 (3.59)	0.85	0.40
Distance to nearest village centre (Km)	7.04 (4.80)	7.16 (5.29)	6.99 (4.59)	-0.21	0.83
Quantity of maize produced by household in current year (Kg)	2,780.49(2,619.20)	2,738.00 (2,828.55)	2,799.12(2,534.81)	1.4	0.89
Quantity of fertilizer used on maize in current year(Kg)	335.17 (280.64)	318.20 (278.02)	342.61(282.68)	0.51	0.61
Number of household members providing labour	4.34 (2.49)	3.74 (2.03)	4.60 (2.64)	1.98	0.05
Area of land under maize (ha)	3.26 (2.17)	2.73 (1.62)	3.50 (2.34)	2.10	0.04
Maize yield in the current year (Kg/ha)	888.47 (484.40)	949.10 (521.81)	861.88 (466.97)	-1.06	0.29

Table 3. Characteristics (%) of women household heads and technology adoption

Characteristic (Variable)	Overall sample (164)	Adopter (50)	Non adopter (114)	P-Value
		Percentages		
Woman engages in running a micro business (dummy)	63.41	76.00	57.89	0.03
Access to a communication gadget like a phone or radio (dummy)	70.73	84.00	64.91	0.01
Membership to a farmers' group/association (dummy)	90.24	92.00	89.47	0.62
Maize selling point regularly used by the woman farmer (dummy)	34.15	0.00	49.12	0.01
Woman farmer's ability to make voluntary savings (dummy)	85.37	84.00	85.96	0.74
Farmer adopted ox plough technology		17.07	82.93	
Farmer agroforestry technology		19.51	80.49	
Woman head of household has access to credit (dummy)	36.59	20.00	43.86	0.003

3.2 Factors Affecting Women Farmers' Awareness of and Knowledge of Farming Technologies

The results of the first stage of the multi-stage logit model Table 4 indicated that as hypothesized engaging in a micro business significantly ($P < .01$) increased the likelihood of a woman's farmer awareness of a farming technology by 9.7 times than when the woman was not engaged in micro business. This is basically because farming is their primary source of income and undertaking a micro business involves moving outside the farm or village to look for merchandise, and in so doing, the woman gets exposure to information on availability of given farming technologies. In addition, many of the extension trainings usually take place in town or village centres where the micro business women usually go to trade. This seems to be in agreement with the findings of [18] that farm operators, whose primary income source is farming, though they may be engaged in other non-agricultural activities in generating income, tend to seek out information on new farming technologies.

A woman farmer's access to a communication gadget such as a phone or radio was also found to significantly ($P < .10$) increase the probability of awareness of farming technologies by about 2.5 times than without the gadget as previously expected Table 4. This is because mobile phones and radios in modern times are used to disseminate information on agricultural prices, inputs and new technologies availability. Imperfect information and knowledge concerning existence of inputs, outputs and new technologies is very prevalent in developing world rural areas, though information technology has potential to reduce the situation [25].

Results also indicated that awareness of farming technologies decreased with increasing education levels. Results Table 4 indicated that every single year increase in education level, leads to about 8% decrease in awareness. This was counter intuitive and counter hypothesis, as education was expected to increase awareness due its effect on decoding information at disposal to the farmer. This is due to the fact that education has a tendency of leading to migration of labour from the agricultural sector into other service and industrial sectors. Therefore, the more women get educated the less they will be enthusiastic about learning about new farming technologies. [26] also emphasised that education and training programmes have favoured the male leaving the rural women in many areas with low education levels and more are illiterate compared to men.

Women farmers were found to be more likely to be aware of farming technologies the farther away they are from the village centre. Every unit increase in the distance away from the village centre could lead to 3.5% increase in awareness Table 4. This is because many of the extension agents usually target those farmers outside the trading centres for training and demonstrations. Furthermore, the farther the women farmers are from town the more they will be entirely involved in farming as a core occupation, hence prompting them to seek knowledge of new and better farming technologies which is solely provided by extension workers who are sometimes overwhelmed by the number of farmers to attend to [27].

3.3 Factors Affecting Women Farmers' Adoption of Ox Plough and Agroforestry Farming Technologies

Results of the second stage of the logit model indicated that access to a communication gadget and total land the woman household head put under cultivation were positive and statistically significant at the conventional levels for probability of adoption of ox plough technology in maize farming. For the same technology, likelihood of adoption was negatively and significantly affected by age (as hypothesised), education level, size of household, distance to the nearest village centre. The signs on coefficients are almost the same for both ox plough and agroforestry apart from age, household size, distance to village centre and total land under cultivation.

A one year increase in age leads to a 6% decrease in the likelihood of adoption of the ox plough technology but increases that of agroforestry by 8.3% Table 5. The main reason for this observation is that ox plough use is an energy demanding task, hence the older the woman farmer, the less they are likely to adopt it unlike agroforestry that is relatively less laborious. [28] noted that the benefits from agroforestry are long term yet the costs of abandonment can be high, this may be one of the reasons why its adoption increases with age.

Collective action, represented by the woman household head being a member of a farmer's group or association has negative impact of adoption of both ox ploughs and agroforestry technologies, which is counter hypothesis. Results in Table 5 indicate that membership to a group reduces the likelihood of ox plough adoption by 5.9% and that of agroforestry by 4.5%. The reason for this is that these technologies are undertaken individually at the farm, and in the case of ox ploughs, there are specialized male attendants who do the ploughing using the oxen [13], hence group work may not be called for, actually a farmer may see it as time wasting.

Results also indicated that the highest price of maize received by the woman farmer decreased the likelihood of adopting both ox plough and agroforestry technology by big percentage points Table 5. This is because when the price is high, women farmers are more likely to expand their production as much as men farmers, which makes them shift from ox plough to tractors and also clear new land rather than agroforestry. [29] highlighted the fact that many women farmers in Africa are increasingly growing staple crops, and hence are affected by any policies like price stabilization policies in the same way as men farmers are affected.

Total land under cultivation by a woman farmer was found to significantly (*P* =.05) increase the likelihood of adoption of ox plough technology (by 29.4% for a unit increase in cultivated land). However, it was also found to reduce the probability of adoption of agroforestry technology by 68.4%. This is basically due to the fact that expanding the land under cultivation requires either a better land preparation technology than a hand hoe, or increasing the manual labour hired. Also when a farmer is to expand production, there is a high chance of foregoing agroforestry and instead open up virgin lands. Noted that access to land by women and its related security of tenure has an effect on technology adoption given its implications for credit and other inputs access.

Education was found to negatively and significantly affect adoption of Ox-plough but it had a positive though not significant effect on adoption of agroforestry technology. This is basically because these technologies are unique in such a way that they require a lot of time to adopt and yet those farmers that are more educated are fond of engaging in other off farm employment which may limit their uptake. However, other studies such as [10,13] found education to be positively associated with technology adoption. It is worth noting that different technologies in different environments are adopted differently depending on the prevailing conditions.

Table 4. Factors affecting women's awareness of a farming technology

Variable	Coefficient	(Std. Err)	Odds Ratio
Access to extension (proxy for technology awareness			
Age of the woman head of household (years)	0.015	0.023	1.015
Woman engages in running a micro business (dummy)	2.274***	0.674	9.722
Access to a communication gadget like a phone or radio (dummy)	0.897*	0.520	2.451
Education level of woman head of household (years in formal school)	-0.083	0.075	0.920
Size of the household (Number of people living in the household)	0.057	0.078	1.059
Distance to the nearest village centre (Km)	0.035	0.056	1.036
Woman being a member of any farmer group or association (dummy)	0.346	0.847	1.413
Maize selling point regularly used by the woman farmer (farm gate=1, otherwise=0)	1.240*	0.654	3.457
Woman farmer's ability to make voluntary savings (dummy)Pseudo R^2 = 0.2025	-0.389	0.698	0.677

*Significant level: * = 10%; ** = 5%; *** = 1%.*

Table 5. Factors affecting women's adoption of two farming technologies

Variable	Ox plough adoption			Agroforestry adoption		
	Coefficient	**(Std.Err)**	**Odds ratio**	**Coefficient**	**(Std. err)**	**Odds ratio**
Age of the woman household head (years)	-0.060*	0.035	0.943	0.083*	0.051	1.087
Access to a communication gadget like a phone or radio (dummy)	1.300*	0.733	4.902	4.217*	2.255	67.842
Education level of woman head of household (years in formal school)	-0.239**	0.109	0.780	-0.153	0.129	0.858
Size of the household (Number of people living in the household)	-0.505*	0.300	0.614	0.419*	0.231	1.521
Distance to the nearest village centre (Km)	-0.005	0.088	0.994	0.092	0.185	1.097
Woman being a member of any farmer group or association (dummy)	-0.593	1.141	0.502	-0.448	1.473	0.639
Income a woman household head earns outside the farm in non-farm related work annually (USD).	0.004	0.003	1.004	-0.005	0.007	0.995
Highest price earned per Kg of maize in current year (USD)	-29.667	20.014	2.840 e^{-13}	-6.070	12.819	0.002
Total land the woman's household put under cultivation in current year (ha)	0.294**	0.145	1.383	-0.684**	0.334	0.505

*Significant level: * = 10%; ** = 5%; *** = 1%*

4. CONCLUSION

Though many factors were found to affect awareness and knowledge of farmers about existing technologies, factors that ensure collective action were quite prominent. Those elements of society that allow women farmers to interact among themselves and other members such as communication through the phones, belonging to a group or association and point of sale of their produce create an atmosphere of sharing knowledge and create awareness about technologies. The implication of increased awareness of technologies is that increased agricultural production and food security will be attained and eventually household incomes will increase.

Adoption of the ox-drawn plough is affected by age of the woman farmer beyond a certain point due to its manpower demands. However, economic status of the woman farmer as measured by the size of land under cultivation promotes ox plough adoption. Therefore, the more women farmers expand in their scale of agricultural operations the more they are likely to adopt better technologies such as shifting from the hand hoe to the ox plough and later to tractors. Hence, it is highly recommended that livestock restocking programs and land tenure systems that ensure increased women access to and expansion of cultivated land be pursued by the authorities to ensure sustainable engagement of women farmers in agriculture.

The benefits from agroforestry are medium to long term. Therefore the likelihood of its adoption by women farmers is promoted by factors related to medium-long term for example age of the woman household head and size of the household (labour). However, expanding scale of production in terms of land put under cultivation negatively affects the likelihood of adoption of the technology. Therefore for sustainable agricultural systems and farmer livelihoods, policy emphasis needs to be focused on the medium to long term agroforestry interventions with the implication that soil productivity will be improved and the negative effects of climate change mitigated.

ACKNOWLEDGEMENTS

This paper is based on earlier research work done by the authors on Women technology adoption in Zambia. The authors are grateful to FK-Norway and the Institute of Economic and Social Research (INESOR) of the University of Zambia for the support that enabled this work. We also wish to acknowledge the women farmers in Zambia who committed their precious time to respond to the questionnaires. However, the views expressed herein do not necessarily represent the position of any of these organizations. All errors in interpretation are entirely attributed to the authors.

COMPETING INTERESTS

The authors declare that we do not have any competing interests whatsoever concerning to this research work.

REFERENCES

1. Adesina AA, Baidu-Forson Z. Farmer's perceptions and adoption of new agricultural technology: Evidence from analysis in Burkina Faso and Guinea, West Africa. Agricultural Economics. 1995;13:1-9.
2. Adesina AA, Mbila D, Nkamleub GB, Endamana D. Econometric analysis of the determinants of adoption of alley farming by farmers in the forest zone of southwest Cameroon. Agriculture, Ecosystems and Environment. 2000;80:255–265.
3. BaumÜller H. Facilitating agricultural technology adoption among the poor: The role of service delivery through mobile phones. ZEF Working Paper Series, No. 93; 2012. Available:http://hdl.handle.net/10419/88382
4. Binswanger HP, von Braun J. Technological change and commercialize-tion in agriculture: The Effect on the Poor. The World Bank Research Observer. 1991;6(1):57-80.
5. Budak DB, Darcan N, Kantar M. Women farmers and extension services in small ruminant production in mountain areas of Turkey. Journal of Arid Environments. 2005;62:507–515.
6. Central Statistical Office (CSO), Ministry of labour and social security. Zambia Labour Force Survey Report; 2013.
7. Doss CR. Twenty-Five years of research on women farmers in Africa: Lessons and implications for agricultural research institutions; with an annotated bibliography. CIMMYT Economics; 1999.
8. Doss CR. Men's crops? Women's crops? The gender patterns of cropping in Ghana.

World Development. 2002;30(11)1987–2000.
9. Doss CR. Designing agricultural technology for African women farmers: Lessons from 25 years of experience. World Development. 2001;29(12):2075-2092.
10. Fernandez_Cornejo J, Hendricks C, Mishra A. Technology adoption and off-farm household income: The case of herbicide tolerant soybeans. Journal of Applied and Agricultural Economics. 2005;37(3):549-563.
11. Giné X, Yang D. Insurance, credit, and technology adoption: Field experimental evidence from Malawi. Journal of Development Economics. 2008;89:1-11.
12. Green WH. Econometric analysis 5th edition. Prentice hall, Upper Saddle River. New Jersey, Prentice-Hall Inc; 2003.
13. Heisey PW, Mwangi W. Fertilizer use and maize production in Sub-Saharan Africa. CIMMYT Economics Working Paper 96-01. Mexico, D.F. CIMMYT; 1996.
14. Karki LB, Bauer S. Technology adoption and household food security. Analyzing factors determining technology Adoption and impact of project intervention. A case of small holder peasants in Nepal. Paper prepared to present in the Deutscher Tropentag. Humboldt-University, Berlin; 2004.
15. Keil A, Zeller M, Franzel S. Improved tree fallows in smallholder maize production in Zambia: Do initial testers adopt the technology? Agroforestry Systems. 2005;64:225-236.
16. Kuhlmann F, Brodersen C. Information technology and farm management: Developments and perspectives. Computers and Electronics in Agriculture. 2001;30:71–83.
17. Kumar SK. Women's role and agricultural technology. Chapter adapted from a presentation at a seminar sponsored by the international service for National agricultural research and the rockefeller foundation, held in Bellagio, Italy March. 1985;135-147.
18. Kuteya A, Kambole C. 2012 Zambian agricultural sector budget analysis. Presented by ACF/IAPRI Budget Breakfast Meeting Pamodzi Hotel, Lusaka November. 2011;30.
19. Lubwama FB. Socio-economic and gender issues affecting the adoption of conservation tillage practices; 1999. Available: http://www.atnesa.org
20. Marenya PP, Barret CB. Household-level determinants of adoption of improved natural resources management practices among smallholder farmers in Western Kenya. Food Policy. 2007;32:515-536.
21. Mason MN, Jayne TS, Myers RJ. The effects of the food reserve agency on smallholder production patterns and maize market prices. Stakeholder's meeting on evidence-based agricultural policy decisions to shape agricultural development for poverty reduction 5 december 2012 Taj Pamodzi Hotel, Lusaka, Zambia. Indaba Agricultural Policy Research Institute (IAPRI); 2012.
22. McBride WD, Daberkow SG. Information and the adoption of precision farming technologies. Journal of Agribusiness. 2003;21(1):21-38.
23. McCulloch AK, Meinzen-Dick R, Hazell P. Property rights, collective action and technologies for natural resource management: A conceptual framework. SP-PRCA Working Paper No. 1. International Food Policy Research Institute, CGIAR System-wide Program on Property Rights and Collective Action; 1998.
24. Mercer DE. Adoption of agroforestry innovations in the tropics: A review. Agroforestry System. 2004;61(1):311-328.
25. Ministry of finance. The next steps, budgetary changes. Budget of Zambia; 2013. Available: www.pwc.com/zm
26. Phiri D, Franzel S, Mafongoya P, Jere I, Katanga R, Phiri S. Who is using the new technology? The association of wealth status and gender with the planting of improved tree fallows in Eastern Province, Zambia. Agricultural Systems. 2004;79:131–144.
27. Pretty J. Hine R. Reducing food poverty with sustainable agriculture: A summary of new evidence. Centre for environment and society, University of essex, Final report from the "SAFE-World" (The Potential of Sustainable Agriculture to Feed the World) Research Project, University of Essex, February; 2001.
28. Smucker GR, White TA, Bannister M. Land tenure and the adoption of agricultural technology in Haiti. Paper presented at the 9th Biennial Conference of the International Association for the Study of Common Property Victoria Falls, Zimbabwe. This

paper was released as CAPRi Working Paper No.6; 2002. Available:http://www.capri.cgiar.org/wp/capriwp06.asp

29. Tembo S, Sitko N. Technical compendium: Descriptive agricultural statistics and analysis for Zambia. Indaba Agricultural Policy Research Institute (IAPRI), Lusaka Zambia, Working Paper 76 2013. Accessed on 25th September 2014. Available:http://www.iapri.org.zm/index.ph?

26

Phosphorus Adsorption Isotherm: A Key Aspect for Soil Phosphorus Fertility Management

Lemma Wogi[1*], J. J. Msaky[2], F. B. R. Rwehumbiza[2] and Kibebew Kibret[1]

[1]School of Natural Resources Management and Environmental Sciences, Haramaya University, Haramaya, Ethiopia.
[2]Department of Soil Science, Sokoine University of Agriculture, P.O. Box 3008, Morogoro, Tanzania.

Authors' contributions

This work was carried out in collaboration between all authors. Author LW designed the study, wrote the protocol, wrote the first draft of the manuscript and managed the literature searches and all laboratory analyses. Author KK edited the data, reviewed and edited the manuscript. Authors JJM and FBRR reviewed and edited the protocol and the manuscript. All authors read and approved the final manuscript.

Editor(s):
(1) Dimka Haytova, Department of Horticulture, Agricultural University of Plovdiv, Bulgaria.
(2) Peter A. Roussos, Lab. Pomology, Agricultural University of Athens, Greece.
Reviewers:
(1) Rusu Teodor, Department of Technical and Soil Sciences, Faculty of Agriculture, University of Agricultural Sciences and Veterinary Medicine Cluj-Napoca, Romania.
(2) Vijay Singh Meena, Crop Production Division, Vivekananda Institute of Hill Agriculture Almora, 263 601, India.

ABSTRACT

Characterization of soils in terms of phosphorus adsorption capacity is fundamental for effective soil phosphorus fertility management and for efficient utilization of phosphorus fertilizers. Thus, this study was conducted to investigate the phosphorus adsorption characteristics of soils of two farms and to elucidate the implication of soil phosphorus adsorption isotherm studies for soil phosphorus fertility management. The two farms, representing the major farming systems of the respective districts, were selected from Adele village in Haramaya district and Bala Langey village in Kersa district in eastern Ethiopia. Soil samples were collected from the crop fields at Adele and Bala Langey farms. Two different P-bearing sources, potassium dihydrogen phosphate (KH_2PO_4) and diammonium phosphate (DAP-$(NH_4)_2HPO4$), were used for the adsorption isotherm studies. The adsorption data were fitted to the linear and Freundlich adsorption isotherm models. Both models revealed that soils of both farms had different P adsorption capacity from the two P sources. Amount of P adsorbed from DAP solution was higher than the amount of P adsorbed

**Corresponding author: E-mail: l.wogi@yahoo.com;*

from KH_2PO_4 solution in soils of both farms. Phosphorus adsorption capacity of Adele farm soils was higher than that of Bala Langey farm soils. Therefore, soils of the two farms should be managed differently for P fertility. Percentages of P adsorbed (% Pa) and P remained in the equilibrium solution (% EC) were also calculated. By plotting the two percentages i.e. % Pa and % EC against the initial concentration of P (IC), two regions were observed. The two regions were described as P intensity and quantity factor windows. Based on the intensity and quantity factor windows, at currently existing soil condition, between 200 and 500 $kgha^{-1}$ P should be applied as fertilizer to soils of Adele at 0-30 cm depth for immediate benefits and soil P fertility maintenance.

Keywords: Phosphorus sources; P-fertility; P- intensity-quantity factor window.

1. INTRODUCTION

Phosphorus is one of the 17 essential elements for plant growth and reproduction. Plants obtain phosphorus from the soil solution which is either from the weathering of the parent materials or from the applied fertilizers. However, accessibility of phosphorus applied with fertilizers to plant is governed by different factors such as phosphorus adsorption- desorption characteristics of soil, soil texture, soil organic matter content and management practices [1,2]. Therefore, understanding interaction of phosphorus with those factors in soil is crucial for soil phosphorus fertility management and for sustaining phosphorus in pedobiochemical cycle.

In the soil system, phosphorus exits in two forms; the labile form which is weakly adsorbed on the surface of soil particles which is considered as not readily accessible to the plant. The other form is a portion of phosphorus in soil solution which is readily accessible to the plant [3]. The labile phosphorus is in equilibrium with phosphorus in soil solution and may be available to plant. But the rate of release is very slow and may not be available to the plant within the short period of crop reproductive cycle. The equilibrium between labile P and solution P will be disturbed when a phosphate fertilizer is applied to soil. This leads to rapid adsorption of P on the surface of soil particles and makes phosphorus to be more firmly held [4]. As a result, the amount of phosphorus adsorbed on the surface of soil particles increases while the quantity of phosphate ions in soil solution decreases.

The amount of phosphorus adsorbed on the surface of soil particles is termed as the quantity factor and the phosphate ions remaining in soil solution is the intensity factor [5]. These factors are very important for soil phosphorus fertility management. Phosphorus adsorption isotherm describes the interdependence of these two factors i.e. the intensity and the quantity factors. In a simple term, phosphorus adsorption isotherm is a plot of quantity factor against intensity factor [6]. The ratio of the quantity factor (Q) to intensity factor (I) is the buffering capacity of the soil [5,7]. This governs soil phosphors supply to plants.

Phosphorus adsorption isotherm is the most useful experimental procedure for studying interaction of phosphate ions with soil constituents [8,4]. It is also a useful parameter to monitor availability of phosphorus to plants. Furthermore, phosphorus adsorption isotherm study can help to describe phosphorus dynamic in the soil system. Characterization of soils in terms of phosphorus adsorption capacity is crucial for effective and efficient utilization of phosphorus fertilizers with respect to quantity, type and placement [9]. Thus, phosphorus adsorption isotherm is a key aspect for soil phosphorus fertility management.

Many workers have investigated phosphorus adsorption characteristic of soils from the environmental sustainability point of views [10,11,12]. Others reported that phosphorus adsorption characteristic of soil is affected by soil clay and organic matter contents [13]. In fact this is directly related to soil fertility. In almost all the studies cited above, KH_2PO_4 was used as phosphorus source for the laboratory studies of soil P adsorption isotherm, which is rarely applied to soil as fertilizer. DAP is commonly being applied as phosphorus fertilizer for soil fertility management in Ethiopia. However, phosphorus adsorption characteristics of soils and impacts of DAP on soil properties when applied as fertilizer have not been studied for soils of eastern Ethiopia.

In general, phosphorus adoption characteristic of soils of the eastern part of Ethiopia is not well investigated either from the environmental or soil fertility management perspectives. Therefore, the objectives of this study were to investigate the P-adsorption characteristics of soils of the study areas using two different P-bearing sources;

potassium dihydrogen phosphate (KH_2PO_4) and DAP, and to elucidate the implication of soil phosphorus adsorption isotherm for soil phosphorus fertility management.

2. MATERIALS AND METHODS

Soil samples for the study were collected from two farms at Adele in Haramaya district and Bala Langey in Kersa district, Oromia Region, Eastern Ethiopia. Both farms were selected as representatives of the respective farming systems of the districts. From each crop field of the farms 16 sampling sites were selected and soil samples collected from 0-30 cm depth. One composite sample was made from soils collected from the 16 representative sampling sites of the crop fields of each farm. Some properties of the soils are presented in Table 1.

Twenty four, 100 ml capacity plastic bottles were prepared and arranged in two rows, each row containing 12 bottles. One gram air dried soil (<2 mm) from Adele farm was placed in every bottle in each row. The rows were labeled as 1st and 2nd. Every bottle within the row was labeled with the P sources, KH_2PO_4 used for P adsorption study in the laboratory and DAP commonly applied as P fertilizer for crop production in the region.

Twenty five ml of 1.0, 5.0, 10, 20, 30, 40, 50, 60, 70, 80, 90 and 100 mgL^{-1} P solution made from KH_2PO_4 and DAP were added to the respective labeled bottles containing the soil. A supporting electrolyte of 10 mM $CaCl_2$ solution was added to all bottles and the contents in the bottles were shaken for 24 hours at 22±2°C on an orbital shaker at 300 rpm [14]. After equilibrating time, the contents of each bottle were filtered through Whatman No. 42 filter paper. Three ml of the filtrates were taken for the color development with ammonium molybdate solution containing potassium antimony tarterate and ascorbic acid [15]. The P contents of the filtrates were read on a spectrophotometer (model T80+) at 880nm after the development of the blue color. Same experiment was repeated for the soil from Bala Langey farm.

Adsorbed P was calculated as the difference between initial and final concentration of P in the equilibrium solution. The adsorption data were fitted to the linear model and Freundlich adsorption isotherm model as described below:

$$\text{Linear Model: } x/m = a + bc$$

$$\text{Freundlich Model: } x/m = kfc^{1}/^{n}$$

Linear form of Freudlich model: logx/m = logkf + 1/nlogc

Where:

- c = Concentration of P in equilibrium solution (EPC) (mgL^{-1})
- x/m = Amount of P adsorbed (mg kg^{-1})
- kf = Proportionality constant for the Freundlich model ($mgkg^{-1}$)
- 1/n = Slope of the curve, when logx/m vs logc was plotted
- a = Y-intercept, when x/m vs c was plotted ($mgkg^{-1}$)
- b = Slope of line, when x/m vs c was plotted. Slope is the buffering capacity of soils with respect to P and clay contents [16].

Percentages of P adsorbed (Pa) and P in the equilibrium solution (EC) were calculated for the adsorption data of DAP solution as:

$$\%Pa = [(IC - EC) \times 100]1/IC$$

$$\%EC = (EC/IC) \times 100$$

Where:

- Pa = Phosphorus adsorbed from initial solution (mgL^{-1}) = Quantity factor
- IC = Initial phosphorus concentration (mgL^{-1})
- EC= Equilibrium phosphorus concentration (mgL^{-1}) = Intensity factor
- %Pa=Percentage of adsorbed phosphorus from initial solution
- %EC=Percentage of phosphorus in equilibrium solution.

Table 1. Some physical and chemical properties of soils from the crop fields of Adele and Bala Langey farms

Sampling sites/crop fields	(%) Sand	(%) Silt	(%) Clay	Textural class	pH	(%) OC	(%) $CaCO_3$	Olsen P(mg/kg)	Total P (g/kg)
Adele farm	58	17	25	SCL	7.24	1.21	0.45	1.56	28.97
Bala Langey farm	58	13	29	SCL	6.53	1.45	0.15	8.74	17.13

**SCL = Sandy clay loam*

Values for the two percentages (%Pa and %EC) were plotted against initial concentration (IC) of P to determine the intensity and quantity factors region or windows for soils of the two farms from DAP adsorption data.

After equilibrating time, pH of the soil suspension to which DAP was added for the adsorption isotherm study was also measured with pH glass electrode. Similarly, pH of soil samples collected from experimental sites established within the crop fields at both farms, where DAP was applied at the rate of 100 $kgha^{-1}$ for two consecutive cropping seasons, was measured. The soil sampling was done before planting in 2012 and 2013, and after harvesting in 2014. The pH was measured for 1:2.5 soil water suspensions with combined pH glass electrode.

3. RESULTS AND DISCUSSION

3.1 Phosphorus Adsorption Characteristics of the Soils

It was found that the adsorption isotherm data best fit to the linear model and Fruendlich adsorption isotherm models. The linear model and Fruendlich adsorption model revealed that soils of the two farms have different P adsorption capacity from KH_2PO_4 and DAP solutions used as P sources for the study (Figs. 1a and b). The adsorption isotherm data showed that amount of P adsorbed by soils from both farms increased with the increased concentration of phosphorus in initial solution of the two P- sources. As shown by Figs.1a and b, at lower concentration amount of P adsorbed from DAP solution is greater than that of P adsorbed from KH_2PO_4 solution by the soils. This can be attributed to the differences in chemical properties of the two compounds used as P-sources. Upon dissolution in water the two compounds yield different phosphate ions as indicate below:

$$KH_2PO_4 + H_2O = K^+ + H_2PO_4^-$$

$$(NH_4)_2HPO_4 + H_2O = 2NH_4^+ + HPO_4^{2-}$$

Thus, one phosphate anion is released from 1 mole of KH_2PO_4 and two phosphate anions from 1 mole of $(NH_4)_2HPO_4$, which are readily absorbable on the soil anion exchange sites.

The P adsorption capacity of Bala Langey farm soils from DAP was not as high as that for Adele farm soils (Figs. 1a and b). Even though, there was a difference between the two soils from the two farms in P adsorption capacity, more P was adsorbed from DAP solution by the soils of the two farms compared to KH_2PO_4 solution (Figs. 1a and b). This suggests that when phosphorus is applied as fertilizer, its accessibility to plants is threatened by the soil factors and the chemical and physical properties of the fertilizers applied. Therefore, studies of soil phosphorus adsorption characteristics as affected by P sources could be very effective for soil P fertility management and for efficient utilization of P fertilizers.

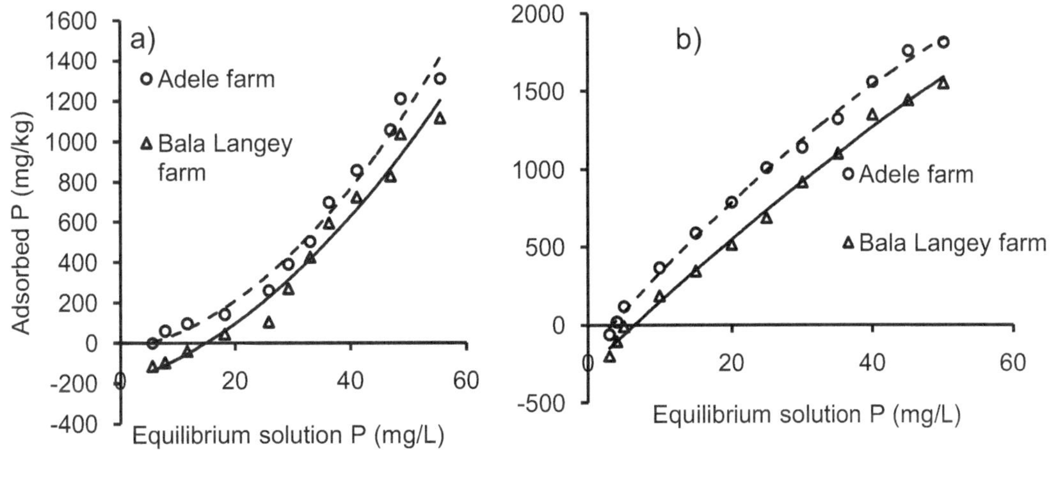

Fig. 1. Phosphorus adsorption characteristics of the soils of the two farms as was affected by different P sources KH_2PO_4 (a) and DAP (b) solutions

Soil organic carbon, clay content and mineralogy, calcium carbonate and soil pH are among the main soil constituents that are responsible for differences in phosphorus adsorption characteristics of soils. It has been reported by several investigators [17,4,1,2,3] that P adsorption is positively correlated with soil clay, calcium carbonate contents and pH, and negatively correlated with soil organic carbon content. Rajput et al. [18] also reported soils with higher available P as well as organic matter content adsorbed less applied P than soils with lower organic matter content. However, soils from Bala Langey farm with relatively higher clay content adsorbed less P than the Adele farm soils (Table 1 and Fig. 1). Therefore, higher P adsorption capacity of Adele farm soils than Bala Langey farm soils can be attributed to its relatively lower organic carbon, higher calcium carbonate and pH value.

The linear and Fruendlich models indicate that soils of Adele farm had higher P adsorption capacity from DAP as well as from KH_2PO_4 solutions than Bala Langey farm soils (Figs. 1a and b). These reveal that soils of the two farms demand different P fertility management strategies to increase the soils productivity. Nevertheless, the soil P fertility management strategy of the extension offices of the two districts is 100 kg DAP ha^{-1} which is a blanket recommendation rate for P fertilizer application. These soils should therefore be managed differently for P fertility.

The slopes of the linear model for the Adele farm soil were 40.27 and 27.32, respectively for both P sources, DAP and KH_2PO_4 (Table 2). This describes that for 1 unit increases in P concentration adsorption increases by 40.27 and 27.32 units respectively for the soil. Slopes of Fruendlich model were also 0.156 and 0.135 for the same soil, respectively for DAP and KH_2PO_4. These again indicate an increase in P adsorption by 0.156 and 0.135 for a unit increase in P solution concentration. Therefore, these clearly show that phosphorus sources have considerable impact on phosphorus adsorption characteristics of soils.

The Y-intercepts of the linear model were -271.30 (KH_2PO_4) and -77.45 (DAP) for soils of Adele farm (Table 2). These suggest that about 271.30 and 77.45 mg kg^{-1} P desorption from the soil when concentration of P is zero in initial solutions. Practically, it is not possible to obtain two observations with such large difference for the same solutions of zero concentration. Therefore, this can be explained as: when the concentration of P in the solutions is very close to zero more P will be desorbed from the soils into KH_2PO_4 solution than into DAP solution. This also indicates soil P-desorption depends on the properties of the P sources applied as fertilizers.

The Y-intercepts for Bala Langey farm soils were also -375.40 (KH_2PO_4) and -230.20 (DAP) (Table 2). As explained above for Adele farm soils, these also show desorption phenomenon at very low concentrations of P. But the Y-intercept values were much higher in magnitude than the Y-intercept values for the Adele farm soils of higher adsorption capacity. This indicates that soils with less adsorption capacity have higher desorption capacity with respect to the phosphorus sources. Therefore, the sorption isotherms clearly depict that phosphorus adsorption/desorption characteristics of soil is a function of soil physico-chemical properties and the chemical and physical properties of fertilizers applied as source of phosphorus.

3.2 Phosphorus Intensity and Quantity Factor Windows of Soils of Adele and Bala Langey farms from Adsorption Isotherm Data of DAP Solution

Phosphorus intensity and Quantity Factor regions (windows) for both farm soils were created from the percentages of adsorption isotherm data i.e. %Pa and %EC. By plotting the

Table 2. Linear model and Freundlich adsorption model parameter for soils of Adele and Bala Langey farms

Soils	P-sources	Linear model parameters			Freundlich model parameters			
		Y-intercept	Slope	R^2	Intercept Log(Kf)	Kf	Slope (1/n)	R^2
Adele farm	KH_2PO_4	-271.30	27.32	0.944	1.79	61.66	0.135	0.95
	DAP	-77.45	40.27	0.990	1.83	67.61	0.156	0.748
Bala Langey farm	KH_2PO_4	-375.40	26.09	0.953	1.78	60.23	0.162	0.868
	DAP	-230.20	37.41	0.993	2.32	208.93	0.109	0.917

two percentages against the initial concentration of P, two regions were observed at the left and right sides of the intersection point in between the line joining the points (Figs. 2a and b).

The region to the left side of the intersection point, with higher percentage of phosphorus in solution represents the intensity factor whereas the region to right side with higher percentage of adsorbed phosphorus represents the quantity factor when P fertilizers are applied to the soil systems. The two regions were described as intensity and quantity factor windows.

The widths of the two windows are interdependent i.e. when the width of intensity factor increases the width of the quantity factor decreases. This exactly demonstrates the P dynamics in the soil systems. The two windows for the study soil revealed the same situation. Adele farm soils with lower available P, higher total P and higher adsorption capacity had narrower intensity factor window and wider quantity factor window (Table 1 and Fig. 2a). Contrary to this, Bala Langey farm soils with relatively higher available P, lower total P and lower adsorption capacity had wider intensity factor window and narrower quantity factor window Table 1 and Fig. 2b.

Furthermore, Figs. 2a and b indicate that the two lines can never cross the Y-axis. This describes the real situation that concentration of P can never be zero in solution and/or in adsorbed form for the materials defined as soil. When the two percentages were plotted against the initial concentration of phosphorus solution, two equilibrium points were also observed. One is at the point where the adsorption line crosses the X-axis and the other is at the intersection of the two lines (Figs. 2a and b). The 1st equilibrium point indicates concentration of P at which adsorption was exactly equal to desorption. The 2nd equilibrium point also indicates concentration of P at which concentration of adsorbed P was equal to concentration of P in equilibrium solution i.e. 50% of the applied P was adsorbed and the remaining 50% was in the soil solution.

The two equilibrium points have an implication for soil P fertility management. For instance, continuous application of P less than the concentration at the 1st equilibrium point would be resulted in the depletion of native soil P. This is in a way explains the reports by the farmers, "fertilizer is killing our soils". Even though, it was not recommended based on studies farmers of the study farms apply P less than the blanket recommendation of 100 kg DAP ha^{-1}, which is less than the quantity at the 1st equilibrium point. Thus, application of P fertilizer at a rate adjusted to the concentration of P between the two equilibrium points based on the crop requirements maintain more P in soil solution for immediate benefits. More application of P greater than the concentration at the 2nd equilibrium point ends with soil P build up. The above therefore, gives a baseline for soil P fertility management.

The adsorption isotherms (Figs. 1a and b) could not indicate quantity of P to be applied as fertilizers to alleviate the problems associated with higher P adsorption by the soils or to replenish depleted P. But the intensity and quantity factor windows were able to show the quantity of P to be applied either to be in the intensity factor window or in the quantity factor window.

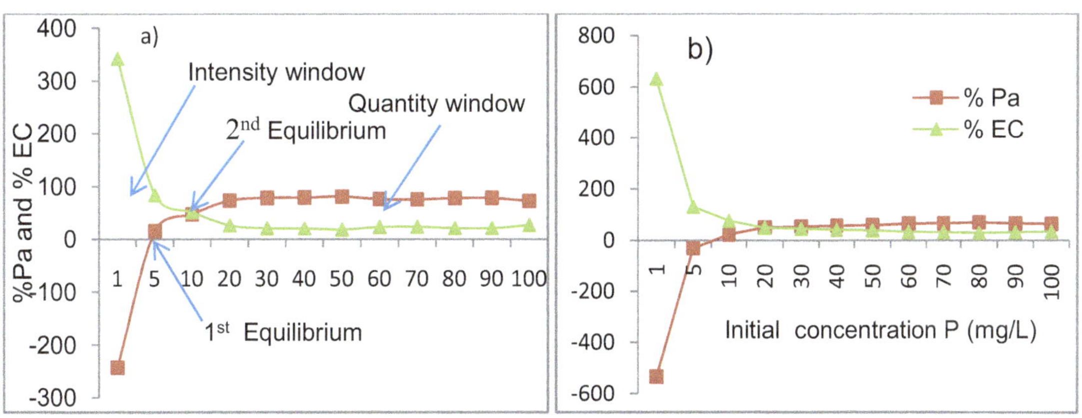

Fig. 2. Intensity and quantity factor windows for soils of Adele (a) and Bala Langey (b) farms from adsorption isotherm data of DAP solution

If the manager of Adele farm is for example, interested in immediate benefits from P application at the currently exiting soils conditions, he/she should apply greater than 2.5 ppm (200 kg Pha^{-1} to the depth of 30 cm) and less than 10 ppm (500 kg Pha^{-1} to the depth of 30 cm). Therefore, P fertilizer application for the immediate benefits should be in between the 1st and 2nd equilibrium points for the 1st season and then be applied based on quantity of P taken up or exported, and crop requirements per cropping seasons.

This seems unachievable from economic points. But there are options, for instance the values of Pha^{-1} presented above were calculated from the soil bulk density values. Therefore, if a farmer is able to reduce soil bulk density through organic matter amendment, regardless of the contributions of organic matter to P availability, then he/she could reduce those values calculated above. Because organic matter addition could reduce P adsorption [19,20], which means organic matter amendment can narrow down the quantity factor window and make wider the intensity factor window.

Thus the soil phosphorus intensity and quantity factor windows help the manager of the farm to make a decision. The manager can decide either to deplete his/her soils by applying less amount of phosphorus up to the point that soil will no longer be productive or maintain the soil phosphorus. Soil P maintenance could be achieved through application of adequate and suitable P fertilizers based on the quantity and intensity factor windows of the soils. Similarly, soil P intensity and quantify factor windows can also be established by studying P sorption isotherm of a particular soil. Therefore, soil phosphorus adsorption isotherm study is a key aspect for soil phosphorus fertility management.

3.3 Impact of Diammonium Phosphate (DAP) Fertilizer on Soil Solution pH

The pH of the equilibrium solution of soils shaken with DAP solution was measured after the equilibrium time to investigate changes in the soil properties induced upon application of DAP to the soil system. Changes in the pH of the equilibrium solution were observed increasing as the concentration of P increased in the initial DAP solution. Changes in pH were rapid for the lower concentration ranges and increased slightly with increased in the P concentration (Fig. 3a). This indicates soil pH buffering capacity against changes in phosphate ions concentration increases with increase in phosphate ions concentrations.

The same was observed for the soil samples collected from the same farms where DAP was applied as P source at the rate of 100 kg ha^{-1} (18 kg N and 46 kg P_2O_5) over two consecutive cropping seasons (Fig. 3b). 100 kg DAP ha^{-1} is the blanket rate for the region. But farmers apply about 4 times less than the recommended rate to plots of selected crops.

The pH increased from 7.28 to 8.01 with 0.75 units for soil from Adele farm and from 6.54 to 7.43 with 0.89 units for soil from Bala Lang farm over the two consecutive cropping seasons (Fig. 3b). Change in pH for Bala Langey farm soils with relatively lower initial pH was higher than changes in pH for the Adele farm soils. This shows, soils of Adele farm with higher P adsorption capacity had relatively higher pH buffering capacity against DAP application.

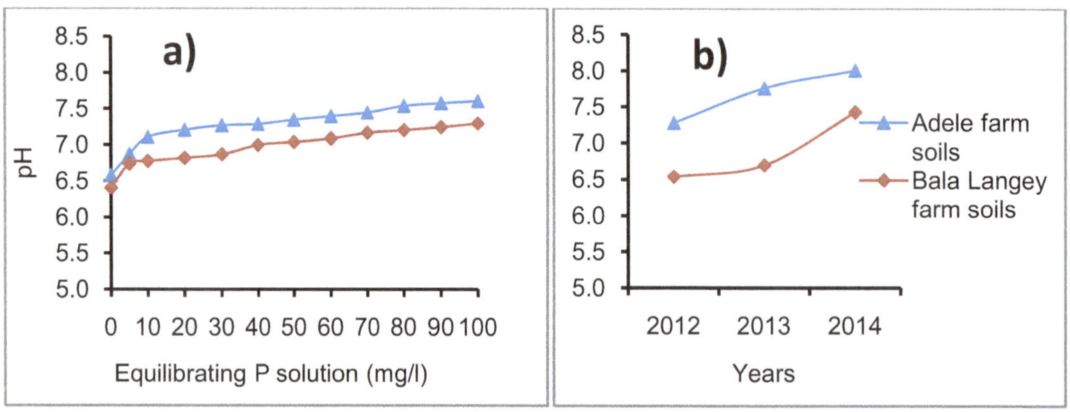

Fig. 3. Changes in pH with increase in P concentration of the equilibrium solution from DAP (a) in laboratory and in soil solution upon application of DAP fertilizer (b) at fields

The pH of equilibrium solutions was also observed increasing as the amount of DAP increased for soil P adsorption characteristics from DAP solution. Similarly, soil pH was also increased upon DAP application at a rate of 100 $kgha^{-1}$ for two consecutive cropping seasons. These changes in soil pH over a short periods i.e. two cropping seasons therefore, put under question the suitability of DAP to those soils if the application doses would be increased.

4. CONCLUSION

- Soils of the two farms had different P adsorption capacity from the two P sources DAP and KH_2PO_4.
- More P was adsorbed from DAP solution than from KH_2PO_4 solution by soils of the two farms
- Phosphorus adsorption capacity of Adele farm soils was higher than that of Bala Langey farm soils.
- The soils of the two farms should be managed differently for P fertility.
- Between 200 and 500 $kgha^{-1}$ P should be applied as fertilizer to soils of Adele farm at 0-30 cm depth for immediate benefits and soil P fertility maintenance.
- Suitability of DAP as source of phosphorus fertilizer for soils of the study areas should be further investigated at field and in laboratory.

ACKNOWLEDGEMENTS

The study was funded by Alliance for Green Revolution in Africa (AGRA) through Sokoine University of Agriculture, AGRA Soil Health programme. AGRA is acknowledged for funding the entire studies and Sokoine University of Agriculture for hosting the programme. Haramaya University is also acknowledged for providing laboratory facilities and technical supports during field and laboratory works.

COMPETING INTERESTS

Authors have declared that no competing interests exist.

REFERENCES

1. Tsado PA, Osunde OA, Igwe CA, Adeboye MK, Lawal BA. Phosphorus sorption characteristics of some selected soil of the Nigerian Guinea Savanna. Int. J. Agri. Sci. 2012;2(7):613-618.
2. Moazed H, Hoseini Y, Naseri AA, Abbasi F. Determining phosphorus adsorption isotherm in soil and its relation to soil characteristics. J. Food Agric. & Enviro. 2010;8(2):1153-1157.
3. Ahmed MF, Kennedy IR, Choudhury AT, Kecske ML, Deaker R. Phosphorus adsorption in some Australian soils and influence of bacteria on the desorption of phosphorus. Commun. Soil Sci. Plant Anal. 2008;39:1269–1294,
4. Amel Idris OA, Sirelkhatim Ahmed H. Phosphorus sorption capacity as a guide for phosphorus availability of selected Sudanese soil series. Afr. Crop Sci. J. 2012;20:59–65.
5. Rashmi I, Biswas AK, Shinogi KC, Subba Rao A. Phosphorus adsorption capacity in soybean growing soils of Madhya Pradesh. Search and Research. 2013;1:15-20
6. Qudrat UK, Jamil K, Saifur R, Sana UK. Impact of equilibrating time on phosphate adsorption and desorption behaviour in some selected saline sodic soils. J. Chem. Soc. P. 2012;34:1.
7. Horta, Torrent. Phosphorus desorption kinetics in relation to phosphorus forms and sorption properties of Portuguese acid soils. Soil Sci. 2007;172(8):038-075X/07/17208-631–638.
8. Ricardo de OM, Ronaldo SB, Ot´avio AC. Phosphorus sorption and redistribution on soil solid phase in a brazilian haplorthox amended with biosolids. Applied and Enviro. Soil Sci; 2011. Article ID 283061, 7 pages. DOI:10.1155/2011/283061.
9. Samadi A. Phosphorus sorption characteristics in relation to soil properties in some calcareous soils of western Azarbaijan province. J. Agric. Sci. Technol. 2006;8:251-264
10. Zhang L, Song H, Jing H, Fuxing G, Yuh-Shan H. Isotherm study of phosphorus uptake from aqueous solution using aluminum oxide. Clean – Soil, Air, Water. 2010;38(9):831–836
11. Cucarella V, Tomasz, Ryszard M. Phosphorus sorption capacity of different types of opoka. Annals of Warsaw University of Life Sciences – SGG Land Reclamation. 2007;38:11–18.
12. Hengpeng Y, Fanzhong Ch, Yanqing S, Guoying S, Jiamo F. Adsorption of phosphate from aqueous solution onto

modified palygorskites. Separation and Purification Technology. 2006;50:283–290
13. Han XZ, Tang C, Song CY, Wang SY, Qiao YF. Phosphorus characteristics correlate with soil fertility of albic luvisols. Plant and Soil. 2005;270:47–56. DOI 10.1007/s11104-004-1083-9.
14. Rowell DL. Soil science methods and applications. Department of Soil Science, Unversity of Reading, Longman Group UK; 1994.
15. Murphy J, Riley JP. A modified single solution method for the determination of phosphate in natural waters. Analytica Chimica Acta. 1962;27:31-36.
16. Hussain A, Abdul G, Ghulam M. Use of models for phosphorus adsorption on some sodic soils of Punjab. Int. J. Agric. Biology. 2006;1560–8530/08–2–241–248.
17. Fisseha H, Heluf G, Kibebew K, Birru Y). Study of phosphorus adsorption and its relationship with soil properties, analyzed with Langmuir and Freundlich models. Agric. Forestry and Fisheries. 2014;3(1):40-51 DOI: 10.11648/j.aff.20140301.18.
18. Rajput A, Panhwar QA, Naher UA, Rajput S, Hossain E, Shamshuddin J. Influence of incubation period, temperature and different phosphate levels on phosphate adsorption in soil. Am. J. Agric. Bio. Sci. 2014;9(2):251-260.
19. Shen J, Lixing Y, Junling Z, Haigang L, Zhaohai B, Xinping C, Weifeng Z, Fusuo Z. Phosphorus dynamics: From soil to plant. Plant Physiology. 2011;156:997–1005.
20. Fadly Hairannoor Y. The relationship between phosphate adsorption and soil organic carbon from organic matter addition. J. Trop. Soils. 2010;15(1):1-10.

In vitro Regeneration of Sweet Potato (*Ipomea batatas* (L.) Lam.) from Node Explants

N. C. Onwubiko[1*], C. Ijeoma Ihezie[2] and M. U. Mozie[1]

[1]Department of Crop Science and Technology, Federal University of Technology, Owerri, Nigeria.
[2]Biotechnology programme, National Root Crops Research Institute, Umudike, Nigeria.

Authors' contributions

This work was carried out in collaboration between all authors. Authors NCO and CII designed the study and wrote the protocol, while. Authors NCO and MUM performed the statistical analysis and wrote the first draft of the manuscript. All the authors read and approved the final manuscript.

Editor(s):
(1) Lanzhuang Chen, Laboratory of Plant Biotechnology, Faculty of Environment and Horticulture, Minami Kyushu University, Miyazaki, Japan.
(2) Daniele De Wrachien, State University of Milan, Italy.
Reviewers:
(1) Anonymous, India.
(2) Anonymous, USA.

ABSTRACT

Investigations on *in vitro* clonal propagation of sweet potato (*Ipomea batatas* (L.) Lam.) were carried out at tissue culture laboratory of National Root Crops Research Institute, Umudike, Abia State, Nigeria. *Ipomea batatas* variety 440293 used was maintained in a culture room at 28 ±2°C under a 16 h photoperiod provided by white florescent tubes (60 µmo/lm-2s-1). Phytohormones- α-naphthaleneactic acid (NAA) at different concentrations and 6-benzylaminopurine (BAP) at 2 mg/l were used for the study. Summarily, data collected on the assessed growth parameters of the explants as tested with F test and t test showed that both the different levels of NAA and the number of weeks of application significantly (P< 0.05) enhanced *in vitro* production of sweet potato.

Keywords: Sweet potato; Ipomea batatas; in vitro regeneration.

**Corresponding author: E-mail: onwubikouche@yahoo.com;*

1. INTRODUCTION

The techniques of tissue culture are widely used in plant regeneration and transformation. *In vitro* vegetative propagation of crop or micropropagation is an important tool for the recovery, conservation of germplasm and embryo rescue [1]. It facilitates genetic manipulation of plants through somatic hybridization which previously were almost impossible, thereby has helped to overcome the limitations posed by basic or convectional breeding operations [2]. It is also effective in maintaining disease-free plants and avoiding genetic instability [3]. Plant tissue culture involves the isolation of cells, tissues and organs (otherwise called explants) from plants and growing them *in vitro* on a culture or nutrient media of known composition under a sterile controlled condition- an environment with optimized physical, nutritional and hormonal factors. [4] listed physical factors that must be maintained within certain limits for efficient *in vitro* vegetative propagation of crops to include temperature, pH, gaseous environment, light (quality and duration), and osmotic pressure. The culture vessel, nutrient or growth media, and the external environment (light, temperature etc) must meet both the physical and chemical needs of the *in vitro* cultured explants for optimal regeneration potential to be achieved.

In the regeneration of sweet potato, correct hormone combination ratio is one of the critical factors to consider in obtaining high regeneration efficiency [5,6]. Plant hormones or plant growth regulators (PGRs) in combination with other nitrogen sources make up the culture medium that supply all the essential mineral ions required for plant growth and development. Plant hormones have a profound effect on the morphology of the tissue developed from the explants as they can enhance not only the growth of some cultured slow growing tissues, but also determine the development pathway of the plant cells. In addition, it has been observed that plant hormones or pytohormones play an important role on the possibility for organogenesis or somatic embryogenesis to occur in culture [7], and the most frequently used PGRs in plant regeneration are auxins and cytokinins or their synthetic analogues. Studies have shown that the rate of proliferation of the root and shoot system are determined by the level of combination of auxins and cytokinins [8], also the type of culture established is determined by the ratio of auxins to cytokinins, and a correct phytohormone constitution ratio is necessary to optimize regeneration of plants [9]. Hence, the need to investigate the effect of combination of different concentrations of α-naphthaleneactic acid (NAA) at 2 mg/l of 6-benzylaminopurine (BAP) application, for optimal regeneration of sweet potato.

2. MATERIALS AND METHODS

Explants were excised from vigorously growing *in vitro Ipomea batatas* cultivar 440293 obtained from the *in vitro* gene bank housed at the Biotechnology programme (Plant tissue culture laboratory) of National Root Crops Research Institute (NRCRI) Umudike, Abia State, Nigeria.

Culture medium containing [10] was used for the study. Sucrose (30 g/l), myo-inositol (100 mg/l), and agar-E (7 g/l) were added to the culture medium.

The first MS medium- the control (CO) contained all the basal medium composition listed above only. The second MS medium contained the control and 6-benzylaminopurine (BAP) at a concentration of 2 mg/l. In third medium, α-naphthaleneactic acid (NAA) was added to the MS basal medium at different concentrations while the concentration of BAP was kept constant at 2 mg/l. The pH was adjusted to 5.8. Agar was dissolved in the three media by heating. 10ml each of the different media were dispensed into 150×25 mm test tubes and autoclaved at 121°C at 1.05kg/cm^2 for 15 minutes.

Explants were seeded singly onto medium in test tubes. The conditions of cultivation were at a temperature of 28±2°C and artificial illumination supplied by white fluorescent tubes that provided an intensity of 60 µmol/m^{-2}s$^{-1.}$ The photoperiod was 16h of light and 8 h of darkness.

The design of the experiment was Completely Randomized Design (CRD). Eight treatments- six levels of NAA ranging from 0.05 to 3.0 mg/l, BAP at 2.0 mg/l, and the control that did contain neither NAA nor BAP- were replicated 25 times. Multivariate analysis of variance (MANOVA) was used to analyze the data collected on number of developed nodes, leaves, roots and plant height, between 1 to 5 weeks after subculture. Specifically the data was tested with F test and Ducan's t test using SPSS package 15.0.

3. RESULTS AND DISCUSSION

The F test result as displayed on Table 1 below shows that there were significant differences on the growth of *I. batatas* as measured by nodes, height, leaves, and roots, with respect to the different levels of NAA applied and the weeks of application at the $\alpha = 0.05$ level of significance.

The result of the Ducan's t test carried out to determine the particular level(s) of NAA and weeks of application that were actually responsible for the observed significant differences on the growth of *I. batatas* are shown in Tables 2 to 5 below. In Table 2, the result on the effect of different levels of NAA application on number of nodes shows that only two levels of NAA- 0.25 and 0.05 were statistically different from one another, while on plant height, only 0.05 level of NAA application was statistically different from all others.

On number of leaves, only 0.30 and 0.05 levels of NAA application were significantly different from each other, the result on number of roots clearly shows that the control and 0.10 NAA level were statistically different from each and all others. Furthermore, 0.30 level of NAA application was statistically different from 0.20, 0.10, and the control (Table 3).

The result on the effect of number of weeks of application of different levels of NAA on the number of developed nodes and plant height shows that significant differences exist among the weeks. No week was statistically the same with another. Again, the means were increasing as the weeks of application increases (see Table 4 below).

Table 5 below shows that the number of weeks of application of NAA significantly ($P<0.05$) improved number of developed leaves and roots. On number of leaves, while weeks 1, 2, 5 were statistically different from one another, week 3 was also not the same with weeks 1, 2 and 5. The f test result on number of roots shows that weeks 1, 4 and 5 were significantly different from each other.

Table 1. F test result on levels of NAA and number of weeks

Source	Dependent variable	Type III sum of squares	df	Mean squares	F	Sig.
NAA levels.	No of nodes	25.610	7	3.659	2.761	.036
	Plant height	50.785	7	7.255	9.832	.000
	No of leaves	68.903	7	9.843	2.933	.032
	No of roots	30.319	7	4.331	7.726	.000
No of Wks	No of nodes	815.103	4	203.776	153.771	.000
	Plant height	497.960	4	124.490	168.708	.000
	No of leaves	540.112	4	135.028	40.239	.000
	No of roots		4	8.232	14.684	.000

Table 2. Ducan's t test result on the effect of different levels NAA applied on number of nodes and plant height

NAA conc.	N	Number of nodes			Plant height		
		Subsets			Subsets		
		1	2	3	1	2	3
0.25	8	5.7120			4.5420		
0.30	8	6.6560	6.6560		5.0760	5.0760	
control	8	6.8680	6.8680		5.2540	5.2540	
0.20	8	6.8880	6.8880			5.7160	
0.10	8	6.9100	6.9100			6.1000	
0.15	8		7.6320	7.6320		6.2180	
0.05	8			8.6320			7.8080
Sig		.175	.267	.206	.192	.051	1.000

Note: mean values in the same column are statistically the same.
N represents sample size.

Table 3. Ducan's t test result on the effect of different levels of NAA applied on number of leaves and roots

NAA levels	Number of leaves			NAA levels	Number of roots				
		Subsets				Subsets			
	N	1	2		N	1	2	3	4
0.30	8	5.2320		Control	5	3.1280			
0.25	8	5.6800		0.10	5		4.0960		
0.20	8	7.2320	7.2320	0.20	5			5.0000	
0.10	8	7.3520	7.3520	0.05	5			5.1520	5.1520
Control	8	7.6240	7.6240	0.25	5			5.5240	5.5240
0.15	8		7.9860	0.15	5			5.7560	5.7560
0.05	8		9.2000	0.30	5				5.9440
Sig.		.071	.135	Sig.		1.000	1.000	.083	.069

Note: mean values in the same column are statistically the same.
N represents the sample size

Table 4. Ducan's t test result on the effect of number of weeks of NAA application on number of developed nodes and plant height

weeks		Number of nodes					Plant height				
		Subsets					Subsets				
	N	1	2	3	4	5	1	2	3	4	5
Week1	5	1.2914					1.0157				
Week 2	5		3.9243					3.5000			
Week 3	5			6.8171					5.6357		
Week 4	5				8.4771					8.0514	
Week 5	5					14.7029					10.8786
Sig.		1.000	1.000	1.000	1.000	1.000	1.000	1.000	1.000	1.000	1.000

Note: mean values in the same column are statistically the same.
N represents sample size

Table 5. Ducan's t test result on the effect of number of weeks of NAA application on number of developed leaves and number of roots

Weeks	N	Number of leaves				Number of roots		
		Subsets				Subsets		
		1	2	3	4	1	2	3
Week1	5	2.0471				4.1257		
Week 2	5		4.8514			4.4229		
Week 3	5			7.6343		4.6114		
Week 4	5			9.2057			5.2914	
Week 5	5				12.1943			6.2629
Sig.		1.000	1.000	.117	1.000	.170	1.000	1.000

Note: mean values in the same column are statistically the same
N represents sample size

In order to improve sweet potato through tissue culture, a reliable and efficient *in vitro* culture technique method that will ensure efficient differentiation- the development of the root and shoot system and whole plant regeneration is essential [11,12,13]. To achieve this, the ratio of the auxins to cytokinins (pytohormones) in the culture medium that supply all the essential mineral ions required for growth and development of *in vitro* cultured sweet potato has to be balanced so as to achieve an optimum culture medium that best supports *in vitro* regeneration of sweet potato. The combination of different levels of NAA at 2 mg/l of BAP, and the number of weeks of application (Tables 1-5) as to achieve an optimum culture medium that will benefit high regeneration efficiency significantly improved the *in vitro* production of sweet potato.

Apparently from the f test result as displayed on Tables 2 and 3, significant differences were observed on the growth parameters measured, with respect to different levels of NAA applied at 2 mg/l of BAP. The application of lower levels or concentration of NAA -0.5 to 0.15 mg/l enhanced the development of nodes, plant height and leaves, and had means that were significantly different unlike higher levels of NAA application. In fact, beyond 0.15 mg/l of NAA application further increase in the quantity of NAA applied did not statistically improve the growth of *Ipomea batatas* variety 440293. This result suggests that *Ipomea batatas* variety 440293 had appreciable quantity of endogenous NAA, hence it was only lower concentration of exogenously applied NAA that statistical improved its growth. The findings of this study is in line with that of [8], who observed that lower levels of auxins supported growth of taller plantlets in *in vitro* propagation of *Dioscorea rotundata* L. (White Yam).

The result on the number of weeks of application of the plant growth regulators on *in vitro* performance of *I. batatas* showed also significant differences on the growth parameters studied (Tables 4-5). Consistently the mean values of number of developed nodes and plant height increased as the number of weeks of application increases (Table 4), and were significantly different between one week and another. At week 5 the mean values of the number of developed nodes and plant height were still increasing considerably, suggesting that the optimum number of weeks of application may not have been attained. Almost a similar result was observed for number of developed leaves and roots (Table 5). On number of developed leaves, weeks 1, 2, and 5 were significantly different from each other, it was only weeks 3 and 4 that were statistically the same. On the other hand, the mean values on number of developed roots from week 1 to 3 were statistically the same. Significant differences were observed between weeks 4 and 5. Again, the mean values on both the number of developed leaves and roots were on the increase as the number of weeks of application increases, suggesting that the number of weeks of application is not adequate to enhance high regeneration efficiency. This is in line with the work of [14].

4. CONCLUSION

Phytohormones like NAA and BAP can enhance the *in vitro* propagation of Sweet potato. However, lower levels or concentration of NAA had more significant effect in the development of roots, leaves, nodes and plant height in sweet potato.

ACKNOWLEDGEMENTS

We are sincerely grateful to staff of tissue culture unit of national root research crop institute Umudike for the use of their facilities in carrying out this experiment.

COMPETING INTERESTS

Authors have declared that no competing interests exist.

REFERENCES

1. Delgado-Sanchez P, Saucedo-Ruiz M, Guzmán-Maldonado SH, Villordo-Pineda, E, González-Chavira M, Fraire-Velázquez S, Acosta-Gallegos JA, Mora-Avilés A. An organogenic plant regeneration system for common bean (*Phaseolus vulgaris* L.). Plant Science. 2006;170(4):822-827.
2. Thottapilly, GLM. Monti, R. Monhon and AW. Moore, Biotechnology: Enhancing research on tropical crops in Africa. CTA-IITA Co-Publication, IITA, Ibadan, Nigeria. 1992;29-30.
3. Long Ming Sheng, China 1989. Chemical regulation of growth of organ formation in *in vitro* tissue cultured sweet Potato. Available: http://www.arc-avrdc.org/pdf
4. Venkatachalam P, Kavipriya V. Efficient method for *in vitro* plant regeneration from cotyledonary node explants of peanut (*Arachis hypogaea* L.). International Conference on Nuclear Energy, Environmental and Biological Sciences (ICNEEBS'2012) September 8-9, 2012 Bangko; 2012.
5. Sihachakr D, Haïcour R, Cavalcante Aves JM, Umboh I, Nzoghé D, Servaes A, Ducreux G. Plant regeneration in sweet potato (*Ipomoea batatas* L., Convolvulaceae). Euphytica. 1997; 96(1):143-152.
6. Gaba VP. Plant growth regulators in plant tissue culture and development. In Trigiano RN, Gray DJ (eds) plant development and biotechnology, Boca Ratom: CRC Press. 2005;87-99.
7. Thorpe TA. Morphogenesis and regeneration. In Thorpe TA, Vasil, IK (eds) Plant Cell and Tissue Culture, Netherlands, Dordrecht, Kluwer Academic Publishers. 1994;17-36.
8. Ezeibekwe IO, Ezenwaka CL, Mbagwu FN, Unamba CIN. Effects of Combination of Different Levels of Auxin (NAA) and Cytokinin (BAP) on *in vitro* propagation of *Dioscorea rotundata* L. (White Yam), New York Science Journal.2009;2(5):1-8.
9. Gubba A, Sivparsad BJ. Development of an efficient plant regeneration protocol for sweet potato (*Ipomoea batatas* L.) cv. Blesbok. African Journal of Biotechnology. 2002;11(84):14982-14987.
10. Murashige T, Skoog F. A revised medium for rapid growth and bioassays with tobacco tissue cultures. Physiol. Plant. 1962;15:473-497.
11. Svetleva D, Velcheva M, Bhowmik G. Biotechnology as a useful tool in common bean (*Phaseolus vulgaris L)* improvement. Euphytica. 2003;131(2):189-200.
12. VARISAI MS, SUNG Jih-Min, JENG Toong-Long, WANG Chang-Sheng. Organogenesis of *Phaseolus angularis* L.: high efficiency of adventitious shoot regeneration from etiolated seedlings in the presence of N6-benzylaminopurine and thidiazuron. Plant Cell Tissue and Organ Culture. 2006;86(2)187-199.
13. Andrés M. Gatica Arias, Jenny Muñoz Valverde, Pilar Ramírez Fonseca, Marta Valdez Melara *In vitro* plant regeneration system for common bean (*Phaseolus vulgaris*): effect of N6-benzylaminopurine and adenine sulphate, Electronic Journal of Biotechnology; 2010. ISSN: 0717-3458
14. Doliński R, Olek A. Micropropagation of sweetpotato (*Ipomoea batatas* (L.) Lam.) from node explants, Acta Sci. Pol., *Hortorum Cultus.* 2013;12(4):117-127.

Permissions

All chapters in this book were first published in AJEA, by SCIENCE DOMAIN International; hereby published with permission under the Creative Commons Attribution License or equivalent. Every chapter published in this book has been scrutinized by our experts. Their significance has been extensively debated. The topics covered herein carry significant findings which will fuel the growth of the discipline. They may even be implemented as practical applications or may be referred to as a beginning point for another development.

The contributors of this book come from diverse backgrounds, making this book a truly international effort. This book will bring forth new frontiers with its revolutionizing research information and detailed analysis of the nascent developments around the world.

We would like to thank all the contributing authors for lending their expertise to make the book truly unique. They have played a crucial role in the development of this book. Without their invaluable contributions this book wouldn't have been possible. They have made vital efforts to compile up to date information on the varied aspects of this subject to make this book a valuable addition to the collection of many professionals and students.

This book was conceptualized with the vision of imparting up-to-date information and advanced data in this field. To ensure the same, a matchless editorial board was set up. Every individual on the board went through rigorous rounds of assessment to prove their worth. After which they invested a large part of their time researching and compiling the most relevant data for our readers.

The editorial board has been involved in producing this book since its inception. They have spent rigorous hours researching and exploring the diverse topics which have resulted in the successful publishing of this book. They have passed on their knowledge of decades through this book. To expedite this challenging task, the publisher supported the team at every step. A small team of assistant editors was also appointed to further simplify the editing procedure and attain best results for the readers.

Apart from the editorial board, the designing team has also invested a significant amount of their time in understanding the subject and creating the most relevant covers. They scrutinized every image to scout for the most suitable representation of the subject and create an appropriate cover for the book.

The publishing team has been an ardent support to the editorial, designing and production team. Their endless efforts to recruit the best for this project, has resulted in the accomplishment of this book. They are a veteran in the field of academics and their pool of knowledge is as vast as their experience in printing. Their expertise and guidance has proved useful at every step. Their uncompromising quality standards have made this book an exceptional effort. Their encouragement from time to time has been an inspiration for everyone.

The publisher and the editorial board hope that this book will prove to be a valuable piece of knowledge for researchers, students, practitioners and scholars across the globe.

List of Contributors

A. E. Agbongiarhuoyi
Cocoa Research Institute of Nigeria (CRIN) P.M.B. 5244 Ibadan. Nigeria

E. O. Uwagboe
Cocoa Research Institute of Nigeria (CRIN) P.M.B. 5244 Ibadan. Nigeria

O. S. Ibiremo
Cocoa Research Institute of Nigeria (CRIN) P.M.B. 5244 Ibadan. Nigeria

F. O. Olasupo
Cocoa Research Institute of Nigeria (CRIN) P.M.B. 5244 Ibadan. Nigeria

E. O. Aigbekaen
Cocoa Research Institute of Nigeria (CRIN) P.M.B. 5244 Ibadan. Nigeria

J. Di Matteo
Faculty of Agricultural Sciences, National University of Mar del Plata, (B7620ZAA), Balcarce, Province of Buenos Aires, Argentina

J. Rattin
Faculty of Agricultural Sciences, National University of Mar del Plata, (B7620ZAA), Balcarce, Province of Buenos Aires, Argentina

A. Di Benedetto
Faculty of Agricultural Sciences, National University of Mar del Plata, (B7620ZAA), Balcarce, Province of Buenos Aires, Argentina
Faculty of Agronomy, University of Buenos Aires, Avenida San Martín 4453 (C1417DSE), Buenos Aires, Argentina

E. F. Fabiyi
Agricultural Economics and Extension Department, College of Agricultural Sciences, Landmark University, Omu-Aran, P.M.B 1001, Kwara State, Nigeria

Valerie Njitat Tsama
Department of Plant Biology (Waste Water Research Unit), Faculty of Science, University of Yaounde I, P.O.Box: 8250 Yaounde, Cameroon

Godwill Mih Chewachong
Department of Plant Biology (Phytopathology Research Unit), Faculty of Science, University of Yaoundé I, P.O.Box 812 Yaoundé Cameroon

Ives Magloire Kengne Noumsi
Department of Plant Biology (Waste Water Research Unit), Faculty of Science, University of Yaounde I, P.O.Box: 8250 Yaounde, Cameroon

Wilfried Arsene Letah Nzouebet
Department of Plant Biology (Waste Water Research Unit), Faculty of Science, University of Yaounde I, P.O.Box: 8250 Yaounde, Cameroon

Nkeze Nyochembeng
Department of Plant Biology (Waste Water Research Unit), Faculty of Science, University of Yaounde I, P.O.Box: 8250 Yaounde, Cameroon

Zachée Ambang
Department of Plant Biology (Phytopathology Research Unit), Faculty of Science, University of Yaoundé I, P.O.Box 812 Yaoundé Cameroon

Geofrey Kawube
National Crops Resources Research Institute, National Agriculture Research Organisation, P.O.Box 7084, Kampala, Uganda
Faculty of Agriculture and Environment, Gulu University, P.O.Box, 166, Gulu, Uganda

Jolly Kabirizi
National Livestock Resources Research Institute, National Agriculture Research Organisation,
P.O.Box 96, Tororo, Uganda

Clementine Namazzi
National Livestock Resources Research Institute, National Agriculture Research Organisation,
P.O.Box 96, Tororo, Uganda

Diaa A. Ibrahim
Department of Scientific Research Center, Faculty of Science, University of Duhok, Iraq

Gharbia H. Danial
Department of Scientific Research Center, Faculty of Science, University of Duhok, Iraq

Vian M. Mosa
Department of Scientific Research Center, Faculty of Science, University of Duhok, Iraq

Belan M. Khalil
Department of Scientific Research Center, Faculty of Science, University of Duhok, Iraq

Ahasanul Hoque
Department of Genetics and Plant Breeding, Bangladesh Agricultural University, Mymensingh-2202, Bangladesh

S. N. Begum
Plant Breeding Division, Bangladesh Institute of Nuclear Agriculture, Mymensingh-2202, Bangladesh

A. H. K. Robin
Department of Genetics and Plant Breeding, Bangladesh Agricultural University, Mymensingh-2202, Bangladesh

Lutful Hassan
Department of Genetics and Plant Breeding, Bangladesh Agricultural University, Mymensingh-2202, Bangladesh

Carla Fabro
Department of Agricultural and Environmental Science, University of Udine, Italy

Stefano Rimoli
Margherita Hack Hummingbirds Centre, Torreano di Martignacco (UD), Italy

Piero A. Susmel
Margherita Hack Hummingbirds Centre, Torreano di Martignacco (UD), Italy

Keshavulu Kunusoth
Department of Seed Science and Technology, Acharya NG Ranga Agricultural University, Hyderabad, TS, India

Krishnasamy Vadivel
Centre for Plant Molecular Biology, Tamil Nadu Agricultural University, Coimbatore, Tamilnadu, India

Raman Meenakshi Sundaram
Crop Improvement, Directorate of Rice Research, Hyderabad, TS, India

Razia Sultana
Department of Seed Science and Technology, Acharya NG Ranga Agricultural University, Hyderabad, TS, India

Passoupathy Rajendrakumar
Crop Improvement, Directorate of Rice Research, Hyderabad, TS, India

Sheshumadhav Maganti
Crop Improvement, Directorate of Rice Research, Hyderabad, TS, India

Lella V. Subbarao
Crop Improvement, Directorate of Rice Research, Hyderabad, TS, India

Sebastian Reyes Chin-Wo
Department of Plant Sciences, University of California, Davis, California, USA

Charles A. Joseph
Institute of Crop Sciences/National Key, Facility for Crop Gene Resources and Genetic Improvement, Chinese Academy of Agricultural Sciences, 12 South Zhong-Guan-Cun St., Beijing 100081, China
Department of Botany, University of Dar es Salaam Box 35091, Dar es Salaam, Tanzania

T. Awio
Department of Agricultural Production, College of Agriculture and Environmental Sciences, Makerere University, P.O.Box 7062, Kampala, Uganda

B. Bua
Department of Agriculture, Faculty of Vocational Studies, Kyambogo University, P.O.Box 1, Kyambogo-Kampala, Uganda

J. Karungi
Department of Agricultural Production, College of Agriculture and Environmental Sciences, Makerere University, P.O.Box 7062, Kampala, Uganda

R. S. Meena
Institute of Agricultural Sciences, Banaras Hindu University, Varanasi (U.P.) - 221 005, India

R. S. Yadav
Swami Keshwan and Rajasthan Agricultural University, Bikaner (Rajasthan) - 334 006, India

M. L. Reager
Swami Keshwan and Rajasthan Agricultural University, Bikaner (Rajasthan) - 334 006, India

Nirmal De
Institute of Agricultural Sciences, Banaras Hindu University, Varanasi (U.P.) - 221 005, India

V. S. Meena
Institute of Agricultural Sciences, Banaras Hindu University, Varanasi (U.P.) - 221 005, India

J. P. Verma
Institute of Environment & Sustainable Development, Banaras Hindu University, Varanasi (U.P.) - 221 005, India

S. K. Verma
Institute of Agricultural Sciences, Banaras Hindu University, Varanasi (U.P.) - 221 005, India

B. C. Kansotia
Swami Keshwan and Rajasthan Agricultural University, Bikaner (Rajasthan) - 334 006, India

David Lomeling
Department of Agricultural Sciences, College of Natural Resources and Environmental Studies (CNRES) University of Juba, P.O.Box 82, Juba, South Sudan

Sebit Mathew Otwari
Department of Agricultural Sciences, College of Natural Resources and Environmental Studies (CNRES) University of Juba, P.O.Box 82, Juba, South Sudan

Yahya Mohammed Khater
Department of Agricultural Sciences, College of Natural Resources and Environmental Studies (CNRES) University of Juba, P.O.Box 82, Juba, South Sudan

Dimitra A. Loka
Department of Crop, Soil and Environmental Sciences, University of Arkansas, Altheimer Laboratory, Fayetteville, AR, USA

Derrick M. Oosterhuis
Department of Crop, Soil and Environmental Sciences, University of Arkansas, Altheimer Laboratory, Fayetteville, AR, USA

Z. Sheykholeslami
Department of Agronomy, Qaemshahr Branch, Islamic Azad University, Qaemshahr, Iran

M. Qasempour Almdari
Department of Agronomy, Qaemshahr Branch, Islamic Azad University, Qaemshahr, Iran

S. Qanbari
Department of Agronomy, Qaemshahr Branch, Islamic Azad University, Qaemshahr, Iran

M. Akbarzadeh
Department of Botany, Mazandaran Agricultural and Natural Resources Research Center, Sari, Iran

U. Mohammed
Niger State Ministry of Agriculture and Rural Development, Minna, Nigeria

R. S. Olaleye
Department of Agricultural Economic and Extension Technology, Federal University of Technology Minna, Niger State, Nigeria

Fabiyi Esther Funsho
Department of Agricultural Economics and Extension, Landmark University, Omu-Aran, Nigeria

Ademiluyi Benson Oluwafemi
Department of Plant Science, Ekiti State University, Ado Ekiti, Nigeria

Abiodun Joseph
Department of Crop and Soil Science, Landmark University, Omu-Aran, Nigeria

L. A. Oppong
Molecular biology section, Biotechnology, Seed and Food Science Division, CSIR-Crops Research Institute, P.O.Box 3785, Fumesua-Kumasi, Ghana

M. D. Quain
Molecular biology section, Biotechnology, Seed and Food Science Division, CSIR-Crops Research Institute, P.O.Box 3785, Fumesua-Kumasi, Ghana

A. Oppong
Plant Health Division, CSIR-Crops Research Institute, P.O.Box 3785 Fumesua-Kumasi, Ghana

H. A. Doku
Molecular biology section, Biotechnology, Seed and Food Science Division, CSIR-Crops Research Institute, P.O.Box 3785, Fumesua-Kumasi, Ghana

A. Agyemang
CSSS Division, CSIR-Crops Research Institute, P.O.Box 3785, Fumesua-Kumasi, Ghana

K. Offei Bonsu
Horticulture Improvement Division, CSIR-Crops Research Institute, P.O.Box 3785, Kwadaso - Kumasi, Ghana

Alexis Mojica
Laboratorio de Investigación en Ingeniería y Ciencias Aplicadas, Centro Experimental de Ingeniería, Universidad Tecnológica de Panam*á, 0819-07289 El Dorado, Panamá*

J. Solís
Laboratorio de Investigación en Ingenieríay Ciencias Aplicadas, Centro Experimental de Ingeniería, Universidad Tecnológica de Panam*á, 0819-07289 El Dorado, Panamá*
Facultad de Ingeniería Civil, Universidad Tecnológica de Panamá, 0819-07289 Zona 6, Panamá

N. Bernal
Facultad de Ingeniería Civil, Universidad Tecnológica de Panamá, 0819-07289 Zona 6, Panamá

F. Vergara
Laboratorio de Análisis Industrialy Ciencias Ambientales, Centro Experimental de Ingeniería,
Universidad Tecnológica de Panam*á. 0819-07289 El Dorado, Panamá*

B. Duarte
Laboratorio de Análisis Industrialy Ciencias Ambientales, Centro Experimental de Ingeniería,
Universidad Tecnológica de Panam*á. 0819-07289 El Dorado, Panamá*

C. A. Ho
Laboratorio de Investigación en Ingenieríay Ciencias Aplicadas, Centro Experimental de Ingeniería, Universidad Tecnológica de Panamá, *0819-07289 El Dorado, Panamá*

J. Ortega
Instituto Nacional de Agricultura, Ministerio de Desarrollo Agropecuario, Divisa, Panamá

L. Pastor
Sorbonne Universités, UPMC Université Paris 06, UMR 7619 METIS, F-75005, Paris, France

David Lace
Department of Biological and Agricultural Engineering, Texas A&M University, College Station, Texas, 303D Scoates Hall, 2117 TAMU, College Station, TX 77843, United States

Jersson Placido
Department of Biological and Agricultural Engineering, Texas A&M University, College Station, Texas, 303D Scoates Hall, 2117 TAMU, College Station, TX 77843, United States

Sergio Capareda
Department of Biological and Agricultural Engineering, Texas A&M University, College Station, Texas, 303D Scoates Hall, 2117 TAMU, College Station, TX 77843, United States

Louis K. Prom
USDA-ARS, Southern Plains Agriculture Research Center, College Station, TX 77845.United States

Ramasamy Perumal
Western Kansas Agricultural Research Center, Kansas State University, Hays, KS 67601, United States

Thomas Isakeit
Department of Plant Pathology and Microbiology, Texas A&M University, College Station, TX 77843, United States

Ghada Radwan
Department of Plant Pathology and Microbiology, Texas A&M University, College Station, TX 77843, United States

William L. Rooney
Department of Soil and Crop Sciences, Texas A&M University, College Station, TX 77843, United States

Clint Magill
Department of Plant Pathology and Microbiology, Texas A&M University, College Station, TX 77843, United States

Louis K. Prom
Crop Germplasm Research Unit, USDA-ARS, Southern Plains Agricultural Research Center, College Station, TX 77845, USA

Ramasamy Perumal
Western Kansas Agricultural Research Center, Kansas State University, Hays, KS 67601, United States

Zheyu Jin
Baicheng City Academy of Agricultural Science, Jilin Province, P.R. of China, 137000, China

Ghada Radwan
Department of Plant Pathology and Microbiology, Texas A&M University, College Station, Texas 77843, United States

Thomas Isakeit
Department of Plant Pathology and Microbiology, Texas A&M University, College Station, Texas 77843, United States

Clint Magill
Department of Plant Pathology and Microbiology, Texas A&M University, College Station, Texas 77843, United States

Filipina J. Krótki
Students Research Society, Witkowski High School, Cracow, Poland

Barbara K. Kościelniak
Students Research Society, College of Medicine, Jagiellonian University, Cracow, Poland

Przemysław J. Tomasik
Tutor of Students Research Society, College of Medicine, Jagiellonian University, Wielicka St. 265, Cracow, Poland

Christopher Sebatta
Department of Agribusiness and Natural Resource Economics, Makerere University, P.O.Box 7062, Kampala, Uganda

Mukata Wamulume
Institute of Economic and Social Research, University of Zambia, P.O.Box 30900, Lusaka, Zambia

Lemma Wogi
School of Natural Resources Management and Environmental Sciences, Haramaya University, Haramaya, Ethiopia

J. J. Msaky
Department of Soil Science, Sokoine University of Agriculture, P.O. Box 3008, Morogoro, Tanzania

F. B. R. Rwehumbiza
Department of Soil Science, Sokoine University of Agriculture, P.O. Box 3008, Morogoro, Tanzania

Kibebew Kibret
School of Natural Resources Management and Environmental Sciences, Haramaya University, Haramaya, Ethiopia

N. C. Onwubiko
Department of Crop Science and Technology, Federal University of Technology, Owerri, Nigeria

C. Ijeoma Ihezie
Biotechnology programme, National Root Crops Research Institute, Umudike, Nigeria

M. U. Mozie
Department of Crop Science and Technology, Federal University of Technology, Owerri, Nigeria